Collision Spectroscopy

Collision Spectroscopy

Edited by

R. G. Cooks
Purdue University
West Lafayette, Indiana

PLENUM PRESS · NEW YORK AND LONDON

Library of Congress Cataloging in Publication Data

Main entry under title:

Collision spectroscopy

Includes bibliographical references and index.
1. Collision spectroscopy. I. Cooks, R. G.
QC454.C634C64   539.7'54   77-10761
ISBN 0-306-31044-9

© 1978 Plenum Press, New York
A Division of Plenum Publishing Corporation
227 West 17th Street, New York, N.Y. 10011

All rights reserved

No part of this book may be reproduced, stored in a retrieval system, or transmitted, in any form or by any means, electronic, mechanical, photocopying, microfilming, recording, or otherwise, without written permission from the Publisher

Printed in the United States of America

Temporary White

Once or twice while walking naked through this empty blue,
I have caught a word that I have understood.
And once or twice another naked being;
Passing through the colours of this walk;
Has heard a word of mine.
These few words are all communication I have had.
Except for one brief moment at my birth
When all of colour was contained in me
When I was still
And was a flash of purifying white.

O.F.C.

Contributors

J. Appell • Laboratoire de Spectrometrie, Rayleigh-Brillouin, U.S.T.L., 34060 Montpellier, France. Formerly, Laboratoire des Collisions Ioniques, associated with the C.N.R.S., Université de Paris-Sud, Orsay, France

J. W. Boring • Department of Nuclear Engineering and Engineering Physics, University of Virginia, Charlottesville, Virginia

R. G. Cooks • Department of Chemistry, Purdue University, West Lafayette, Indiana

T. R. Govers • Laboratoire de Résonnance Electronique et Ionique (part of the Laboratoire de Physico-chimie des Rayonnements, associated with the C.N.R.S.), Université de Paris-Sud, Orsay, France

R. E. Johnson • Department of Nuclear Engineering and Engineering Physics, University of Virginia, Charlottesville, Virginia

D. L. Kemp • Department of Chemistry, Purdue University, West Lafayette, Indiana

Q. C. Kessel • Department of Physics and The Institute of Materials Science, University of Connecticut, Storrs, Connecticut

J. Los • F.O.M. Instituut voor Atoom- en Molecuulfysica, Amsterdam, The Netherlands

J. T. Park • Department of Physics, University of Missouri, Rolla, Missouri

E. Pollack • Department of Physics, University of Connecticut, Storrs, Connecticut

W. W. Smith • Department of Physics, University of Connecticut, Storrs, Connecticut

Contents

Introduction
R. G. Cooks

1. The Subject	1
2. Metastable Ions	3
3. Types of Collisions	3
4. Collision Cross Sections	5
5. Energy Interconversions	6
6. Energy Resolution and Angular Resolution	7
7. Laboratory and Center-of-Mass Coordinate Systems	8
8. Guideposts to the Contents	14
References	16

1 Collisional Excitation of Simple Systems
John T. Park

1. Introduction	19
1.1 Types of Collisions	20
1.2. Elements of Collision Theory	21
1.3. Experimental Methods	24
2. Determination of Excitation Cross Sections from Ion Energy-Loss Spectrometry	26
2.1. Apparatus	26
2.2. Analysis of the Energy-Loss Spectrometric Method	30
3. Determination of Excitation Cross Sections from Photon Emission Data	35
3.1. Incident Particle Flux Determination	38

3.2. Target Density Determination 40
  3.3. Photon Flux Determination 42
4. The Excitation of Atomic Hydrogen 44
  4.1. Modulated Crossed-Beam Technique 44
  4.2. Energy-Loss Spectra of Atomic Hydrogen 49
5. Excitation of Helium 53
  5.1. Energy-Loss Spectra of Helium 53
  5.2. Optical Studies of the Excitation of Helium 54
6. Differential Cross-Section Measurements 63
7. Excitation of Simple Molecules 71
  7.1. Excitation of Molecular Hydrogen 73
  7.2. Excitation of Molecular Nitrogen 75
  7.3. Excitation of Molecular Oxygen 79
  7.4. Excitation of Singlet–Triplet Transitions 81
8. Discussion 84
  References 87

2 Charge Transfer in Atomic Systems
R. E. Johnson and J. W. Boring

1. Introduction 91
2. Theory 96
  2.1. Introduction 96
  2.2. Semiclassical Cross Sections Differential in Angle and Energy 99
  2.3. Transition Probabilities and Energy-Loss Spectra . . . 101
  2.4. General Considerations 107
3. Experimental Methods 109
  3.1. Introduction 109
  3.2. Pressure Measurement 110
  3.3. Detectors 111
  3.4. Beam Production 113
  3.5. Cross-Section Measurements 114
4. Experimental Data and Interpretation 119
  4.1. Introduction 119
  4.2. Integrated Cross Sections 119
  4.3. Energy-Loss Spectra with Doubly Charged Ions 124
  4.4. Angular Differential Cross Sections 131
  4.5. Cross Sections Differential in Angle and Energy 138
5. Summary 142
  References 143

3 Inelastic Energy Loss: Newer Experimental Techniques and Molecular Orbital Theory
Q. C. Kessel, E. Pollack, and W. W. Smith

1. Introduction 147
2. The Molecular Orbital Model 149
   2.1. Molecular Orbitals for Collision Processes 149
   2.2. Transitions between Molecular Orbitals 156
3. Experimental Techniques and Results 163
   3.1. Energy-Loss Measurements 166
      3.1.1. A Typical Scattered-Particle Apparatus 166
      3.1.2. Time-of-Flight Measurements 177
      3.1.3. The Use of Molecular Targets 189
      3.1.4. Inner-Shell Excitations 193
   3.2. Ion–Photon and Ion–Electron Coincidence Measurements 198
      3.2.1 Introduction 198
      3.2.2. Experimental Principles 200
      3.2.3. Charge-Exchange Measurements Using Coincidence Techniques 203
      3.2.4. Stueckelberg Oscillations 206
      3.2.5. Rosenthal Oscillations and Long-Range Couplings 211
      3.2.6. Polarization and Angular Distribution Measurements in Coincidence 214
      3.2.7. Ion–Molecule Coincidence Measurements 217
4. Summary and Conclusion 219
   References 221

4 Double Electron Transfer and Related Reactions
J. Appell

1. Introduction 227
2. Cross Sections 229
3. Experimental Techniques of Double-Charge-Transfer Spectroscopy 231
   3.1. The Instrument 231
   3.2. The Energy-Loss Scale 232
   3.3. An Apparatus with Higher Energy Resolution 233
   3.4. Target Pressure Dependences of the Negative Ion Intensities 234
4. (H^+, M) Spectra and the Study of the States of the Target Species 235

4.1. The Spin Conservation Rule		236
4.2. The Applicability of the Franck–Condon Principle		238
4.3. Selection Rules for the Symmetry of the Molecular States		240
4.4. Observation of Double-Charge-Transfer Processes at Different Scattering Angles		242
4.5. Determination of the Double-Ionization Potentials of Some Molecules		243
4.6. Observation of Radiative Transitions without Detection of the Photon		247
5. (A^+, Ar) Spectra: An Insight into Future Possible Developments		248
5.1. Determination of the State Composition of the A^+ Beam		249
5.2. Determination of the Geometric Structure of a Polyatomic A^+ or A^- Ion		250
5.3. The Electron Affinity of the Projectile Species A		250
5.4. Dissociative Double-Charge-Transfer Spectra		251
Appendix: Energy Loss of the Projectile in an Inelastic Collision		252
References		255

5 Ionic Collisions as the Basis for New Types of Mass Spectra
D. L. Kemp and R. G. Cooks

1. Introduction	257
2. $2E$ Mass Spectra	260
3. $-E$ Mass Spectra	267
4. $E/2$ Mass Spectra	272
5. $+E$ Mass Spectra	281
6. Other Types of Spectra	284
References	288

6 Collision-Induced Dissociation of Diatomic Ions
J. Los and T. R. Govers

1. Introduction	289
2. Dynamics of Dissociation	291
2.1. The Two-Step Model	294
2.2. Electronic Excitation	296
2.3. Vibrational/Rotational Dissociation	300
2.4. Predissociation	302
3. Experimental Techniques	305
3.1. The Aston Band Method	306
3.2. Initial-State Preparation	314

	3.3. Final-State Analysis	316
	3.4. The Apparatus Function	317
4.	Direct Dissociation in Heavy-Particle Collisions	321
	4.1. The Dissociation of H_2^+	322
	4.2. The Dissociation of HeH^+	331
	4.3. Discussion	335
5.	Translational Spectroscopy	335
	5.1. Collisional Dissociation of 4–10 keV N_2^+ Ions	336
	5.2. Collisional and Unimolecular Dissociation of 10-keV HeH^+ Ions	340
	5.3. Photodissociation of H_2^+	347
6.	Concluding Remarks	351
	References	353

7 Collision-Induced Dissociation of Polyatomic Ions
R. G. Cooks

1.	Introduction	357
	1.1. Scope of Chapter	357
	1.2. Comparison with Metastable Ions	358
	1.3. Development of CID Studies	359
2.	The Reaction	363
	2.1. The Basic Phenomenon and Its Experimental Characterization	363
	2.2. Mechanism	367
	2.3. Cross Section	380
	2.4. Effects of Experimental Variables	385
3.	Experimental Procedures	395
	3.1. Instrumentation and Scanning Methods	395
	3.2. Energy Resolution and Kinetic Energy Measurement	399
	3.3. Other Considerations	401
4.	Applications	402
	4.1. Ion Structure Determination	402
	4.2. Thermochemical Determinations	413
	4.3. Fragmentation Mechanisms	417
	4.4. Molecular Structure Determination	422
	4.5. Analysis of Mixtures and Isotope Incorporation	425
	4.6. Product Characterization in Ion–Molecule Reactions and Other Applications	427
5.	Related Reactions	430
	5.1. Negative Ions	431
	5.2. Dissociative Charge Transfer	434
	5.3. Related Ion–Surface Reactions	441

5.4. Related Photodissociations	442
5.5. CID Studied by Ion–Cyclotron Resonance Spectrometry	443
5.6. CID of Neutral Molecules	444
6. Prospects	445
References	446
Index	451

Introduction

R. G. Cooks

This introduction has three purposes: (a) to summarize some of the chief features of energy spectrometry of ions and to sketch in a little of the background to this subject, (b) to present some simple facts about collision processes which one skilled in, say, mass spectrometry but innocent of any knowledge of bimolecular collisions might find of value, and (c) to indicate the scope and content of the volume.

1. The Subject

This book takes as its subject, ion–molecule and ion–atom reactions occurring at high energies. It emphasizes the study of inelastic reactions at high energy through measurements of translational energy. The investigation of these reactions using other procedures has been important in the cases of the simpler systems. In particular, the emitted radiation has been investigated and this subject is therefore discussed where appropriate. For more complex species, however, there is little information available other than from energy spectra.

The defining characteristic of the energy range of interest is that momentum transfer to the neutral target is negligible for small scattering angles. The result of this apparently bland condition is a welcome simplicity in the interpretation of the results of what appears to be developing into a

---

R. G. Cooks • Department of Chemistry, Purdue University, West Lafayette, Indiana 47907.

unique form of spectroscopy. The names ion kinetic-energy spectrometry, translational energy spectrometry, collision spectroscopy, and energy-loss spectrometry have all been used to describe this subject (cf. Section 5). Hallmarks of collision spectroscopy are that it treats isolated and usually readily defined systems with the ability, because of velocity amplification (which is discussed further in Section 5), to make highly refined measurements on the energetics of the system. Laboratory translational energies of several hundred volts and above satisfy the momentum transfer condition just noted. The energy range of interest is bounded at high energies by the onset of nuclear reactions although more mundane experimental limitations are chiefly responsible for the fact that most work has been concentrated near the low end of the energy range of interest. It is also of note that valence electrons are chiefly involved in the reactions in the lower range of energies (low-kilovolt range) and that this is therefore where the reactions of most chemical interest occur. At higher energies, especially above 100 keV, inner-shell excitations occur and emphasis in such experiments has been placed on the behavior of atomic systems.

The high-energy ion–molecule reactions of interest here share a common genesis with their counterparts occurring at energies near or below chemical bond energies. Both can be traced to Thomson's experiments with the parabola mass spectrograph. Low-energy ion–molecule reactions showed spectacular growth in the 1950s and 1960s, stimulated by their relationship to natural atmospheric phenomena; by instrumental developments including tandem mass spectrometers, ion cyclotron resonance, and chemical ionization; and by their similarity to chemical reactions occurring in solution.

Some of the history of high-energy ion–molecule reactions is traced in Chapter 7, Section 1. The subject is deserving of more attention than it has received until recently. The rapid growth of collision physics in the last decade has contributed greatly to our understanding of the interactions of ions, electrons, and neutral atoms. A parallel development in chemistry has occurred during the same period where experiments on molecular beams have emphasized interactions between molecules at low, chemically interesting, translational energies.

It is an apparent paradox that the high-energy reactions of concern in this book can also provide detailed information on reaction mechanisms, energetics, and dynamics and on the structures of the species involved. It is an important objective of this work to substantiate this claim. The paradox can be briefly accounted for if it is noted that the high kinetic energies refer to nuclear kinetic energies. The ion–molecule reactions of interest involve electronic excitations and electron transfer and the internal energy changes associated with these processes appear as changes in translational energy in these isolated systems. Thus, the nuclei serve as templates which register energy changes associated with changes in electronic configuration.

2. Metastable Ions

The ionic collisions which form the subject of this book yield fast product ions with particular distributions of kinetic energy. The unimolecular dissociation of metastable ions gives products with the same characteristics.

The two phenomena can be studied using the same instrumentation, and the interpretation of the widths of peaks in the two sets of energy spectra in terms of kinetic energy release is analogous. In recent years metastable ions have taken on increased importance in mass spectrometry and they are now frequently studied using instruments which are modified to the point where they are more accurately termed energy spectrometers than mass spectrometers.

A few years ago a monograph on metastable ions appeared.[1] This emphasized the unity of the subject of energy spectrometry and, although covering primarily unimolecular (metastable ion) reactions, it also treated high-energy ion–molecule reactions. The present volume may be read as a companion to the first. In particular, treatments of instrumentation for energy spectrometry, of the theory of ionic fragmentations, and of the relationships between kinetic energy release and peak shape given there are not reproduced at length here.

3. Types of Collisions

Because this work is intended for readers having different backgrounds, including those with no previous experience in collision processes, some fundamental concepts are covered here. The reader is referred to other texts[2-6] and to other sections of this one for more detail.

Elementary collision reactions may be exothermic, endothermic, or thermoneutral. *Elastic collisions* are those in which there is no change in the internal energy states of any species. In these reactions there may be and usually are changes in the kinetic energies, momenta, and directions of motion of the collision partners. *Inelastic reactions* include all those in which there is any change in the internal state of the system. These internal energy changes may balance each other so that there is no *net* interconversion of translational and internal energy. In this case the internal energy change will be zero and a condition of resonance may be said to obtain. More generally inelastic reactions will involve nonzero energy changes.

Inelastic reactions which are exothermic are often termed *superelastic* reactions. These processes lead, at zero scattering angles, to a fast product which emerges from the collision with a higher velocity than the fast reactant.

Another type of inelastic reaction is that which leads to a change in chemical bonding. Such processes are said to be *reactive collisions*. The chemical reorganization occurs on the time scale of the collision process. Reactive collisions are of little importance in the energy range of concern here where nuclear motion is slow relative to the time scale of interest so that electronic motion may be considered independently of it (the Born–Oppenheimer approximation). Fragmentation may follow the collision, however, and this phenomenon of collision-induced dissociation is covered (Chapters 6 and 7).

The essential nature of the collisions of interest here is that they involve electron-cloud interactions with relatively small momenta exchange. These have been referred to as soft collisions. Hard collisions, in which there occur large exchanges of momenta between the collision partners, are of less interest. A few cases of such collisions leading to dissociation have been observed but they have very low cross sections. Certainly, complete momentum interchange, as occurs in low-energy ion–molecule reactions which proceed via collision complexes, does not occur. The distinction between ion–molecule reactions which occur at low translational energy and those that occur at high translational energy, therefore, has a basis in the dynamics of the processes.

The usual situation in collision experiments is that a beam of high-energy particles interacts with a target or collision gas confined to a collision or scattering chamber at or near ambient temperatures. In this book, the fast reactant is almost always an ion beam (see, however, Chapter 3, Section 3.1) and the kinetic energies of interest are in the kilovolt range. This makes the assumption of a stationary target an excellent approximation. Energy can be converted from one form to another in collisions, and in particular the translational energy of the fast particle—the laboratory kinetic energy—may in part be converted into internal energy of the system, or vice versa. It is a feature of reactions carried out at high laboratory energies and small scattering angles that any such interconversion of energy *is registered in the energy of the fast particle*. This is an excellent approximation for small scattering angles and for energies in the kilovolt range, and makes the determination of the thermochemistry of the isolated reaction very simple. The endo- or exothermicity of the reaction is determined directly from the translational *energy change* in the fast species. Since many reactions are endothermic, there is often a net conversion of translational into internal energy, the heat of reaction being supplied from the translational energy of the fast particle. This has led to the term *energy-loss spectra* to describe experiments of this type. Exothermic reactions leading to an increase in the translational energy of the fast species are also encountered and the more general term *energy-change spectra* is also appropriate. The distinction between the measured kinetic energy change and the actual exo- or endoergicity of the process should be borne in mind even though the two quantities may be equal to a good approximation under the conditions of interest here. The heat of reaction is given as Q and the experimental quantity, the energy change, is given as ΔE. The validity of the approximation $\Delta E \cong Q$ is determined by the kinetic energy of the fast particle, the masses of the species, and the scattering angle. It may be expressed as a disproportionation factor, that is, the ratio in which the energy change is associated with the fast particle and the target. This factor, as shown in Chapter 4, Appendix, is given at zero scattering angle by Eq. (3.1) where m_1 and m_2 are the masses of the projectile and target, respectively,

$$D = \frac{4m_2 E_0}{m_1 Q} \quad (3.1)$$

and E_0 is the initial laboratory translational energy of the fast species.

4. Collision Cross Sections

The extent to which a given process occurs on collision is conventionally expressed as its cross section. This is a measure of the probability that interaction will lead to the specified products, the nature of the interaction and particularly the impact parameter (see later) being determined statistically. Cross sections are given in units of area per atom or molecule.

A cross section is perhaps best thought of as an absorptivity leading to a specified final state of the system. The analog of the Beer–Lambert law of optical absorption holds and can be employed in measuring cross sections from experimentally accessible parameters. For a beam of intensity I passing through a gas of number density n, the decrement dI in beam strength is proportional to the number density, the distance dl, and the cross section σ, i.e.,

$$-dI = In\sigma \, dl \quad (4.1)$$

integrating

$$-(\log I - \log I_0) = n\sigma l \quad (4.2)$$

or

$$\frac{I}{I_0} = e^{-n\sigma l} \quad (4.3)$$

where I_0 is the intensity at $l = 0$.

The factors which control reaction cross sections occupy much of the discussion herein. One simple generalization is that collisions which involve the interconversion of large amounts of internal and translational energy are relatively less likely. The closer the approach to resonance the higher the reaction cross section will tend to be.

Related to the concept of cross section is that of the impact parameter—the closest distance of approach of the collision partners where each is considered as a mass point (see Chapter 3, Fig. 8 for a precise definition). Statistically, collisions with large impact parameters are more likely than those with small impact parameters. The correlation between cross section and impact parameter follows because processes with large cross section (and small energy defects) can occur at large impact parameters. Expressed differently, reactions that require only soft collisions have larger cross sections.

In plotting collision data reduced coordinates are often used. Of particular importance is the reduced scattering angle τ, defined by

$$\tau = E_0 \theta \tag{4.4}$$

where E_0 is the laboratory ion energy and θ is the laboratory scattering angle. At small scattering angles, τ is primarily a function of the impact parameter b.

5. Energy Interconversions

The central role played by measurements on the kinetic or translational energies of the species involved in high-energy ion–molecule or ion–atom collisions underlines the fact that we are dealing with a form of energy spectrometry. Energy loss or more generally energy change spectra (Section 3) emphasize the interconversion of translational and internal energy on collision. Another phenomenon, the results of which also show up indelibly in an energy spectrum, is the dissociation of a fast particle with release of some of its internal energy as relative translational energy of the fragments. This process often follows a collision and it is worth emphasizing the differences between the interconversion of translational and internal energy (a) on fragmentation and (b) in an inelastic collision.

In the latter case the energy change is associated almost entirely with the fast particle and occurs in approximately the original direction of motion. (Large-angle scattering of the fast particle requires considerable momentum transfer to the target and is of less interest here.) In the former case, the energy change is associated with two fragments. This energy change is *amplified* in the laboratory system so that broad peaks can be observed corresponding to small energy releases in the center of mass.

The kinetic energy release T is given[1] for the generalized reaction $m_1^+ \to m_2^+ + m_3$ by Eq. (5.1), where Ve is the kinetic energy of m_1^+, m_2 and m_3 are the masses of the products m_2^+ and m_3, and $\delta E/E$ is the fractional width of the energy peak. E can be measured in any convenient units:

$$T = \frac{m_2 Ve}{16 m_3} \left(\frac{\delta E}{E} \right)^2 \tag{5.1}$$

The range of measured ion energies δE is greater than the kinetic energy release T which causes it, by an amplification factor A given by

$$A = 4 \frac{(m_2 m_3)^{1/2}}{m_1} \left(\frac{Ve}{T} \right)^{1/2} \tag{5.2}$$

The large size of this factor is the underlying reason for the precision with which kinetic energy releases can be measured (cf. Chapter 7).

Historically, the study of energy spectra to determine kinetic energy releases began with metastable ions. Diffuse peaks observed in mass spectra (which are actually momentum spectra) have their characteristic shapes because of kinetic energy release. More recently such processes have been explicitly examined by determining directly the energy spectra of the product ions. The name ion kinetic energy spectra came to be associated with this field of study. Fragmentation reactions in which the dissociating species is activated by collision have characteristics analogous to those of metastable ion reactions and so the subject matter of ion kinetic energy spectrometry has widened to include ion–molecule reactions of these types as well as charge exchange processes. Meanwhile the subject of collision physics was changing to emphasize energy loss spectra, and then to encompass fragmentation reactions with the introduction of diatomic and polyatomic ions and targets. The term translational energy spectroscopy came to be associated with such studies. A single field of study is clearly emerging from these two streams and the term collision spectroscopy describes it adequately.

6. *Energy Resolution and Angular Resolution*

Measurements of both the angular and the kinetic energy distributions of the products of collisions are necessary to define completely a collision in the general case. In the simplest measurements, data may be taken for a small range of scattering angles distributed about zero and the experiments interpreted in terms of the associated energy spectra. Data may also be taken as a function of scattering angle, and such processes are discussed in Chapters 2 and 3.

In addition to possible changes in the dynamics of the collision as a function of scattering angle, there is a more practical reason why this is of concern. Translational energy spectra are generally measured using an energy analyzer which measures only the component of energy in a plane of analysis. Kinetic energies which are associated with particles moving in three dimensions are measured using two-dimensional analyzers. To the extent that ions having different energies may have the same component in a particular plane, energy resolution is degraded by the angular acceptance of the instrument. Although the laboratory scattering angles of interest in much of this work are small because of the high reactant-ion energies and the interest in glancing interactions, high-energy resolution such as is required to observe vibrational fine structure requires that the angular resolution also be high. In particular, ions moving at large angles relative to the plane of analysis must be rejected. Ions traveling within the plane of analysis but at an angle relative to the central ray will be correctly analyzed (to first order) by an electric sector if collisions occur at the point of focus of the sector. To the extent that collisions are not localized some degree of angular discrimination in the plane of analysis will also be necessary.

7. Laboratory and Center-of-Mass Coordinate Systems

Many of the reactions of interest in this volume can be described in the laboratory system. The center-of-mass system is introduced here, however, because it is more useful for measurements differential in scattering angle. It is also used to clarify the dynamical consequences of conversion of internal to translational energy, and vice versa, both for inelastic collisions and for fragmentations of fast-moving particles.

Given a system in which m_1 and m_2 undergo a collision, the origin of the laboratory frame of reference may be taken as the collision point. The velocities of the collision partners are v_1 and v_2. Prior to the collision the laboratory coordinates of m_1 and m_2 are given by the vectors $\mathbf{R}_1$ and $\mathbf{R}_2$. The center of mass of this isolated system moves at a constant velocity relative to the laboratory (along a line given by the path of m_1 for the case of chief interest here, where $v_2 = 0$). Relative to the center of mass, which is at some distance R_{cm} from the laboratory origin at unit time prior to collision, the coordinates $\mathbf{r}_1$ and $\mathbf{r}_2$ of m_1 and m_2, respectively, are vectors

$$\mathbf{r}_1 = \frac{m_2}{m_1 + m_2}(\mathbf{R}_1 - \mathbf{R}_2) \qquad (7.1)$$

and

$$\mathbf{r}_2 = \frac{-m_1}{(m_1 + m_2)}(\mathbf{R}_1 - \mathbf{R}_2) \qquad (7.2)$$

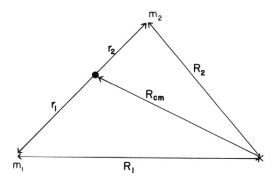

Fig. 1. Laboratory and center-of-mass coordinate systems of m_1 and m_2 prior to collision (which will occur at the laboratory origin $*$). $\mathbf{R}_{cm}$, the coordinate of the center of mass in the laboratory system, is equal to $-\mathbf{r}_{lab}$, the coordinate of the laboratory in the c.m. system. The c.m. origin is shown as ●.

Hence, by the vector addition shown in Fig. 1, in the laboratory frame, we have

$$\mathbf{R}_1 = \mathbf{R}_{cm} + \mathbf{r}_1 \tag{7.3}$$

and

$$\mathbf{R}_2 = \mathbf{R}_{cm} + \mathbf{r}_2 \tag{7.4}$$

Equations (7.1) and (7.2) can be written as the corresponding forms (7.5) and (7.6) involving center-of-mass (c.m.) velocity vectors $\mathbf{u}_1$ and $\mathbf{u}_2$ by time differentiation:

$$\mathbf{u}_1 = \frac{m_2}{m_1 + m_2}(\mathbf{v}_1 - \mathbf{v}_2) \tag{7.5}$$

$$\mathbf{u}_2 = \frac{-m_1}{(m_1 + m_2)}(\mathbf{v}_1 - \mathbf{v}_2) \tag{7.6}$$

By definition the sum of the c.m. momenta of m_1 and m_2 must be zero:

$$\left(\frac{m_1 m_2}{m_1 + m_2}\right)(\mathbf{v}_1 - \mathbf{v}_2) - \left(\frac{m_1 m_2}{m_1 + m_2}\right)(\mathbf{v}_1 - \mathbf{v}_2) = 0 \tag{7.7}$$

Because of conservation of momentum, the total laboratory kinetic energy of the fast particle undergoing collision with a stationary target is not available for conversion into internal energy of the products. The available energy is termed the relative kinetic energy. It is equal to the total kinetic energy in the c.m. reference frame since total linear momentum in the c.m. is zero. From Eqs. (7.5) and (7.6) the sum of the c.m. kinetic energies of m_1 and m_2 is

$$\frac{1}{2}m_1\left(\frac{m_2}{m_1+m_2}\right)^2(g^2)+\frac{1}{2}m_2\left(\frac{m_1}{m_1+m_2}\right)^2(g^2)$$

$$=\frac{1}{2}\frac{m_1 m_2}{m_1+m_2}g^2\left(\frac{m_2}{m_1+m_2}+\frac{m_1}{m_1+m_2}\right)$$

$$=\tfrac{1}{2}\mu g^2$$

where μ is the reduced mass and g is the relative velocity of m_1 and m_2, i.e., $\mathbf{v}_1-\mathbf{v}_2$. If m_2 is stationary, then $g=\mathbf{v}_1$ and the relative kinetic energy is $\tfrac{1}{2}\mu v_1^2$.

The limitations which relative kinetic energies place upon energy interconversions in collisions are most important in low-energy ion–molecule reactions. In the systems of interest here, the energy converted from one form to another seldom exceeds 1% of the laboratory energy.

We can now illustrate, with the aid of two-dimensional Newton diagrams of the type shown in Fig. 2, the consequences for the laboratory kinetic energies of the products when internal energy is converted into translational energy (a) upon collision at high relative velocity and (b) upon spontaneous fragmentation.

For an elastic collision translational energy is conserved so we have in lab coordinates

$$\tfrac{1}{2}\mu g^2 = \tfrac{1}{2}\mu g'^2 \tag{7.8}$$

or

$$|\mathbf{v}_1-\mathbf{v}_2|=|\mathbf{v}_1-\mathbf{v}_2|' \tag{7.9}$$

where the prime refers to final states of the system. That is, relative velocity is conserved in an elastic collision and from (7.5) and (7.6), the c.m. velocities $\mathbf{u}_1$ and $\mathbf{u}_2$ are also conserved. For an elastic collision the products must lie on circles of radii $\mathbf{u}_1$ and $\mathbf{u}_2$ in the velocity vector diagram, Fig. 2. Measurement of just one laboratory scattering angle defines the behavior of the system since m_1 and m_2 are scattered through the same c.m. angle Θ.

In the case of particular interest the target is stationary in the laboratory system, i.e., $\mathbf{v}_2=0$, and elastic scattering is described by the simpler Newton diagram, Fig. 3. This also illustrates the second characteristic with which we are chiefly concerned, that is, small laboratory scattering of the fast species. It will be seen that this results in small c.m. scattering angles for m_1 and m_2 and a large laboratory scattering angle for the target m_2.

Consider now an inelastic collision between a high-velocity species m_1 and a stationary target m_2 where the processes of experimental interest (a) lead to small laboratory scattering angles and (b) involve energy interconversions which represent a small fraction of the laboratory kinetic energy of

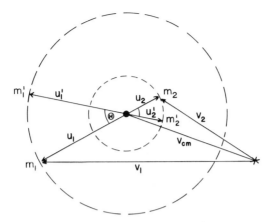

Fig. 2. Newton diagram for elastic scattering in the general case. Subject to constraints inherent in the potential energy function and to angular momentum considerations, the products m_1 and m_2 may occur at any points on the large and small circles joined by a line passing through the center of mass. One particular result is given showing velocity vectors before and after collision (primes for final state of the system).

the fast particle. The reaction is accompanied by the conversion of Q of internal energy into translational energy. In the case illustrated, Q is positive and the reaction is of the superelastic type, but the results for kinetic energy loss are analogous.

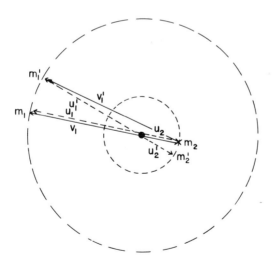

Fig. 3. Newton diagram for elastic scattering for the stationary target case. The c.m. scattering angle Θ is that subtended by $\mathbf{u}_1$ and $\mathbf{u}_1'$ whereas the laboratory scattering angle θ of the fast species is that subtended by $\mathbf{v}_1$ and $\mathbf{v}_1'$.

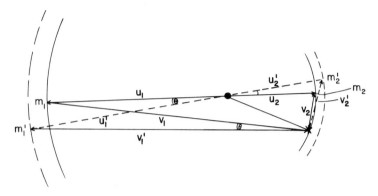

Fig. 4. Newton diagram for inelastic scattering at small laboratory angles (θ) of a fast ion m_1 off a stationary target m_2. The amount of energy converted from internal to translational is a small fraction of the initial translational energy. The full circles shown are for the product velocity vectors in the c.m. Their radii are therefore different from those (dashed circles) for the elastic collision.

Conservation of energy requires that the translational energy (c.m.) prior to collision equal that after collision plus the internal energy change, i.e.,

$$\tfrac{1}{2}m_1 u_1^2 + \tfrac{1}{2}m_2 u_2^2 = \tfrac{1}{2}\mu g^2 = \tfrac{1}{2}\mu v_1^2$$
$$= Q + \tfrac{1}{2}m_1(u_1')^2 + \tfrac{1}{2}m_2(u_2')^2 = \tfrac{1}{2}\mu(g')^2 \qquad (7.10)$$

Equation (7.10), together with the fact of small laboratory scattering angles, means that the changes in $\mathbf{u}_1$ and $\mathbf{u}_2$ on going to $\mathbf{u}_1'$ and $\mathbf{u}_2'$ must be small. This leads to the situation shown in Fig. 4 for inelastic scattering.

It is interesting to compare the situation shown in Fig. 4, in which internal energy is released as translational energy in a collision, with that illustrated in Fig. 5, in which internal energy is released as translational

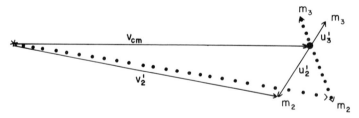

Fig. 5. Newton diagram for fragmentation of a fast particle m_1 accompanied by kinetic energy release T, illustrating two arbitrary angles at which fragmentation can occur and the resulting variation in the laboratory velocities of the fragments m_2 and m_3 and velocity amplification of T. Laboratory origin (∗) is taken as the point of fragmentation; the c.m. origin is denoted as in Fig. 1 by ●.

energy upon fragmentation of a single species. The kinematics of these two cases are quite different: In the latter case the isotropic energy release in the c.m. system causes a wide range of laboratory energies for the two fragments. In the former case, it is chiefly the laboratory energy of the fast fragment which is changed, the change being equal to a good approximation to the energy release.

Let us examine the ion–molecule reaction (Fig. 4) a little more closely. The change in translational energy represented by the changed relative velocity vector is greatly exaggerated for clarity (experimentally the energy change is <1% of the laboratory energy), as is the initial laboratory velocity of the target m_2.

Since momentum is conserved in the c.m. we have

$$m_1\mathbf{u}_1 = -m_2\mathbf{u}_2 \quad \text{and} \quad m_1\mathbf{u}_1' = -m_2\mathbf{u}_2'$$

This fixes the changes in $\mathbf{u}_1$ and $\mathbf{u}_2$ as the difference in the radii of the full and broken circle segments.

Limiting consideration to small laboratory angle scattering, it is apparent from the figure that if $\tfrac{1}{2}m_2v_2^2$ is negligible, so too is $\tfrac{1}{2}m_2(v_2')^2$. Hence, after collision the laboratory energy of the system is given to a good approximation by

$$\tfrac{1}{2}m_1(v_1')^2 = \tfrac{1}{2}m_1(v_2)^2 + Q \tag{7.11}$$

Since the energy change (ΔE) in the fast particle is defined by

$$\Delta E = \tfrac{1}{2}m_1(v_1')^2 - \tfrac{1}{2}m_1(v_1')^2 \tag{7.12}$$

one sees that

$$\Delta E \approx Q \tag{7.13}$$

Figure 5 shows fragmentation occurring in two randomly chosen directions for a species (m_1) moving with velocity $\mathbf{v}_{cm} = \mathbf{v}_1$ in the laboratory system. Because fragmentation can occur in all directions, the resulting laboratory velocities $\mathbf{v}_2'$ and $\mathbf{v}_3'$ of the fragments will clearly cover a wide range and show up as a broad peak in a measured velocity spectrum. A further important point relating to fragmentation of the fast species m_1 concerns velocity amplification of the energy release. Conservation of energy in the center-of-mass system means that the kinetic energy release represents the only postfragmentation c.m. energy

$$T = \tfrac{1}{2}m_2(\mathbf{u}_2')^2 + \tfrac{1}{2}m_3(\mathbf{u}_3')^2 \tag{7.14}$$

Moreover, the velocity of the center of mass in the laboratory system is, as shown in Fig. 5, very large in the cases of interest. Hence, for forward and backward scattering (defined with respect to the direction of motion of the center of mass), the laboratory velocities of the products are given by

$$\mathbf{v}'_2 = \mathbf{v}_{cm} \pm \mathbf{u}'_2 \tag{7.15}$$

$$\mathbf{v}'_3 = \mathbf{v}_{cm} \pm \mathbf{u}'_3 \tag{7.16}$$

Thus, although the range of c.m. energies of m_2^+ arising from the fact of isotropic fragmentation is a function of $\mathbf{u}'_2$, the corresponding range of laboratory energies is a function of $\mathbf{v}_{cm} + \mathbf{u}'_2$. This increased energy on changing coordinate systems constitutes velocity amplification (cf. Section 5).

8. *Guideposts to the Contents*

To assist the reader in locating material of particular interest, this section briefly describes the coverage of each chapter. Broadly, the sequence followed is that of increasing complexity—atomic, diatomic, and polyatomic systems and collisions involving excitation, charge change, and dissociative excitation.

Chapter 1, on collisional excitation of simple systems, covers electronic excitation of the ion or the target in elastic collisions in the kilovolt energy range (up to 200 keV). Charge exchange, ionizing collisions, and dissociations are not treated. The emphasis is on the simplest atom systems, comparisons of the experimental results with theory being emphasized. Section 1.2 contains a concise description of the elements of collision theory. The excitation of some diatomic molecules is also covered and brief coverage of the proton excitation of organic targets is provided. The chapter covers two experimental techniques for the study of excitation—translational energy spectroscopy in the form of energy-loss spectra and photon emission measurements. The former, for which cross-section measurements are rather simply made, provides direct information on the collision itself but is limited in terms of energy resolution. Cross-section data for the emission measurements are subject to more systematic uncertainties but the energy resolution is high.

Chapter 2 covers charge transfer in atomic systems from both a theoretical and an experimental point of view. Single electron transfers between singly and doubly charged ions and atomic targets are emphasized. Information on potential energy functions for these reactions is experimentally accessible by measuring cross sections which are differential in both scattering angle and kinetic energy. The theoretical treatment emphasizes the use of crossings in diabatic potential functions to account for small impact parameter reactions. Coupling between reactant and product states at larger impact parameters is shown to provide an alternative noncrossing mechanism of charge transfer.

Chapter 3 deals with both charge exchange and excitation reactions, chiefly for atomic species. It emphasizes the value of translational energy measurements as a source of data on the primary collision event in contrast to measurements on emitted radiation which refer to subsequent deexcitation processes. The chapter describes a molecular orbital theory of inelastic collisions which seeks to construct time-variable MOs for the dynamic system and emphasizes the role of diabatic states in these phenomena. The system may exit on the same MO or, for an inelastic collision, on a different one. Curve crossing probability will depend on the energy between the states and the relative velocity of the particles. In addition to nonadiabatic electronic transitions between states of the same symmetry in the vicinity of avoided crossings various coupling mechanisms may promote transitions from one state to another. In particular, radial coupling due to the relative orbital angular momentum of the collision partners and rotational coupling occur.

The chapter also details the time-of-flight approach which has been employed to study the fast neutral products of inelastic collisions. It is shown that this represents a form of translational energy spectrometry which complements measurements on ionic products and so greatly aids in the elucidation of the behavior of many systems. An important section of the chapter deals with ion–photon and photon–photon coincidence techniques. These methods are experimentally demanding but may provide the most complete information yet available on collision dynamics.

Chapter 4 deals with the charge inversion of positively charged ions. The reaction is characterized in terms of the kinetic energy spectra of the product anions. Much of the work describes the use of proton beams with the selection rules for the electron-transfer process and the final states of the doubly charged target being of interest. Applications of this measurement to the determination of the double-ionization potentials of molecules, the state composition of ion beams, and the structures of polyatomic ions are described. A valuable appendix describes the dynamics underlying energy-loss measurements.

Chapter 5 shows how several modified types of mass spectra, which depend on the occurrence of ion–molecule reactions, can be recorded. Ions which undergo a particular type of charge-changing reaction are distinguished from all others by their unique energy-to-charge ratios. By appropriate adjustment of the electrostatic analyzer of a mass spectrometer it can readily be arranged that only these ions will be transmitted. The resulting spectra are a unique source of information on the ionic chemistry of polyatomic ions, including information on ion structures.

Chapter 6, on collision-induced dissociation of diatomic ions, emphasizes the mechanisms of this process and provides examples of the use of translational energy measurements in determining energies of ionic states with an accuracy which approaches that characteristic of spectroscopic measurements on neutral molecules. A detailed understanding of the collisional excitation phenomenon is sought through the close study of a few very simple molecular ions, especially H_2^+, HHe^+, and N_2^+. Much of the information obtained promises to enhance our understanding of the behavior of more complex ions. Electronic and vibrational/rotational excitations are treated and special emphasis is given to predissociation phenomena. Anisotropic effects in the collision are shown to provide a route to detailed dynamical information. Photodissociation of fast ions, which is intimately related to collision-induced dissociation, is shown to possess unique advantages. The relationship of the kinetic energy methodology developed in this chapter to several modern techniques in molecular photochemistry and photophysics is noted.

Chapter 7 treats those ion–molecule reactions of polyatomic ions which result in fragmentation. It traces something of the history of studies on ion–molecule reactions and treats qualitatively the various mechanisms which can lead to dissociation. Emphasis is given to the effects of experimental variables on the nature of the cross sections of collision-induced dissociation. Experimental methods for optimizing this type of study are presented. Applications are cited involving the study of isomerization in nonfragmenting ions, ion structure elucidation, qualitative analysis, and thermochemical determinations. Related reactions involving collisions at surfaces and photodissociation are briefly treated as are processes in which both fragmentation and charge inversion occur.

ACKNOWLEDGMENTS

The support of the National Science Foundation of the work done at Purdue University is gratefully acknowledged. So too are the numerous contributions of other individuals in this laboratory to the development of this subject. It has been a special privilege to have been associated with Professor J. H. Beynon.

References

1. R. G. Cooks, J. H. Beynon, R. M. Caprioli, and G. R. Lester, *Metastable Ions*, Elsevier, Amsterdam, 1973.
2. J. Hasted, *Physics of Atomic Collisions*, American Elsevier, New York, 1972.
3. H. S. W. Massey and H. B. Gilbody, *Electronic and Ionic Impact Phenomena*, 2nd ed., Oxford University Press, London, 1974.

4. E. W. McDaniel, V. Cermák, A. Dalgarno, E. E. Ferguson, and L. Friedman, *Ion–Molecule Reactions*, Wiley-Interscience, New York, 1970.
5. E. W. Thomas, *Excitation in Heavy Particle Collisions*, Wiley-Interscience, New York, 1972.
6. E. W. McDaniel, *Collision Phenomena in Ionized Gases*, Wiley, New York, 1964.

1

Collisional Excitation of Simple Systems

John T. Park

1. Introduction

In any encounter between an ion and an atom, an exchange of energy can occur. Because the energy required to excite a ground-state atom to a higher state is usually small compared to the energy of an ion from an accelerator or in a plasma, the excitation of the target and/or the projectile is the most common inelastic process in most collisions. Emission from the excited target gas is a common indicator of a collisional process. In this chapter we are primarily concerned with the characterization of the direct excitation process in the simplest ion–atom collisions.

Although the excitation process is very common, even the simplest of ion–atom collisions provides difficulties in interpretation. Just to identify completely the processes which occur, the state of both the projectile and atom before and after the collision, the trajectory of both, and the state and trajectory of any ejected particles must be known. The experiments should be arranged so that the precollision conditions are known. However, after the collision the state and trajectories of the particles are generally unknown. Direct excitation is only one of the many possible results of the collisional process. The determination of the direct excitation cross section from experimental measurements is complicated by the large variety of

---

John T. Park • Department of Physics, University of Missouri, Rolla, Missouri 65401.

possible configurations of final states. The experimental data must be carefully analyzed to eliminate the effects of contributions to the detected signals from these competing processes.

The discussion in this chapter is limited to ion–atom collisions in the kiloelectron volt range. The adiabatic criterion of Massey[1] indicates that a cross section will be small if the ion velocity is much less than $a|Q|/h$, where a is the radius of the first Bohr orbit of the hydrogen atom, Q is the energy defect of the process, and h is Planck's constant. At lower kinetic energies, the ions have velocities that are small compared to the orbital velocity of the target atom's electrons. The ion–atom system is therefore able to accommodate itself adiabatically to the relatively gradual change in separation. At high impact energies, the Born approximation provides a satisfactory interpretation of the collision. The energy region addressed in this chapter covers the ion energies between these extremes. In this energy range, the excitation cross section normally reaches its maximum.

1.1. Types of Collisions

The notation of Hasted[2] with only slight modifications is employed in the discussion. The symbols for the participants in the collisional process are listed in Table I. The convention resulting from Hasted's notation indicates the charge and state of excitation of the participants both before and after the collision. The conditions before and after the collision are separated by a solidus. The incident particle is listed first, and the order remains the same on both sides of the solidus. Any new particles resulting from the collision are listed after the original particles. We can identify the cross section for a particular process using the shorthand notation to indicate the process involved. As an example, the cross section for target excitation to the j state, $X^+ + Y \to X^+ + Y_j^*$, $(10/10_j^*)$ is given by $\sigma(10/10_j^*)$. This notation clearly identifies the cross section involved.

Table I. List of Symbols

Φ_{ij}	Photon from a transition from the state i to the state j
e	Electron
0	Neutral atom
(00)	Neutral molecule
0^*	Excited atom; state not specified
0_j^*	Excited atom in state j
1	Singly ionized ion
1_j^*	Excited singly ionized ion in the state j
2, 3, ...	Multiply ionized ion
$\bar{1}$	Negative ion
(01)	Singly ionized molecule
$(00)_v^*$	Vibrationally excited molecule

The formation of excited atoms or ions can result from several processes; a partial list is given in Table II. The dominant mechanism in most collisional systems is direct excitation. The other processes included in Table II must of course be considered, because they provide alternate channels for excitation, which can easily result in misinterpretation of the observed emission. Cascade processes and charge-transfer excitation are most important from the view of avoiding errors in interpretation.

1.2. Elements of Collision Theory

Theoretical studies of the collisional excitation of atoms by ions have met with serious difficulty. The success achieved in electron–atom collisions has not been matched in ion–atom collisional excitation. The incident ion can capture the orbital electrons from an atom. The ion itself has structure and may have bound electrons which can be exchanged with those of the

Table II. Excitation Mechanisms

Notation		Process
$(10/10^*)$	$X^+ + Y \to X^+ + Y^*$	Direct excitation of the target
$(10/1^*0)$	$X^+ + Y \to (X^+)^* + Y$	Direct excitation of the ion
$(10/1^*0^*)$	$X^+ + Y \to (X^+)^* + Y^*$	Direct simultaneous excitation of both target and incident ion
$(10/10_k^*/10_j^*\Phi_{kj})$	$X^+ + Y \to X^+ + Y_k^* \to X^+ + Y_j^* + \Phi_{kj}$	Excitation cascade
$(10/11^*e)$	$X^+ + Y \to X^+ + (Y^+)^* + e$	Simultaneous excitation and ionization of the target
$(10/1^*1e)$	$X^+ + Y \to (X^+)^* + Y^+ + e$	Simultaneous excitation of the ion and ionization of the target
$(10/1^*1^*e)$	$X^+ + Y \to (X^+)^* + (Y^+)^* + e$	Simultaneous excitation of both target and ion with ionization of the target
$(10/0^*1)$	$X^+ + Y \to X^* + Y^+$	Charge-transfer into an excited state
$(10/01^*)$	$X^+ + Y \to X + (Y^+)^*$	Charge-transfer with simultaneous target excitation
$(10/0^*1^*)$	$X^+ + Y \to X^* + (Y^+)^*$	Charge-transfer with simultaneous excitation of both target and projectile
$(1(00)/1(00)^*)$	$X^+ + YZ \to X^+ + YZ^*$	Excitation of a molecule
$(1(00)/10^*0)$ $((10)0/10^*0)$	$\left\{ \begin{array}{l} X^+ + YZ \to X^+ + Y^* + Z \\ XY^+ + Z \to X^+ + Y^* + Z \end{array} \right\}$	Dissociative excitation
$(1(00)/110^*e)$	$X^+ + YZ \to X + Y^+ + Z^* + e$	Dissociative excitation with simultaneous ionization
$(1(00)/1(00)^*/10^*0)$	$X^+ + YZ \to X^+ + YZ^* \to X^+ + Y^* + Z$	Excitation into a dissociating state

atom. These additional possibilities must be incorporated into any theory which represents the collisional process.

The theoretical techniques applied are discussed in texts on scattering theory. See, for example, R. G. Newton *Scattering Theory of Waves and Particles*.[3] The basic problem in a theoretical study of any collisional process is to describe the transition between the initial conditions, where an incoming ion strikes a stationary target atom, and the final conditions, where an outgoing ion leaves an excited atom. Theoretically, the problem involves the calculation of a transition matrix (T-matrix) for the purpose of describing the collisional process. In the simplest case, this calculation includes an integral over all space and involves the wave function that describes the initial system, the wave function that describes the final system, and the interaction potential between the projectile and the target. The incident ion is initially assumed to be a free particle with kinetic energy and momentum; hence the wave function describing the incident ion is usually taken to be a plane wave. If the wave function representing the outgoing ion and target atom system could be given exactly, the problem could be solved; however, the wave function representing the final system now contains information about the interaction.

To proceed further, some simplifying approximation is required. The most frequently applied theoretical approach involves the first Born approximation. The Born approximation requires that a plane wave, which has the final momentum and kinetic energy of the scattered ion, be substituted for the more complex wave function, which includes the information about the interaction and truly represents the final system of the scattered ion and the excited target. The final kinetic energy must equal the initial kinetic energy minus the energy required for any excitation or other inelastic process. Because the scattered ion is treated like a free particle when it is represented by a plane wave, the Born approximation can only be expected to be valid when the incident kinetic energy is very large compared to any interaction potential. This limits the validity of the first Born approximation to collisions with ions of high incident kinetic energy.

It is obvious that if a collision takes place, the wave function representing the scattered ion must include the effects of the interaction potential to be a true solution to the Schrödinger equation for the collisional process. The scattered wave function can be expanded in a series which represents successively higher order corrections to the plane wave function. In the second Born approximation, the first correction term is retained as well as the plane wave. Each additional term of the series expansion, which is retained, gives a better representation of the scattered ion, but the intergrals become more and more difficult. Also, this Born series appears to be an asymptotic approximation, and the retention of successively higher terms does not rapidly improve the results of the calculation.

Another technique, which is used to incorporate the effects of the interaction potential into the wave function representing the scattered ion, is the eikonal approximation. The plane wave is modified or distorted to take into account the fact that the scattered ion is influenced by the interaction potential both as it approaches and leaves the target atom. Both the incident and scattered wave functions are no longer plane waves; hence it is expected that the eikonal approximation should be valid at lower impact energies than the first Born approximation.

The eikonal approximation is applied with several different forms of the potential that are used to distort the plane waves. The Glauber approximation is usually employed when the incident ion has no electronic structure, for example, a proton or α particle. In the Glauber approximation, the distorting potential is the interaction potential itself. The projectile ion is assumed to travel in straight-line paths, and the momentum transferred is assumed to be perpendicular to the projectile's path; thus momentum is not conserved. This restricts the application of the Glauber approximation to certain substates of the target; however, the results of the Glauber approximation compare favorably with experimental results. Other forms of the eikonal approximation that have been developed allow for conservation of momentum. These forms usually employ an average potential to account for the distortion of the plane wave. The results of these approximations also compare favorably with experimental results.

In any ion–atom collision, because many different possible reactions occur, the scattered wave function should be a composite of the various wave functions which represent the different reactions. The coefficients for a particular wave function are related to the probability that the collision results in the particular reaction represented by that wave function. This approach is called the close coupling approximation. In the simplest case, it is assumed that there are only two possible reactions: the elastic scattering of the projectile and the particular reaction of interest. This yields a 2×2 matrix which represents the potential interaction. This results in a set of coupled equations which, when solved approximately, is called the distortion approximation. The more general close coupling studies involve several reactions beyond the one of particular interest. The sets of coupled equations result in a potential interaction matrix in which each element is related to the probability of the collision undergoing one of the possible interactions. When the equations are solved numerically in a computer, the cross sections for all of the possible reactions are obtained simultaneously. Because of the complexity of the potential matrix when several reactions are included, the number of reactions that can be handled is limited by the ambition of the investigator and the availability of computer time.

The close coupling approximation is usually applied with an additional approximation to assist in solving the coupled equations. The projectile ion is assumed to follow a straight-line trajectory, which passes a fixed distance (impact parameter) from the target. The differential equations are solved for a given impact parameter. In this way, a solution is found as a function of the impact parameter, which is related to the total cross section. This impact parameter approach provides results that are in favorable agreement with experimental results if enough reactions can be included in the potential interaction matrix that is used to solve the coupled differential equations.

Most of the theoretical effort has been applied to proton–atomic hydrogen collisions. For this type of collisional system, the theory is almost as well developed as the experiment. In other collisional systems, the theoretical attempts are quite limited. Nevertheless, considering the difficulties, notable progress has been made. Of the possible collisional processes which can be studied, direct excitation involves fewer complications than processes such as ionization. Absolute experimental data in ion–atom collisional excitation are needed to guide theoretical development in this difficult field.

1.3. Experimental Methods

There are two major experimental methods used for obtaining information on excitation processes during collisions. The two are distinguished by the particle detected. In emission experiments, a secondary particle, usually the emitted photon, is detected. In energy-loss experiments, the primary particle itself is detected, and the desired information is obtained from its energy distribution.

The most frequently employed method involves detection of the photons emitted from the excited states resulting from the exciting collision. This technique has been employed for nearly 50 years, but few quantitative results were obtained until recently when high-quality photomultipliers became available. The technique has the advantage of high resolution and sensitivity. Its major disadvantage is that it allows one to look only at the emission process not at the excitation process directly. It is therefore difficult to study excitation of metastable states. If the system is complex, it is often impossible from the observed emission to identify unambiguously the process responsible for the excitation, because direct excitation with the cross section $\sigma(10/10_j^*)$ represents only one of the many possibilities that leave the target atom in an excited state after the collision. Even in the case of direct excitation, the detected photon comes from a two-step process $(10/10_j^*/10_i^*\Phi_{ji})$, and frequently the emission detected is from only one of several possible transitions from the excited state.

The basic optical detection technique has been modified to solve special problems. The use of modulated crossed-beam techniques has expanded the application of the basic technique to unstable and reactive targets, particularly atomic hydrogen targets. The recent use of coincidence techniques to establish a definite relationship between the emitted photon and the projectile or other particle greatly assists in determining the specific collision process which produces the excitation. The angular dependence of the cross section for the excitation process can also be obtained by using coincidence techniques. The coincidence techniques are included in the general grouping with the optical techniques, even when both the projectile ion and photon are detected. This is because the cross-section measurement still depends on the absolute calibration of the photon detector rather than on the energy resolution of the particle detector.

The second major method involves the observation of the energy lost by the incident projectile itself. This technique has been employed for several years, although it has only very recently been extended into the intermediate (15–200 keV) energy range. The advantage of this technique is the unambiguous identification of the excitation process. The measurement provides cross sections for excitation directly but gives no information on the emission process. The technique can simultaneously provide additional information on the scattering of the ions. Because the energy-loss technique is not dependent on the optical selection rules involved in emission, the technique can also be used to study the excitation to metastable levels whose decay transitions are optically forbidden. The disadvantage of the technique is its lack of resolution and sensitivity. Energy resolutions between 0.2 and 2 eV are typical. This permits the study only of the excitation of dominant states that are energetically isolated from other states.

The purpose of the experiments in which either technique is used is to relate the detected signals to the fundamental parameters of the collisional process. The cross section σ for a specific process is the constant of proportionality that relates the number of interactions, n_i, to the product of the number of incident particles, N, the number density of the target gas, n, and the path length L: $\sigma = n_i/NnL$. In an actual experiment, all the terms in this equation provide difficulties in measurement. The detected events giving n_i may well not be due to the interaction under study but may be the result of some competing process. The number N of incident particles is depleted by charge and state changes as the beam passes through the gas. The number density n of target atoms is also depleted by collisional processes. If N and n are kept sufficiently small, these multiple collision effects may be made negligibly small. The conditions in which n_i is linearly dependent on n and N are called single-collision conditions. Demonstration of the linearity is required of all experiments unless a more sophisticated analysis of the data is applied.

The effects of multiple collisional processes and competing processes are not demonstrated in the same way in both techniques, and hence the two techniques can complement each other. The two techniques together can provide a rather complete picture of the excitation process. The energy-loss technique unambiguously identifies the excitation mechanism, whereas the optical techniques clearly identify the emitting state.

2. Determination of Excitation Cross Sections from Ion Energy-Loss Spectrometry

Energy-loss spectrometry was one of the earliest methods applied to collisional processes. The Frank–Hertz experiment is an example. The application of the technique to obtain cross sections for intermediate-energy (15–200 keV) ions is, however, a recent development. Even though the general idea is very easily understood, considerable effort is required to obtain an adequate energy resolution for the purpose of determining the cross sections from energy-loss spectra.

The technique can be used to study a wide range of ion–atom collisional processes. It has been applied to the study of the excitation of atoms, ions, and molecules, ionization, total inelastic cross sections, ionic stopping powers, and ejected electron spectra. It is especially useful for studying collisional systems that do not result in the emission of a single particle. The excitation of a molecule to a dissociating state, the excitation of a metastable state in the target, and the simultaneous excitation of both target and projectile are examples of such systems which have been studied with the technique.

2.1. Apparatus

An ion energy-loss spectrometer consists of a source of monoenergetic ions, an ion accelerator, a collisional region, and an energy analyzer. The information on the collision is contained in the energy spectrum of the scattered ion.

The heavy ion energy-loss spectrometer employed by Moore[4] is similar in design to a modern electron spectrometer. The design emphasis in this apparatus is directed toward high resolution to permit identification and diagnosis of molecular states rather than toward absolute cross-section measurements. The system currently can obtain a resolution of 0.150 eV. Moore obtains absolute cross sections with reduced resolution by using a modification of this apparatus.[5] Figure 1 shows a schematic of the apparatus. Mass-analyzed ions are decelerated to 10–30 eV before entering the hemispherical electrostatic monochromator. The electron optics are based on a design of Kuyatt and Simpson[6] that eliminates the need for real apertures in

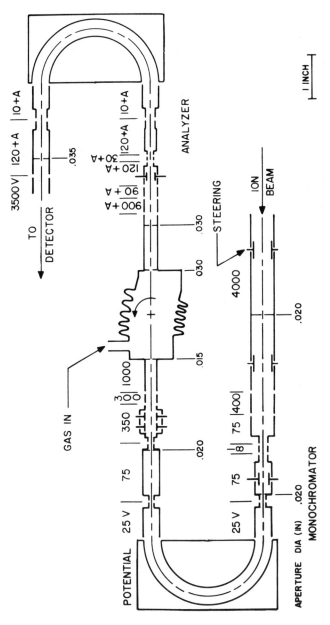

Fig. 1. Schematic of the ion spectrometer used by Moore.[4] Aperture sizes and typical values of the (negative) voltages applied to the lens elements in order to achieve 0.3-eV FWHM resolution at an ion-impact energy of 1 keV are shown. Lens element dimensions are shown to scale. Lens gaps are 0.1 times the corresponding lens diameters. A is the spectral sweep voltage and is numerically equal to the energy lost by ions which will be transmitted through the analyzer to the detector.

the immediate vicinity of the monochromator. Ions leaving the monochromator are accelerated to the desired energy and focused on the entrance aperture of the collision chamber. Ions scattered by gas in the chamber into the acceptance cone of the analyzer are decelerated and focused on the focusing plane of the second hemispherical electrostatic analyzer. Ions with the correct energy for transmission through the analyzer are accelerated into an electron multiplier. The hemispherical electrostatic analyzer is set to transmit ions of one energy. The energy-loss spectra are obtained by adding back energy to the ions which are scattered into the analyzer system. The potential difference between the collision chamber and the analyzer system is varied by applying a voltage across the elements of an "adder lens" whose properties remain essentially unchanged when the potential across it is changed. This provides essentially constant transmission of the analyzer system throughout the energy scan.

Moore's apparatus has been applied for ion energies to 5 keV. At higher energies, the uncertainty in the best available acceleration and deceleration power supply voltages becomes very large compared to the desired resolution. The apparatus shown in Fig. 2 has been designed to overcome this problem and permits the use of the technique to 200-keV energies.[7] The beam extracted from the ion source is momentum analyzed. The analyzed ion beam passes through an einzel lens into the main accelerating column. The ion source is connected to the positive terminal of a

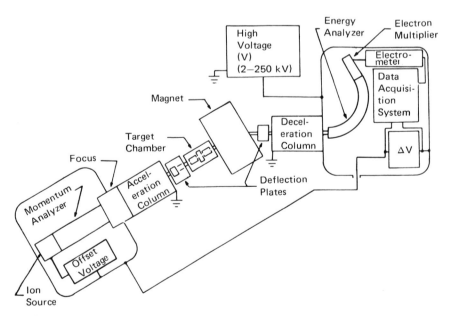

Fig. 2. Schematic drawing of an ion energy-loss spectrometer for use at high ion energies.

precision power supply, which provides a positive offset voltage, V_0. The negative terminal of this supply is connected to a variable power supply which supplies an additional smaller voltage ΔV. The high-voltage power supply, which provides a voltage V, is connected to the common of the ΔV power supply and to the decelerator cage. Therefore, the ions are accelerated through a total voltage of $V + V_0 + \Delta V$. After passing through the collision chamber, the beam is magnetically analyzed to remove the undesired ions and is then decelerated through voltage V. The beam now enters the electrostatic analyzer, which is located in the decelerator cage. The analyzer detects ions having the correct energy for the setting of its electrostatic field.

With no gas in the collision chamber, the voltage ΔV is set to zero, and the ion beam is focused on the slits of the electrostatic analyzer. The electrostatic analyzer voltage is then set to maximize the detected-ion beam current. The analyzer is set to detect an ion reaching it with an energy $qV_0 + E_p$, in which E_p is the most probable kinetic energy of an ion in the ion source and q is the charge of the ion. The electrostatic analyzer voltage, the voltage on the einzel lens, the offset voltage V_0, the high voltage V, the deflecting magnet current, and the electrostatic deflection voltages are not adjusted again during an experimental measurement. With all other parameters held fixed, the detected current is measured as a function of ΔV. Because the analyzer has a finite resolving power, ions with energies slightly different from the desired energy have a probability of being detected. The apparatus resolution function represents the dispersive effects of the entire apparatus on the ion source energy distribution, including the analyzer resolution. The zero energy-loss distribution obtained in this measurement is therefore a convolution of the true ion source energy distribution and the apparatus function.

When gas is present in the collision chamber, the ion beam loses energy as a result of collisions. An ion which is formed in the ion source with the most probable energy E_p but has lost energy ΔE from the collisions. is detected by the analyzer electron multiplier if $\Delta E = q \, \Delta V$, i.e., if it arrives at the analyzer with kinetic energy equal to $qV_0 + E_p$. To be detected, such an ion must enter the decelerating column with the energy $E_p + q(V + V_0)$ and reach the analyzer with the energy $E_p + qV_0$ because

$$E_p + q(V + V_0 + \Delta V) - \Delta E - qV = E_p + qV_0 \quad \text{when} \quad \Delta E = q \, \Delta V$$

By varying ΔV, the energy-loss spectrum can be obtained. The actual current detected as a function of ΔV is a convolution of the spectrum just indicated with the ion-source energy distribution and the apparatus function.

This method eliminates the difficulties of determining the effect of changing deceleration conditions. It assures that all the ions being detected pass through the magnet and enter the decelerating column and the analyzer with the same energy regardless of the energy lost by the ions in the stopping cell. The only change is an increase in the acceleration voltage by ΔV.

The high voltage, V, is connected first to the decelerator and then through the ΔV and V_0 power supplies to the accelerator. Any fluctuations in the high-voltage power supply are added to both acceleration and deceleration voltages. Small high-frequency fluctuations in the high voltage are efficiently filtered by the high-voltage power supply, and large catastrophic events are infrequent and easily detected. Fluctuations with risetimes that are long compared to the time of flight of the ions in the system cancel out of the equation. These fluctuations only contribute to the dependence of the energy-loss spectrum on total energy. If this energy-loss spectrum is a slowly varying function of total ion energy and the fluctuations are small, the effects of the fluctuations in the high-voltage supply on the energy-loss spectrum are too small to be detected.

2.2. Analysis of the Energy-Loss Spectrometric Method

The ion beam entering the collision chamber is nearly monoenergetic. When the projectile ions impinge on the target particles, a large variety of inelastic processes can occur. The projectile ions not only lose energy in the collision but also change charge state. Within the incident ion beam, those ions which have undergone the various charge-changing interactions and the various energy-loss processes constitute an array of smaller beams or partial beams denoted by charge state and energy loss. These ions in having undergone different energy losses can be ideally viewed individually by analyzing the energy of the beam that emerges from the scattering region. A diagrammatic representation of a helium ion beam as it passes through a collision chamber containing helium gas is shown in Fig. 3. In this figure, the energy loss is indicated by a decrease in the vertical position of a partial beam within the hatched regions corresponding to the various charge states. The upper hatched region corresponds to doubly charged ions and is labeled I_2. The middle hatched region corresponds to singly charged ions, I_1, and the lower hatched region corresponds to the neutral atomic beam, I_0. The beam entering the collision region from the left is entirely in the singly charged state and is monoenergetic.

Stripping, electron capture, excitation, and target ionization are the dominant processes with respect to an analysis of the beam composition. Stripping $(10/20e)$ is indicated in Fig. 3 by a transfer of part of the beam in the singly charged monoenergetic beam into the doubly charged state with an energy loss corresponding to the energy required to remove the second electron from the singly charged ion. Capture $(10/01)$ is represented by a transfer of part of the singly charged monoenergetic beam into the neutral

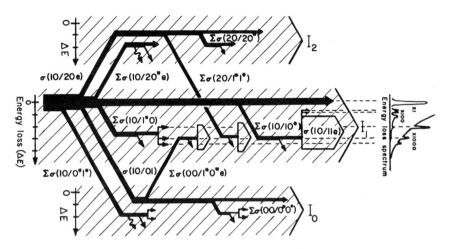

Fig. 3. Schematic representation of the partial beam generation and loss contributions in the scattering region. Only the singly charged component I_1 is observed in energy-loss spectrometry as applied here. The example shown is for helium-ion bombardment of a helium-gas target.

beam with an energy loss corresponding to the energy required to accelerate an electron to the projectile velocity plus any difference in the energy levels between the initial and final states of the electron. In either case, the projectile may be left in an excited state and radiate this energy. This radiation is indicated by "dangling" vectors with a single sinusoidal oscillation. Deexcitation of the beam by radiative transitions does not represent a loss of beam from a particular charge-state but indicates that projectiles in those partial beams can change excitation states by radiative transitions. Because the kinetic energy required for excitation will already have been lost during the original collision, these radiative transitions do not affect the energy-loss data and are subsequently ignored.

Because the method as applied here involves the analysis of the singly charged beam component, multiply charged or neutral projectiles only represent losses in a singly charged ion beam current. If multiply charged projectiles return to the singly charged beam by capture or if neutral projectiles return to the singly charged beam by stripping, they will lose energy at both steps in the capture–stripping cycle. Energy analysis will then provide identification of these ions. They may of course be further identified because of the quadratic dependence of multiple collision processes on target density.

Of particular interest are the $\sigma(10/1^*0)$ and $\sigma(10/10^*)$ interactions between the incident beam and the target particles (Fig. 3). The former represents excitation transitions in the projectile ions, and the latter represents inelastic transitions in the target particles. Direct excitation processes of either the projectile or the target cause the projectile to lose an amount of energy that corresponds to the difference between the initial and final quantum states. The energy-loss spectrum is simply the singly charged ion current detected as a function of the energy lost. The singly charged ion current, which has lost that energy required to excite either the target or the projectile into a definite state, is a direct measure of the probability of exciting that state in a collision.

From each of the partial beams, there are losses that result from the occurrence of additional inelastic and charge-transfer collision processes before the partial beam particles emerge from the scattering region. These losses are indicated in Fig. 3 by straight dangling vectors. Those loss components, which do take an active role in the analysis of the experiment, are elaborated upon in more detail in the following discussion.

The coincidence of energy loss for a partial beam generated by a single discrete transition with that for a partial beam generated by the resultant of multiple discrete energy-loss transitions would be largely accidental. However, a commonly occurring situation involves the superposition of a transition into a discrete state and transitions into a continuum at a given energy loss. Another source of ambiguity in partial beam identification arises from stripping collisions between fast neutralized projectiles and target particles. [See transitions $\sum \sigma(00/1^*0^*e)$ in Fig. 3.] These contributions also appear in the form of a continuum. If the total contributions to the continuum result in a background that is continuous and slowly varying in the vicinity of a superimposed discrete transition, it is shown that the cross section for the discrete transition can be determined by suppressing this background.[8]

Figure 3 is schematic and does not include all possible inelastic processes. The critical point is that inelastic processes involving the positive ion partial beam are directly correlated with features on the energy-loss spectrum. In many cases, there is only one inelastic process corresponding to a resolved structure in the energy-loss spectrum. In this large set of cases, the cross section for an excitation may be determined in a very straightforward manner from the energy-loss spectrum. All the charge exchange processes can be considered as resulting from a single cross section σ_c, accompanied by a decrease in the singly charged components. Inelastic processes, including the one causing the peak under study, can be grouped into a single cross section σ_j, which represents a decrease in the monoenergetic component of the singly charged beam.

In the case of a resolved inelastic transition in the target particles, a simplified model with sufficiently accurate first-order corrections is shown in Fig. 4 for the transition $\sigma_k \equiv \sigma(10/10_k^*)$. By temporarily limiting the discussion to inelastic transitions with energy losses less than the ionization potential, the contributions resulting from $\sum \sigma(00/10^*e)$ interactions can be

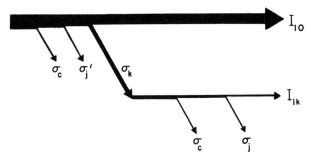

Fig. 4. Simplified partial beam model for target transitions.

neglected. In Fig. 4, the cross section for transitions to all states other than k is given by $\sigma'_j = \sigma_j - \sigma_k$.

The differential equations describing the model in Fig. 4 are

$$dI_{10} = -I_{10}(\sigma_c + \sigma_j)n\, dx \tag{2.1a}$$

and

$$dI_{1k} = I_{10}\sigma_k n\, dx - I_{1k}(\sigma_c + \sigma_j)n\, dx \tag{2.1b}$$

The term I_{10} represents the singly charged zero energy-loss partial beam and I_{1k} the singly charged partial beam that has an energy loss corresponding to the transition being studied. In this derivation, σ_c and σ_j are considered to have the same values for both I_{10} and I_{1k}. This justifiably ignores the negligible difference in cross sections produced by the slight loss of ion beam energy resulting from the σ_k transition.

With the boundary conditions $I_{10} = (I_{10})_i$ and $I_{1k} = 0$ when $x = 0$, the solutions at $x = L$ for these equations are

$$(I_{10})_f = (I_{10})_i \exp[-(\sigma_c + \sigma_j)nL] \tag{2.2a}$$

and

$$(I_{1k})_f = (I_{10})_i \sigma_k nL \exp[-(\sigma_c + \sigma_j)nL] \tag{2.2b}$$

The term $(I_{10})_f$ respresents the singly charged zero energy-loss partial beam, which exits the collision chamber after having traveled a distance L in the gas, and $(I_{1k})_f$ represents the singly charged partial beam leaving the collision chamber with an energy loss corresponding to the transition being studied.

A combination of Eqs. (2.2a) and (2.2b) gives

$$\sigma_k = \left(\frac{1}{nL}\right)\frac{(I_{1k})_f}{(I_{10})_f} \tag{2.3}$$

In spite of the simple form of Eq. (2.3), it is nevertheless an exact solution to the differential equations. It does not require single-collision conditions; thus, it technically permits the use of high target pressures. However, the ratio of the current $(I_{1k})_f$ to the total incident beam current reaches a maximum when the pressure is such that $nL = 1/(\sigma_c + \sigma_i)$. This corresponds to pressures on the order of 0.01 torr for a 1-cm scattering chamber. The practical requirement of obtaining a large signal thus limits the range of pressures actually employed.

Interpretation errors may result if the energy loss corresponding to one partial beam should coincide (within the limit of resolution of the apparatus) with the energy loss of another partial beam. This is precisely the situation when a discrete transition is superimposed on a continuum, such as that introduced by ionization, $(10/11e)$. As a result of the additive uncoupled nature of the results obtained when the differential equations are solved for these cases, the continuum in the energy-loss spectrum can be suppressed to expose the isolated discrete transition for evaluation.[8] Capture–loss cycling, which also appears as a continuum, and energy-loss transitions resulting from double scattering may also be superimposed on the ionic transitions; however, these various responses are also additive.

Another interpretative error may occur if the energy-loss beam resulting from two successive inelastic collisions should coincide with that of a singly scattered partial beam. The solution in this case varies quadratically with the reduced pressure. Least-squares fitting of experimental data can be used to isolate the square-law dependency of the double scattering from the linear dependence of the single scattering. The latter can then be used for cross-sectional evaluation.

In considering cross sections for transitions in the projectile, the inelastic-collision losses and the charge-changing losses for the partial beam are not necessarily identical to those for the incident zero energy-loss beam because some of the projectiles remain in an excited state after the collision throughout the remainder of the scattering region. In these cases, Eq. (2.3) is still valid if $[(\sigma_c + \sigma_i) - (\sigma_c^* + \sigma_i^*)]nL \ll 1$.[8] The term σ_c^* represents the total cross section for charge-exchange processes for the excited projectile, and σ_i^* represents the total cross section for inelastic processes for the excited projectile. It is clear that the differences between the total cross sections for the ground state projectile and the excited projectile must be quite large before this restriction becomes significant.

The largest single uncertainty in the experiment is due to the pressure measurement. This is discussed in more detail in the following section. Other systematic errors include errors in determining the effective path length in the gas, the gas temperature, nonlinearities in particle multiplier response, and nonlinearities in electrometer response. Absolute cross-sectional data must be shown to be free of systematic errors resulting from the ion beam being scattered outside of the acceptance geometry of the analyzer. An error resulting from such a situation presents a serious problem even if the actual current lost is small because the ions have a high probability of a relatively large energy loss. This error also has a strong possibility of being energy dependent because the scattering angle increases with decreasing ion energy. The problem becomes more severe as the resolution is increased because the modifications required to improve resolution also tend to decrease the acceptance angle of the analyzer.

3. Determination of Excitation Cross Sections from Photon Emission Data

The principal advantage of the optical technique is its high resolution. This permits unique identification of both the radiating and final states. The intensity of the emission is directly proportional to the number of atoms in the excited state. The polarization of the radiation provides information on the populations of the magnetic sublevels of the excited state. The basic optical instrumentation required to detect and analyze the emission is well understood. The instruments are commercially available, capable of high resolution, and measure accurately with high sensitivity.

Problems arise in relating the detected photons to the fundamental quantity, i.e., the excitation cross section. The emission is related to the population of the emitting state. This population in turn is related to all the processes that populate and depopulate the state, not just the direct excitation process. These various processes may depend strongly on both target gas density and experimental geometry. In many cases, it is not possible to derive the cross section for specific states from emission measurements.

The most severe experimental problems are related to the absolute accuracy of the calibration of the quantum detection efficiency of the optical system. The calibration is a complicated procedure fraught with traps for the unwary. As a result, large systematic differences frequently occur between the cross sections measured by different, highly qualified workers.

Detailed information on the experimental procedures employed in optical studies of excitation processes are given in Refs. 9 and 10. Tabulations of the results of the optical experiments and an estimate of the errors involved are indicated in Ref. 10. A brief review of the basic methods is included here with a short discussion of their advantages and limitations.

Figure 5 is a schematic of a typical apparatus. Ions formed in an ion source, hopefully in the ground state, are accelerated to the collisional velocity. A magnetic momentum analyzer is usually employed to isolate the desired ion. The mass-selected ion beam is directed through a differential pumping region into a collision chamber. The differential pumping region can also include a charge-exchange cell to neutralize part of the beam and

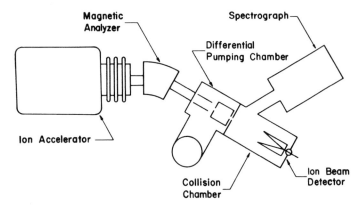

Fig. 5. Schematic representation of a typical emission cross-section apparatus.

aid in the assessment of possible contributions to the detected optical signal from neutrals. Ions and neutrals entering the collision chamber may collide with the target gas. The resulting emitted radiation, if scattered into the solid angle subtended by the spectrograph at the spectrometer's transmission wavelength, will reach its detector.

In an emission experiment, the number of interactions n_j and ultimately the excitation cross section must be related to the number of photons detected by the experimental apparatus. The number of photons emitted is directly proportional to the population of atoms in the excited state corresponding to the upper level of the transition. If we are interested in the number of atoms in a given state, n_i, after a beam of ions passes through a target, processes in addition to the direct excitation process must be included. If processes representing the alternate excitation channels are included as well as loss mechanisms, the differential equation representing the population of the state i is

$$\frac{dn_i}{dt} = \left(\frac{dN}{dt}\right)\left\{n\left[\sigma(10/10_i^*) + \sum_j \sigma_j\right]L\right\} + \sum_{k>i} n_k A_{ki} - n_i \sum_{l<i} A_{il} + C_i(n, N, v) \tag{3.1}$$

where N is the number of incident ions, and n is the number of target atoms in the collision volume. The term $\sum_j \sigma_j$ rather crudely lumps together competitive excitation processes that can also populate the state i. The term A_{ki} represents the transition probability from states higher than the state i; therefore, the term $\sum_{k>i} n_k A_{ki}$ represents cascade. The term $n_i \sum_{l<i} A_{il}$ indicates the radiative depopulation of the ith state. The radiation, i.e., the number of photons per second at a particular frequency, is related to the number of atoms in the state n_i by its radiative transition probability A_{ij}.

The excited target atom cannot be assumed to be thermal. A considerable amount of kinetic energy can be transferred to the target atom in the collision. The excited target particle may undergo a second collision with another atom in the target chamber before radiating. Such a collision could easily result in a change in the target particle's state of excitation. Secondary collisions can lead to both population and depopulation of the excited state i. Secondary collisions are particularly important in the depopulation of long-lived or metastable states. Following the notation used by Thomas and Bent,[11] the term $C_i(n, N, v)$ is used to represent secondary collisional processes, such as the collisional depopulation of the state as well as the alternate excitation channels. This term should have a nonlinear dependence on the target gas density. Its precise form depends on the target gas and the incident ion, which in turn determines which of the secondary processes are dominant. If the data are accumulated under single-collision conditions, this term is negligible. Frequently, it is not possible to obtain an adequate photon count under single-collision conditions. In these cases, the data taken at various pressures are sometimes plotted against target gas pressure and extrapolated to zero pressure to remove the effects of secondary processes.

The operational definition of an emission cross section is

$$\sigma(\Phi_{ij}) = \frac{n_{ij}}{nNL} = \frac{n_i A_{ij}}{n(dN/dt)L} \qquad (3.2)$$

where n_{ij} is the number of photons emitted in the transition $i \to j$ from the element of the target of length L. This in turn is equal to the number of atoms in the state i multiplied by the transition probability for the transition between the states i and j. Under steady state conditions, $dn_i/dt = 0$. If the competing production processes are negligible, i.e., $\sum_j \sigma_j \approx 0$, the differential equation (3.1) yields

$$\sigma(10/10_i^*) = \frac{n_i \sum_{l<i} A_{il}}{(dN/dt)nL} - \frac{\sum_{k>i} n_k A_{ki}}{(dN/dt)nL} - \frac{C_i(n, N, v)}{n(dN/dt)L} \qquad (3.3)$$

The excitation cross section can now be related to the emission cross section by combining Eqs. (3.2) and (3.3):

$$\sigma(10/10_i^*) = \sum_{l<i} \sigma(\Phi_{il}) - \sum_{k>i} \sigma(\Phi_{ki}) - \frac{C_i(n, N, v)}{n(dN/dt)L} \qquad (3.4)$$

Because the intensity in the i–j transition is related to the total photon emission

$$\frac{n_{ij}}{\sum_{m<i} n_{im}} = \frac{A_{ij}}{\sum_{m<i} A_{im}} = \frac{\sigma(\Phi_{ij})}{\sum_{m<i} \sigma(\Phi_{im})}$$

If the transition probabilities are known, Eq. (3.4) can be written[11]

$$\sigma(10/10_i^*) = \frac{\sigma(\Phi_{ij})}{A_{ij}} \sum_{l<i} A_{il} - \sum_{k>i} \left[\frac{\sigma(\Phi_{km})}{A_{km}}\right] A_{ki} - \frac{C_i(n, N, v)}{n(dN/dt)L} \quad (3.5)$$

Thus, it is not necessary to determine all emission cross sections for a level if the transition probabilities are known. It is necessary, however, to know at least one emission cross section $\sigma(\Phi_{km})$, from each of the higher levels to some lower level m in order to make a correction for cascade processes.

As can be seen from this discussion, the determination of an excitation cross section by optical techniques is the result of a series of experiments. The measurement of the signal corresponding to a particular transition must be supplemented by accurate measurements of the quantum yield of the detector, the polarization of the radiation, the isotropic nature of the radiation, the emission cross sections from higher levels, the transition probabilities, and the effects of competing processes, such as excitation by the neutral beam component. The basic experimental measurements involve the determination of particle flux, the determination of target particle density, and the measurement of photon flux.

3.1. Incident Particle Flux Determination

Early measurements occasionally involved some uncertainty in the ion species. This type of contamination is unlikely in modern experiments. A few ion species can provide special problems, such as the problem of removing H_2^+ contamination from a He^{2+} beam. The major contamination of the ion beam results from charge state changes, which occur as a result of collisions with either the background gas or the target gas. With modern high-vacuum techniques, the effects of the background gas can be made negligible. The target gas obviously is necessary, and separate experiments are required to evaluate the contribution to the observed signal from projectiles that have undergone charge state changes in collisions with target gas. The definition of an excitation cross section is only meaningful if the projectile responsible for the excitation is clearly identified. In many cases, the excitation cross section for a specific state is much smaller than the total cross section for charge exchange. The most obvious check for a homogeneous beam requires a demonstration of the linear relationship between the target pressure and photon signal. This test is not always adequate. An example given by Thomas[10] concerns the case in which the incident beam reaches charge state equilibrium, i.e.,

$$N = N_1 + N_0$$

and

$$N_0 \sigma(00/10e)nL \simeq N_1 \sigma(10/01)nL$$

In this case,

$$n_i = N_1\sigma(10/10_i^*)nL + N_0\sigma(00/00_i^*)nL$$

This condition can cause a linear dependence of the excitation with pressures (over a narrow range) far exceeding those of single-collision conditions. The careful experimenter makes a determination of either $\sigma(00/00_i^*)$ and/or the charge-exchange cross sections to assist in determining single-collision conditions.

Although the effects of charge state changes can be determined with relatively straightforward tests, there is a severe problem in evaluating the effects of excited states in the projectile beam. Excited ions are produced in the ion source and by collisions with the background and target gases. Excited-state ion contamination of the ion beam can strongly influence the experimental results. In some cases, optically forbidden states which cannot be excited by ground-state ion projectiles can be excited by metastable ion collisions.[12] Whereas, in a few cases, the excited state ions can be eliminated from the ground state beam, there is no general method for separating ions in excited states from ground-state ions. Because there is no general technique for removing the ions in the excited states, considerable effort is expended to avoid producing them. Electron bombardment or plasma ion sources are operated so that the available energy is just sufficient to produce ground-state ions. Frequently, this effort is negated by applying extraction fields high enough to cause excitation while the ion is still in the source. Because ion beam currents must be quite large in order to permit detection of the emitted photons, it is often impossible to satisfy the minimum ion beam requirement and simultaneously reduce extraction fields and electron bombardment energies. The excited state content is presumably dependent on ion-source operating conditions. It is important that the experimenter demonstrate that the detected signals are independent of source operating conditions. However, this test does not give adequate proof that the excited state content can be neglected. In evaluating available literature, it is often impossible to determine if adequate attention has been applied to evaluating the problems resulting from ion excited states. Future experimenters will need to use great care to remove the effects of this problem. (Compare the related discussion in Chapter 3, Section 3.1.1.)

Charged ion beams are usually detected by a Faraday cup, which consists of a deep cup preceded by an aperture plate. The cup is connected to an electrometer. The aperture is biased to prevent secondary electrons from escaping from the cup and providing an erroneous beam current reading. Ejected secondary ions are attracted to the aperture plate, but the coefficient for ejection of secondary ions is quite small, and hence this process does not seriously affect the current measurement. Reflection of the projectiles can be inhibited by the orientation of the surfaces of the Faraday cup and by the introduction of an aperture, which confines the beam in the cup.

The Faraday cup is regarded as an absolute device, and frequently the only test of the Faraday cup operation is the measurement of ion beam current as a function of electron-suppression voltage. The possibility of having a significant error resulting from reflections and secondary ion emission is probably small; however, the assumption that the Faraday cup is an absolute detector is too often left unchecked.

Neutral beams are usually detected by measuring the current of secondary electrons ejected from a surface bombarded by the neutral beam. The secondary electron-emission coefficients of a metal surface vary with the material, the condition, and history of the surface as well as with the energy, mass, charge state, and state of excitation of the incident projectile. The assumption that the secondary electron coefficients for the charged and neutral projectile of the same mass and energy are equal is not always valid.[13] Calibrations of neutral detectors that are determined by calculating the ratio of the secondary emitting current to the incident ion beam current and by assuming that it is the same for neutral projectiles are not dependable. Because the secondary emission coefficients are dependent on the condition, history, and orientation of the surface, the ratios of neutral and ion secondary emission coefficients obtained by other experimenters cannot be easily applied. Neutral detectors, which are dependent only on the energy deposited in the detector, should be independent of the surface secondary emission coefficient. Thermal detectors, such as thermocouples, thermopiles, or thermistors, can be used to determine the energy deposited by the neutral beam by noting the change in temperature of the detector. The calibration of the detector with known positive currents is straightforward and free of the problems involved in the calibration of secondary emission detectors. The problem with thermal detectors is in their low sensitivity and slow response.

3.2. Target Density Determination

Problems in target pressure measurements are common to both the optical and energy-loss techniques. Low-pressure measurement is a very difficult procedure. Many serious systematic errors can be traced to inadequate attention to accurate pressure measurement. The McLeod gauge has traditionally been used as the standard device for such measurements, but there are many practical problems associated with its use. Because the mercury tends to stick in the capillary tubes of the gauge, their surfaces must be extremely clean; condensable gases cannot be used, measurement is slow and painstaking, and there is the ever present threat of mercury-vapor contamination of the gas sample. These problems can be overcome with considerable effort; however, the necessity of a cold trap between the McLeod gauge and the target chamber to prevent mercury vapor contamination of the gas introduces a serious potential for error. This error arises from two effects: thermal transpiration and mercury vapor pumping. Thermal transpiration results from equilibrium pressure differences in two chambers at different temperatures. The effect of thermal transpiration is a strong function of the pressure and the connecting tube diameters.[14] Mercury vapor pumping occurs when a continuous stream of mercury vapor flows from the McLeod gauge to the cold trap. The magnitude of the error introduced by this pumping effect is a function of the temperatures of the cold trap and the McLeod gauge, the diameters of the connecting tubes, and the mass of the gas in the system. This effect can produce very large errors for heavy gases.[15]

Because of the great care required in the use of the McLeod gauge and the long time required to make a single measurement, the usual experimental procedure has been to use the gauge to calibrate a thermocouple vacuum gauge, ionization gauge, or discharge gauge *in situ* for each gas being studied. Although this procedure makes it possible to determine the gas pressure more rapidly, it simply transfers any errors that are made by the McLeod gauge in the pressure determination to the gauge being calibrated as well as introducing additional possibilities for error.

Except in the few cases in which published information provides unusually complete details of the geometry of the connections to the McLeod gauge, it is impossible to make retroactive corrections to the data. There is also some question as to the absolute magnitudes of the corrections, although the validity of the transpiration and mercury-vapor pumping effects is universally accepted. Because much of the published data has been obtained from experiments in which a McLeod gauge was used, errors resulting from its use must be considered if these data are to be employed.

Recently, experimenters have employed a differential capacitance manometer that consists of two chambers separated by a thin metal diaphragm. A differential pressure produces a small deflection of the diaphragm, which also forms the common plate of two capacitors. The change in the capacitance is proportional to the deflection. Commercially available devices are calibrated by using a dead-weight technique at relatively high pressures, and the calibration is extrapolated to lower pressures. The validity of this calibration at low pressure is somewhat an act of faith in the case of a particular instrument, although tests performed with several of these manometers have proven the reliability of the calibration of those tested.[16] The capacitance manometer is rapidly becoming the accepted device for determining pressures. It provides continuous readings, introduces no mercury-vapor contamination, can be used with condensable vapors, is free of errors resulting from thermal transpiration and mercury vapor pumping, and is much easier to use. Even though the errors inherent in the use of the McLeod gauge can be overcome, the effort hardly seems worthwhile when the capacitance manometer provides such an attractive alternative.

In practice, the actual purity of the gas is less likely to produce serious errors than the absolute pressure determination. The spectrum of each substance provides a unique fingerprint, and significant contamination is unlikely to pass undetected.

3.3. Photon Flux Determination

The calibration of the quantum detection efficiency of an optical system is a difficult and complicated procedure. The accuracy achieved, even with the expenditure of great care and considerable effort, is often poor. The detection system consists of an optical system to collect light, an analyzing device, such as a monochromator, and a photon detector. Each part of this system can be calibrated separately; however, the most dependable technique involves *in situ* calibration of the entire detection system.

The number of photons, n_{ij}, detected depends on the geometry of the system. In addition to geometrical effects, the detector has a certain efficiency depending on the wavelength λ, which is expressed in the quantum yield, $k(\lambda)$, of the detector. The quantum yield is a function of polarization as well as wavelength.

Experimentally, the quantum yield $k(\lambda)$ of the photomultiplier is combined with losses at the refractive and reflective surfaces to produce a product term labeled $K(\lambda)$. This term is determined by measuring the signal from a source of known emission. If a standard tungsten strip filament lamp is positioned at the point previously occupied by the strip of emission from the ion beam, the effective quantum efficiency $K(\lambda)$ can be accurately determined. In many experiments, the standard emission source has a very different geometry or location than that occupied by the emitting atoms. If the field of view and the solid angle change markedly, it is very difficult to ensure that the response is identical for the calibration and data measurements. The problem is eliminated by a standard emission source whose geometry and location duplicate the region of the target excited by the beam.

The emissive power $E(\lambda)$ of the standard lamp at a wavelength λ is expressed in terms of the number of photons per second per angstrom bandwidth emitted from a unit area into one steradian solid angle. If the area viewed by the optical system is wL, i.e., w wide by L long, the number of photons per second emitted by this area is $E(\lambda)wL$. The observed signal from the standard lamp is then $S_\lambda = K(\lambda)E(\lambda)wL\omega$ (photons/sec), in which ω is the effective solid angle of the detector. When observing the radiation from the excited atoms, the number of emitted photons n_{ij} is related to the detected signal S_{ij} (photons/sec) as $S_{ij} = [dn_{ij}/dt][K(\lambda)/4\pi]\omega$, in which it is assumed that the solid angle ω is not dependent on position over the length of the observation region, L. The number of photons per second is then given by

$$n_i A_{ij} = \frac{dn_{ij}}{dt} = 4\pi E(\lambda)\left(\frac{S_{ij}}{S_\lambda}\right) wL$$

If the standard lamp is placed in the same position as the emitting region, L is the same in both calibration and experiment. The emission cross section for isotropic radiation is then

$$\sigma(\Phi_{ij}) = \frac{4\pi w E(\lambda) S_{ij}}{S_\lambda n (dN/dt)}$$

The substitution technique does not require the separate measurement of either the solid angle ω or the length of the observation region, L.

The emissive power of the standard lamp filament can be determined from tables based on the work of de Vos[17] and Larrabee.[18] An optical pyrometer is usually used to determine the temperature of the filament. This procedure could be increased in accuracy by making a direct comparison between the standard lamp and a blackbody.[10]

A disadvantage of simulating the ion beam with a standard lamp is the large difference between the calibration count rate and the signal count rate. Hasselkamp et al.[19] used neutral density filters to reduce the count rate of the calibration to the same order of magnitude as the experimental count rate. The introduction of neutral density filters reduces possible saturation and scattered light problems but introduces another possible source of systematic error. The use of this technique is probably wise, especially if all the data are taken in a narrow range of the spectrum so that only one set of neutral density filters is required.

Considerable care must be exercised during the calibration to eliminate any scattered light. Gray light from the very intense emission in the visible regions of the spectrum can completely obscure low-intensity emissions in the ultraviolet.

When an ion beam strikes a surface, secondary particles and photons are ejected from the point of impact. To avoid spurious signals from bombarded surfaces, the target chamber and ion-beam detection system must be designed so that the photon detection system does not view any surfaces subject to projectile beam impact. Further, the design must prevent the secondary particles from entering the collision region.

Although measurement of polarization is often not an objective of the experiment, it is always necessary to determine the polarization in order to correct the data for anisotropy in emission. By using a Polaroid analyzer, the difference between the spectrometer transmission of light polarized parallel and perpendicular to the spectrometer slit can be used to estimate the polarization fraction. If the radiation is not isotropic, then it is necessary to integrate the detected current over the entire solid angle to obtain n_{ij}. Polarization corrections require correction of the effective quantum yield, $K(\lambda)$, at each angle before integration.

The analysis to this point assumes that the target atoms radiate before leaving the region of the target chamber observed by the detection system. The optical technique cannot be easily applied to measurements involving long-lived or metastable states that will leave the field of view before they decay with the emission of radiation. In some special cases, optical techniques can be applied to metastable states by utilizing quenching fields. One example, the excitation to the $H^*(2s)$ state, is discussed in Section 4.1.

The actual optical arrangement employed must be analyzed to determine the effects of its dispersion, slit dimensions, and/or filter transmission. The properties of some instruments change dramatically as the wavelength changes, and calibration is required at each wavelength employed. Because the techniques and apparatus vary so greatly, the discussion of the effects of specific instrumentation covering particular wavelengths and of the various modifications to the basic technique is reserved until specific experiments are discussed.

4. The Excitation of Atomic Hydrogen

The simplest ion–atom collision is a collision of a proton with an atomic hydrogen atom. This system has been intensively studied by theorists. Calculations in which many different approximations have been used are available; however, the range of validity of the various theoretical approaches is still uncertain. The number of experimental measurements is very sparse in spite of the obvious interest in the cross section. The problem of obtaining an atomic hydrogen target has been solved by dissociating hydrogen in a furnace. The furnace must be heated to almost 2700 K before the hydrogen gas is 90% dissociated. In order to use optical techniques, the photon detector must be optically shielded from the white hot furnace.

4.1. Modulated Crossed-Beam Technique

The basic optical technique can be modified by using modulated crossed-beam techniques. The basic modulated crossed-beam apparatus is shown schematically in Fig. 6. The atomic beam is usually produced in a furnace. The beam issuing from a hole in the side of the furnace is collimated by apertures which also serve to isolate the differential pumping region

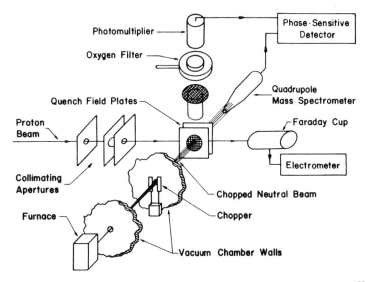

Fig. 6. Schematic drawing of the crossed-beam apparatus used by Morgan et al.[20]

around the furnace. The atomic beam is mechanically chopped before arriving at the interaction region. The ion beam collides with the atomic beam in the small area defined by the overlap of the ion and atomic beams. The region of interaction is viewed by the optical detection equipment. The density of the atomic beam is usually comparable to the density of the background gas even though fast high-vacuum pumps are employed throughout the system. The observed optical signal represents the sum of the modulated radiation from the modulated atomic beam and the constant radiation from the background gas. The desired signals from the modulated atomic beam are separated from the background signals by a phase-sensitive detector (lock-in amplifier), which accepts only signals of the correct frequency and phase. The required phase and frequency are derived from the beam chopper.

The recent paper of Morgan et al.[20] covers four of the excitation processes that can result from a proton–atomic hydrogen collision:

$$H^+ + H(1s) \rightarrow H(2p) + H^+$$
$$\rightarrow H(2s) + H^+$$

electron capture into an excited state (10/0*1)

$$H^+ + H(1s) \rightarrow H^+ + H(2p)$$
$$\rightarrow H^+ + H(2s)$$

direct excitation (10/10*)

Their data were taken for proton energies between 2 and 26 keV. The crossed-beam apparatus they used is as just described (see Fig. 6). The beam of highly dissociated hydrogen is produced in a tungsten tube furnace heated to 2700 K. The intersecting beams are carefully aligned. The diameter of the neutral hydrogen beam is large enough to cover the proton beam completely. The background pressure is maintained at 2×10^{-7} torr. The Lyman α radiation arising from the beam-intersection region is recorded by an open-ended multiplier, which is provided with an oxygen filter cell. The ends of the oxygen filter are lithium fluoride, which provides a short-wavelength cutoff. A nonreflective shield confines the field of view to the interaction region to reduce signals from radiation reflected from nearby surfaces. The entire Lyman α detector can be rotated about the target beam in the plane containing the proton beam.

A quench field of 600 V/cm^{-1} can be applied to the beam interaction region by supplying voltage to a pair of metal plates. In the absence of this field, the Lyman α signals arise from spontaneous decay of H(2p) atoms formed in the beam interaction region. When the field is applied, metastable H(2s) atoms are quenched by Stark mixing of the 2s and 2p states. The Lyman α signal with the quench field applied is due to both H(2p) and H(2s) atoms.

A quadrupole mass spectrometer in line with the neutral hydrogen beam is used to determine the H/H$_2$ fraction in the beam.

The Lyman α radiation arising from charge transfer excitation of the fast beam must be separated from Lyman α radiation arising from direct excitation of the relatively slow-moving target beam. This is accomplished by making use of the extremely narrow transmission window of the oxygen filter. This transmission window is narrow enough so that Lyman α radiation emitted at 54.7° by the fast-moving excited atoms is Doppler shifted to such an extent that it causes much greater attenuation in the oxygen filter than the Lyman α from the slow-moving target atoms. Lyman α radiation emitted at 90° to the beam direction is not Doppler shifted, and radiation from all processes is attenuated equally by the oxygen filter.

Six separate measurements were made, with the detector at 54.7 and 90°, with the oxygen filter in and out, and with the quench field on and off. The transmission of the oxygen filter for unshifted Lyman α radiation and the transmission of the oxygen filter for Doppler shifted Lyman α radiation were both measured. Because the transmission of the filter for the Doppler shifted Lyman α is a function of the proton-beam velocity, it was evaluated at each energy. The relative fraction of the fast H(2p) atoms recorded in the absence of the quenching field and the fractions of fast H(2p) plus fast H(2s) atoms recorded by the detector with the quenching field applied were measured.

To obtain absolute cross sections, the Lyman α detector had to be calibrated. The data were normalized to the 2p capture cross sections for 18 keV H$^+$+H$_2$ collisions measured by Birely and McNeal.[21] Their measurements in turn were normalized to absolute cross sections for Lyman α production in $H^+ + Ne$ collisions measured by Andreev et al.[22]

For electric dipole radiation, the photon yield at 54.7° to the axis of quantization is independent of the polarization fraction of the emitted radiation. The effects of polarization are thus eliminated in cross-section determinations taken entirely at 54.7°. Cross sections determined with measurements taken at both 54.7 and 90° are dependent on polarization; however, Morgan et al.[20] noted that the two methods yield the same results within the experimental error, implying negligible polarization. This is consistent with the measurements of Kauppila et al.[23] which show that the polarization of Lyman α radiation produced by direct excitation of hydrogen by protons is small in this energy range. The calculation of Crandall and Jaecks[24] was used to show that the effects of polarization of the radiation from the quenched $H(2s)$ would produce a change of less than 5% in their $H(2s)$ formation cross sections.

Cascade contributions to the detected photon emission were estimated by using Born approximation calculations (Section 1.2) to determine the populations of the higher states. These estimates indicate that the cascade contribution is negligible in the case of the measured capture cross sections but can contribute 10% to the direct excitation. The cross sections obtained by Morgan et al.[20] are probably too large by this amount.

The cross section for $H^+ + H(1s) \rightarrow H^+ + H(2p)$ is shown in Fig. 7. The data are those of Morgan et al.,[20] Kondow et al.,[25] and the earlier work of Stebbings et al.[26]

The cross sections in the work of Stebbings et al. were normalized to the cross sections for the Lyman α radiation production in electron bombardment reported by Fite and Brackman.[27] Their measurements in turn were normalized to the cross section for impact excitation of the $H(2p)$ state by 300 eV electrons, which were calculated by means of the Born approximation. The data as plotted were corrected for a pressure dependence in the oxygen absorption coefficients by Young et al.[28]

Kondow et al.[25] used a standard crossed-beam apparatus with addition of cryopumping to reduce the background gas pressure. The relative cross sections were normalized to the cross section for Lyman α production in e-H collisions for 6 keV electrons. Data were accumulated at both 54.7 and 90° to allow for the separation of photons emitted from fast-moving and stationary atomic hydrogen. They did not apply a quenching field to the collision region and hence did not measure cross sections for excitation of the $2s$ states. The experiment was conducted with high precision, and uncertainties of only 6 to 11% are mentioned.

The three sets of data are in very good agreement. A peak in the cross section curve located at 6 keV is observed in the measurements of Morgan et al. and Kondow et al. This peak is not observed in the earlier measurements of Stebbings *et al.*; however, the relatively few data points in the early work made resolution of such a peak unlikely. Overall, the general agreement with the pioneering work of Stebbings is good.

The cross section for $H^+ + H(1s) \rightarrow H^+ + H(2s)$ is given in Fig. 8. In this case, the measurements of Morgan *et al.*[20] are the only ones available.

Also shown on the figures are various theoretical calculations. The Born approximation calculations in general exhibit a peak which overestimates the observed cross sections and which occurs too low in energy. The Born calculations (curve B)[29] include only the initial and final states of the hydrogen electron and thus neglect coupling between other states. The distortion calculations (curve D)[30] have been applied to the direct excitation processes. These calculations include matrix elements that cause distortion of the atomic wave functions. The effect is to reduce the magnitude of the derived cross section near the peak in the curve and to increase the energy at which the peak occurs. This results in a net improvement in the agreement when compared to Born approximation calculations, but the agreement is still poor. The best agreement with the data is obtained by the close coupling calculations based on the impact parameter methods.[31-33] The improvement in the agreement is marked when the charge-transfer channels are included (curve C4). Further improvement is obtained with the inclusion of couplings to the $3s$, $3p_0$, and $3p_{\pm 1}$ levels[32] (curve C7). The best overall agreement is

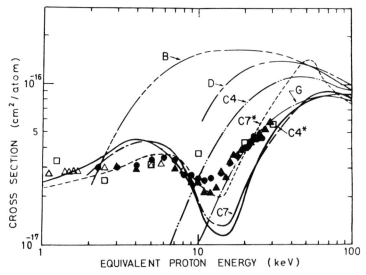

Fig. 7. Cross sections for the process $H^+ + H(1s) \rightarrow H^+ + H(2p)$: ●, Morgan *et al.*[20]; △, Kondow *et al.*[25]; □, Stebbings *et al.*[26,28] as reported in Ref. 10. Curves: B, Born[29]; C4, four-state close coupling without exchange[31]; C4*, four-state close coupling with exchange[33]; C7, seven-state close coupling with exchange[32]; C7*, seven-state close coupling with pseudostates[33]; D, distortion[30]; G, Glauber.[34]

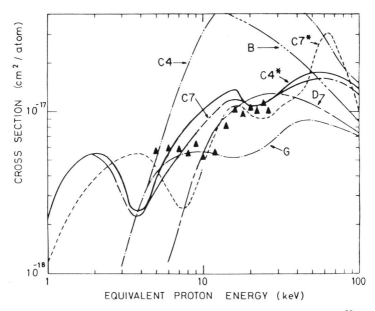

Fig. 8. Cross-sections for the process $H^+ + H(1s) \rightarrow H^+ + H(2s)$: ▲, Morgan et al.[20] Curves: B, Born[29]; C4, four-state close coupling without exchange[31]; C4*, four-state close coupling with exchange[33]; C7, seven-state close coupling with exchange[32]; C7*, seven-state close coupling with pseudostates[33]; D, distortion[30]; G, Glauber.[34]

obtained by the seven-state close coupling calculation of Cheshire et al.[33] (curve C7*). In this calculation, exchange and excitation are treated simultaneously. The problem of representing the intermediate state of He^+ formed at small proton separations is accomplished by choosing three pseudostates. This calculation reproduces the peak in the cross section curve at 6 keV and accurately locates the minimum at 12 keV. Curve G is a Glauber approximation calculation by Franco and Thomas.[34] This calculation gives surprisingly good results at the higher end of the energy range and is expected to reproduce the close coupling approximation at high energies.

4.2. Energy-Loss Spectra of Atomic Hydrogen

The cross section for excitation to the $H(n = 2)$ states has been obtained with heavy-ion energy-loss spectrometry.[35] The apparatus and general method employed in measuring the cross section have been discussed previously. In this case, the target furnace is constructed of tungsten tubes. Current flows coaxially along the furnace wall and returns through an adjacent coaxial shield. The furnace is Joule heated to 2700 K to dissociate the hydrogen target. The coaxial design minimizes any effect from magnetic fields produced by the currents in the furnace.

With the target furnace cold, the energy-loss spectrum of molecular hydrogen is obtained when hydrogen gas is introduced into the target cell. As the furnace is heated, the spectrum begins to change. A peak at 10.2 eV energy loss appears and increases while the molecular peak at 12.5 eV energy loss decreases.

The energy-loss spectrum of atomic hydrogen that is obtained with a hot furnace is shown in Fig. 9. Clearly present is a peak at 10.2 eV energy loss that corresponds to the excitation to the $n = 2$ state of atomic hydrogen. A secondary peak at 12.7 eV corresponds to the excitation to the $n = 3$ state.

The determination of the cross section for the excitation to the H$(n = 2)$ state is not dependent on the complete dissociation of the molecular hydrogen, because the 10.2-eV atomic peak is resolved from the molecular peak at 12.5 eV. The molecular fraction is estimated to be less than 3% during the data-acquisition period. This fraction is determined not only from pressure–temperature curves but also from plots of the ratio of the ion currents at 10.2 and 12.5 eV energy loss taken as a function of furnace temperature. At low temperatures, the peak at 12.5 eV contains contributions from collisions that excite both the $n = 3$ states of the atomic hydrogen in the furnace and the Lyman α bands of the molecular hydrogen. As the furnace temperature increases, the molecular hydrogen is depleted, and the peak at 12.5 eV decreases. The ratio of the peak at 10.2 eV to the peak at 12.5 eV increases with temperature until it reaches a plateau. Higher furnace temperatures do not make any further changes in spectral shape, indicating that the molecular hydrogen no longer makes a significant contribution to the spectrum.

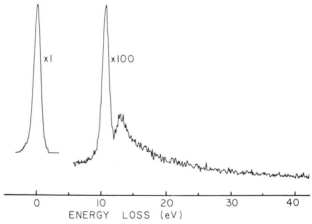

Fig. 9. Energy-loss spectrum for 50-keV protons incident on atomic hydrogen.

A relative cross section can be obtained directly from the data.[8] Energy-loss spectra are measured in series. Consecutive spectral measurements are taken at various energies from 200 keV down to 15 keV. During each series of spectral measurements, the pressure in the chamber is held fixed to permit normalization to a single cross section at 200 keV. The cross sections obtained for the excitation to the $n = 2$ state of atomic hydrogen are shown in Fig. 10. Averages of all the available data are given by the triangles. The averaged data are normalized at 200 keV to the Born approximation calculation for proton excitation to the $H(n = 2)$ state ($\sigma = 6.637 \times 10^{-17}$ cm^2).[29]

At low energies (15–30 keV), the data obtained with an energy-loss spectrometer can be compared to the crossed-beam measurements. To obtain the cross section for the excitation to the $n = 2$ state of atomic hydrogen, the cross sections for both the $H(2s)$ and $H(2p)$ states have to be included. In the case of Stebbings et al.[26,28] and Kondow et al.,[25] the $H(2s)$ excitation cross sections from Morgan et al.[20] are added to the measured $H(2p)$ excitation cross sections to give the excitation cross section to the

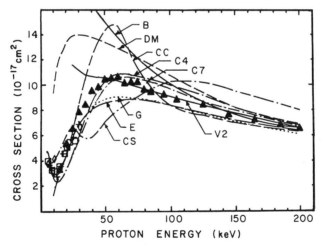

Fig. 10. The cross section for excitation of the $n = 2$ states of atomic hydrogen by protons. Theory and experiment: ▲, Park et al.[35]; □, Morgan et al.[20]; ◇, Stebbings et al.[26,28]; +, Kondow et al.[25] The $H(2s)$ excitation cross sections of Morgan et al.[20] are added to the $H(2p)$ excitation cross sections measured by Stebbings et al.[26,28] and Kondow et al.[25] to yield the cross sections for excitation to the $H(n = 2)$ state. Curves: B, Born approximation calculation[29]; CC, seven-state close coupling with orthogonal pseudostates[33]; C4, four-state close coupling[33,32]; C7, seven-state close coupling[33,32]; CS, coupled-state calculation in the Sturmian representation[46]; DM, 20-state diagonalization method[47]; V2, second-order potentials calculation[45]; E, distorted-wave eikonal[48]; G, Glauber calculation.[34]

H($n = 2$) levels. Considering the major differences in technique and normalization, the agreement between the data from the energy-loss experiments and data from these crossed-beam experiments is unexpectedly good.

The data from the energy-loss experiment cover the energy range in which the cross section reaches its maximum value. This energy range had for some time been a source of difficulty for theoreticians. Neither approximations suitable for very high ion velocities nor those suitable for low ion velocities are completely satisfactory in this energy range.

Examples of the various theoretical calculations are also shown in Fig. 10. The Born approximation calculations[29] (curve B) in general exhibit a peak that overestimates the observed cross sections and occurs at too low an energy. The results of the second Born calculations for the direct excitation to the H*($n = 2$) state do not markedly improve the theoretical fit to the data over the first Born approximation calculations.[36-38] Impact parameter formulations of the Born approximation similarly produce negligible improvement[39,40] in the theoretical fit to the data.

The distortion approximation[30,41] agreement with the experimental measurements is poor but is better than for the Born approximation.

Impact parameter coupled-state calculations have been undertaken by several groups. Unless exchange channels are included, the theoretical values are not in much better agreement with the experimental data than the distortion-approximation results. The differences between the various four-state calculations are noticeable,[31,32,42] but the overall fit to the experimental data is roughly equivalent. The data of Rapp and Dinwiddie[32] are shown for both four-state (curve C4) and seven-state (curve C7) calculations. It is noted that the inclusion of the additional states does not produce any dramatic changes. The agreement between theory and experiment is very good.

It was observed that the best overall agreement at energies less than 20 keV is obtained by the seven-state close coupling calculation of Cheshire et al.[33] (curve CC). This calculation includes exchange and uses pseudostates to represent coupling to the higher states. Measurements indicate that Cheshire's calculation tends to overestimate the cross section at the peak of the curve. Kondow et al.[25] also noted that at their highest energies the cross sections of Cheshire et al.[33] were larger than their experimental measurements.

Bransden and Coleman[43-45] have developed a technique in which they use second-order potentials to make allowance for states omitted in truncating the close coupling expansion. The four-channel calculation that uses second-order potentials does not, however, provide a good fit to the data below 100 keV (curve V2). The coupled-state calculations of Gallaher and Wilets,[46] who used a Sturmian representation to form a complete basis set, are shown in curve CS. The agreement with experiment is not good. No structure was detected in the experimental data that would correspond to the minimum in the cross section, which was obtained in this calculation at ~35 keV.

Baye and Heenen[47] have recently applied a diagonalization method which includes 20 states. The cross sections obtained do not provide a good fit to the data at energies below 100 keV (see curve DM).

The Glauber approximation calculated by Franco and Thomas[34] (curve G) gives surprisingly good results in the energy range under study. The distorted-wave eikonal calculation of Joachain and Vanderpoorten[48] (curve E) is only slightly different than the Glauber approximation and fits the data equally well. The Glauber approximation is lower than the experimental measurements at the maximum in the curve of the cross section. The agreement with the crossed-beam measurements is not satisfactory below 10 keV. Nevertheless, the agreement is quite good over a large energy range, especially if one considers the relative simplicity of the Glauber approximation.

5. Excitation of Helium

The bombardment of helium by protons is the most thoroughly studied of all the collision systems. Experimentally, helium is easy to handle. The availability of accurate theoretical transition probabilities for helium permits the emission cross sections to be converted to excitation cross sections. Cross sections for excitation by protons of all the helium singlet states through $n = 6$ have been measured. Excitation of the triplet states from the $1\,^1S$ ground state requires a change of spin. If the Wigner spin conservation rule is obeyed, proton excitation of the triplet state is forbidden. The triplet states of helium can be excited in electron exchange collisions by a projectile carrying a bound electron. Observation of triplet state excitation is sometimes an indication of neutral contamination in the proton beam.

5.1. Energy-Loss Spectra of Helium

An example of cross-section measurements obtained from ion energy-loss spectrometry for protons incident on a gas target is shown in Fig. 11. The data, which are for the excitation of the helium ground state to the $n = 2$ level, $H^+ + He(1\,^1S) \rightarrow H^+ + He(2\,^1S + 2\,^1P)$,[49] are absolute; the resolution is 2 eV.

The cross section for excitation of the $2\,^1P$ state obtained by energy-loss spectrometry is shown in Fig. 12. The structure in the energy-loss spectra corresponding to the $He(1\,^1S \rightarrow 2\,^1S)$ excitation was not completely resolved from the larger peak in the spectra corresponding to the $He(1\,^1S \rightarrow 2\,^1P)$

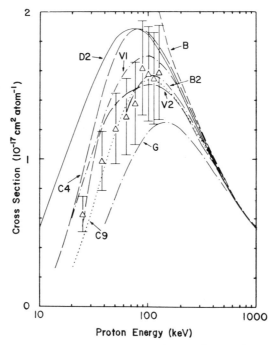

Fig. 11. Excitation cross sections for the process $H^+ + He(1\ ^1S) \rightarrow H^+ + He(2\ ^1S + 2\ ^1P)$: △, Park and Schowengerdt.[49] Curves: B, Born[52]; B2, second Born[55]; C4, four-state close coupling[56]; C9, nine-state close coupling[57]; D2, second-order diagonalization[59]; V1, first-order potential[58]; V2, second-order potential[60]; G, Glauber.[58]

excitation and the uncertainty in the $He(1\ ^1S \rightarrow 2\ ^1P)$ excitation cross section is large because of the poor resolution.

5.2. Optical Studies of the Excitation of Helium

Figure 12 also displays the data of Hippler and Schartner.[50] These data were obtained from optical experiments in which a vacuum ultraviolet monochromator was used as the detector (see Fig. 13). The light was observed at a 54.5° angle with respect to the proton beam and with the monochromator rotated 45° around its optical axis. This arrangement eliminates difficulties that result from the polarization of the emitted light.[51] The apparatus was calibrated with photons, which were emitted from the collision region when a 1-keV electron beam was substituted for the proton beam. The normalization technique was justified by the good agreement between the theory and the experiment for the 1-keV electrons that bombarded the helium. It has been shown that the technique provides excellent agreement with calibrations when a tungsten strip lamp is used to

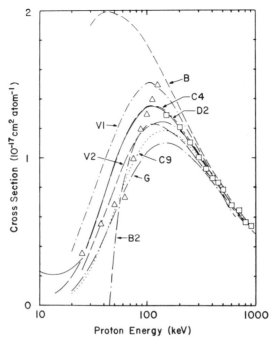

Fig. 12. Cross sections for the process $H^+ + He(1\,^1S) \rightarrow H^+ + He(2\,^1P)$: △, Park and Schowengerdt[49]; □, Hippler and Schartner.[50] Curves: B, Born[52]; B2, second Born[55]; C4, four-state close coupling[56]; C9, nine-state close coupling[57]; D2, second-order diagonalization[59]; V1, first-order potential[58]; V2, second-order potential[60]; G, Glauber.[58]

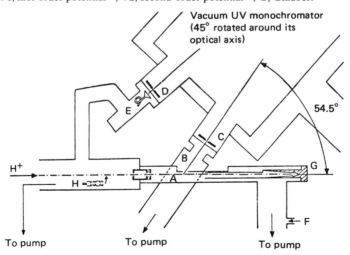

Fig. 13. Schematic drawing of the apparatus used for the excitation of helium by Hippler and Schartner.[50] A, Collision chamber; B, differential pumping chamber; C, entrance slit to the vacuum monochromator; D, exit slit for the vacuum monochromator; E, channel electron multiplier; F, gas inlet; G, Faraday cage; H, electron gun.

replace the ion beam.[19] The data were corrected for cascade contributions by using theoretical excitation cross sections to determine the population of the higher excited states. The agreement between the optical and the energy-loss measurements is satisfactory considering the uncertainties involved: 10% for Hippler and Schartner,[50] 25% for Park and Schowengerdt.[49]

The experimental data do not merge perfectly; however, a smooth curve can easily be drawn through the error bars. Both theory and experiment indicate that the maximum in the cross section appears at ~ 135 keV for protons. It is unfortunate that neither of the data sets extends over the peak in the cross-section curve.

Several theoretical calculations of the $H^+ + He \rightarrow H^+ + He^*(2\,^1P)$ excitation cross sections are available. Some of these calculations are also shown in Figs. 11 and 12. The accurate Born approximation calculation of Bell et al.[52] and that of Kim and Inokuti[53] show a peak which appears at too low a proton energy and overestimates the cross section. This situation is only slightly improved in a distortion approximation calculation.[54] The second Born approximation calculation of Holt et al.[55] provides a reasonable agreement at proton energies greater than 150 keV. The peak in the theoretical curve appears to be at too low an energy, and the theoretical cross section underestimates the experimental data at low energies.

Impact-parameter close-coupling techniques have been applied to proton–helium collisions. A four-state calculation by Flannery[56] is in good agreement with the experiments, although this calculation tends to be larger than the experiments at low energies. The nine-state calculation of Van den Bos[57] provides very good agreement over the entire range of available data. The calculation is slightly lower than the data near the peak in the cross-section curve but well within the estimated error of the data.

The uncertainty in the data does not permit a sound choice between these close coupling calculations and the recent calculations using other approaches. Although their first-order potential calculation results are significantly higher than the data at low impact energies, the Glauber approximation calculation of Joachain and Vanderpoorten[58] is in very good agreement. The second-order diagonalization calculation of Baye and Heenen[59] produces results very much like those obtained when the close-coupling technique is used. The second-order potential calculation of Begum et al.[60] provides equally good agreement for most of the proton energies.

The contribution of the $2\,^1S$ cross section affects the theoretical calculations differently. The Glauber calculation for $2\,^1S$ and $2\,^1P$ is markedly lower than the measurement (see Fig. 11), whereas the agreement was quite good when only the $2\,^1P$ cross section was compared with theory. The exact opposite effect is obtained when the same comparison is made for the second-order diagonalization calculation and the four-state close-coupling calculation. These calculations differ much more with respect to the cross section for the $2\,^1S$ excitation than they do for the $2\,^1P$ excitation cross section.

Several optical studies are available for the higher excited states. The experiment of Thomas and Bent[11] provides a good example of these studies. In this experiment the source of 0.15–1.0 MeV protons was a Van de Graff positive ion accelerator. The incident proton beam energy was determined to within 2 keV by a 90° deflection in a magnetic analyzer. Beam currents were typically 6 µA. A schematic of the apparatus is shown in Fig. 14. The collimated beam with a diameter of approximately $\frac{1}{16}$ in. enters the collision region through an orifice whose diameter is large enough to transmit all of the beam. This reduces the possibility of having secondary particles or secondary electrons enter the observation region while still providing adequate gas containment for the collision chamber. The ion beam was monitored on a deep parallel plate assembly with an inclined backplate. Voltages were applied to the plates to assure complete suppression of secondary electrons. Helium gas, which was passed through a cold trap to remove condensable materials, leaked into the collision chamber. The target gas pressure was measured with a McLeod gauge. The McLeod gauge pressure was not corrected for errors introduced by the pumping effect of mercury vapor moving from the reservoir to the cold trap, and an overall limit of accuracy of the pressure measured was estimated to be 5%.

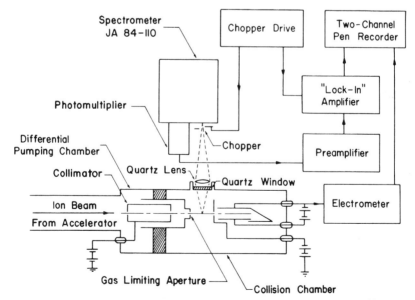

Fig. 14. Schematic drawing of the apparatus used by Thomas and Bent.[11]

Radiation from the collision region passed through a quartz window in the side of the collision chamber and was focused on the entrance slit of a Jarrel Ash 84-110 scanning spectrometer by a quartz lens. An ac detection technique was employed to reduce the effects of noise in the photomultiplier detector (EMI6256S). A tuning-fork chopper modulated the light flux entering the spectrometer at 100 Hz. The modulated signal from the photomultiplier was fed into a "lock-in" phase-sensitive detector. The signal identified by the modulating frequency and phase was extracted from the background noise signal, which was independent of the modulation.

Thomas and Bent found that the polarization fraction did not exceed 8% in the worst case and therefore introduced an error of less than 4%. Their estimate of polarization fraction, however, is at variance with the measurements of other workers in the case of the $4\,^1D \to 2\,^1P$ emissions.[10] The error might be as large as 9% at the extremes of the energy range used.

Experimental runs were normally taken by varying the impact energy while holding the target pressure constant. Preliminary data runs were employed in an effort to establish a region of pressure independence, i.e., single-collision conditions. Such a pressure-independent region could not be established, and it was necessary to extrapolate the apparent cross section to zero pressure. The spectrometer was set manually to the wavelength under study, and the data were recorded for an adequate interval to assure a good statistical average.

The measurements were repeated at different pressures. The detection efficiency of the system was determined at frequent intervals, and the data were used to calculate emission cross sections. The emission cross sections obtained at different "single-collision" pressures were consistent to within ±3%. The emission cross sections were used along with ratios of transition probabilities from the work of Gabriel and Heddle[61] to obtain excitation cross sections.

The cross-section data for the excitation of the He($3\,^1P$) state by protons are shown in Fig. 15. The data of Hasselkamp et al.[19] are almost indistinguishable from the data of Hippler and Schartner.[50] These data and those of Thomas and Bent[11] are in excellent agreement, even though the calibration techniques differ markedly. Hasselkamp et al. focused the light from the collision region on the entrance slit of a Leiss monochromator. This optical arrangement permitted the introduction of neutral density filters between the collision region and the monochromator when the tungsten strip was switched into the beam position. This technique overcomes the large difference between the calibration count rate and the signal count rate. The calibration was also made by using the Born approximation cross section for the electron-bombardment excitation of He($3\,^1P$) to calibrate the detection system when a 400-eV electron beam was substituted for the proton beam. The two techniques proved to be in excellent agreement.

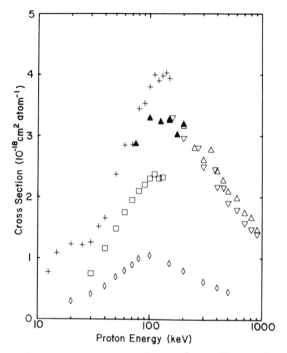

Fig. 15. Cross-section data for the process $H^+ + He(1\ ^1S) \rightarrow H^+ + He(3\ ^1P)$: △, Thomas and Bent[11]; ▲, Thomas and Bent,[11] data using deuterium ions; ▽, Hasselkamp et al.[19]; □, Dodd and Hughes[63]; ◇, Denis et al.[62]; +, Van den Bos et al.[64,10]

The data of other experimenters[62-64] are not in such good agreement. Much of this disagreement is due to the calibration techniques used in the various experiments. If the data are renormalized to a common curve, as demonstrated in Fig. 16, only the data of Denis et al.[62] deviate from it appreciably. Possible explanations for this deviation are discussed by Thomas.[10]

Figure 16 also shows several theoretical calculations for $H^+ + He(1\ ^1S) \rightarrow H^+ + He^*(3\ ^1P)$ excitation. The Born approximation[52] characteristically overestimates the measured cross section and exhibits a maximum at too low a proton energy. The impact parameter–close coupling calculation of Van den Bos[57] provides much better agreement. The calculation of Bransden and Issa[65] who used the second-order potential method[43] involves a nine-state approximation including transitions via the $2\ ^1P$ level. This calculation provides good agreement with the normalized experimental data. The calculation is not expected to give agreement below 100 keV and can be seen to begin to deviate from the experiment for 200-keV protons. The second-order diagonalization calculation of Baye and Heenen[59] shows excellent agreement for proton energies between 30 and 1000 keV.

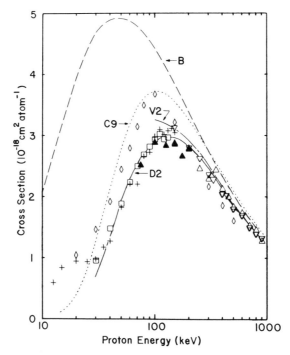

Fig. 16. Normalized cross-section data and theory for the process $H^+ + He(1\,^1S) \to H^+ + He(3\,^1P)$: △, Thomas and Bent[11]; ▲, Thomas and Bent,[11] data using deuterium ions; ▽, Hasselkamp et al.[19]; □, Dodd and Hughes[63]; ◇, Denis et al.[62]; +, Van den Bos et al.[64,10] Curves: B, Born[52]; C9, nine-state coupled state[57]; D2, second-order diagonalization[59]; V2, second-order potentials.[65]

Some of the results of the optical experiments for the excitation of the $He(4\,^1S)$ state are shown in Fig. 17. The data of Thomas and Bent,[11] Robinson and Gilbody,[66] Van den Bos et al.,[64] Dodd and Hughes,[63] Denis et al.,[62] and Hasselkamp et al.[19] are shown. The emission functions from the $4\,^1S \to 2\,^1P$ (5047 Å) line are used. There are no known secondary processes which would significantly populate or depopulate the $He(4\,^1S)$ state at the pressures employed. Cascade into the ^1S levels can occur through transitions from the higher ^1P levels. As noted, it is not necessary to measure cross sections for the precise transition involved in the cascade if the transition probabilities are known and if a transition from the higher state to some lower state is known. This permits the use of the $n\,^1P \to 2\,^1S$ transition to provide the emission cross sections for levels up to $n = 6$. For higher levels, the assumption that the cross section for the formation of a level decreases as n^{-3} provides an estimate of the cascade contribution (see Refs. 64 and 67). The cascade contribution is found to be less than 2%. The data of Denis et al.[62] show some deviation from those of the other experimenters.

This may be due to inadequate correction for the effects of neutral hydrogen in the incident beam.

Incident deuterons appear to produce the same excitation as protons of the same velocity. Thomas and Bent extended their data to lower velocities by using deuterons.

The data of various experimenters show good agreement in regard to curve shape. The agreement in absolute magnitude is not as impressive. Polarization of the $4\ ^1S \rightarrow 2\ ^1P$ transition is zero; hence the major sources of systematic errors in the experiments are probably to be found in the determination of the target density and in the determination of the detection efficiency of the optical system.

It is not possible to state objectively which of the measured values for this cross section is correct, because each is an independent measurement. The differences between the measurements are larger than the estimates on the limits of accuracy of the various measurements. The shape of the cross-section curve is not affected by the problems of making absolute pressure and detector efficiency determinations and hence may be compared

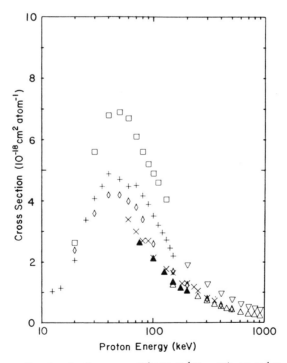

Fig. 17. Cross section data for the process $H^+ + He(1\ ^1S) \rightarrow H^+ + He(4\ ^1S)$: △, Thomas and Bent[11]; ▲, Thomas and Bent,[11] data using deuterium; ×, Robinson and Gilbody[66]; +, Van den Bos et al.[64,10]; □, Dodd and Hughes[63]; ◇, Denis et al.[62]; ▽, Hasselkamp et al.[19]

with theory. The $1\,^1S \rightarrow 4\,^1S$ excitation transition is optically forbidden and is expected to have the characteristics of a quadrupole transition. Typically, the cross-section curves for quadrupole transitions exhibit a sharper peak at lower ion energy than does the cross-section curve of an optically allowed transition with the same transition energy.

In order to make a comparison with theory, the curves have been normalized in Fig. 18 to the theory of Baye and Heenen.[59] The high-energy measurements have been normalized to the theory at 200 keV and the low energy measurements at 100 keV. The measurements of Hasselkamp *et al.*[19] have not been renormalized. The latter used two separate techniques to calibrate their measurements. They used a tungsten strip lamp and placed it in the same geometrical position on the beam as did Thomas and Bent.[11] They also normalized to the Born approximation calculation for 400-eV electron bombardment excitation of the $3\,^1P$ level and followed this by determining the cross-section ratios for $\sigma(3\,^1P)/\sigma(4\,^1S)$. At energies above 1 MeV, the experimental value for this ratio is the same as that obtained from Born approximation calculations.

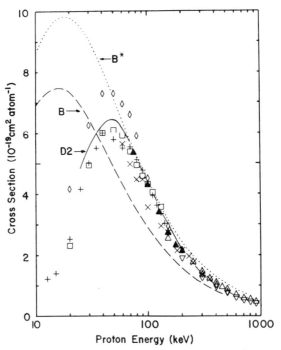

Fig. 18. Normalized cross-section data and theory for $H^+ + He(1\,^1S) \rightarrow H^+ + He(4\,^1S)$: △, Thomas and Bent[11]; ▲, Thomas and Bent,[11] data using deuterium ions; ×, Robinson and Gilbody[66]; +, Van den Bos *et al.*[64,10]; □, Dodd and Hughes[63]; ◇, Denis *et al.*[62]; ▽, Hasselkamp *et al.*[19] Curves: B, Born[52]; B*, Born[68]; D2, second-order diagonalization.[59]

The agreement of the normalized data with the measurements of Hasselkamp et al.[19] and with the theoretical calculation of Baye and Heenen[59] is very good. Baye and Heenen employ a second-order diagonalization method in an impact parameter treatment. The general agreement in curve shape indicates, as noted above, that the energy dependence in the experimental measurements is much more dependable than the absolute magnitudes. The normalization technique used by Hasselkamp et al. provides a method for circumventing the problem of absolute calibration of the optical detectors but also limits a serious comparison with theory. It should be noted that Hasselkamp et al. report excellent agreement between their tungsten calibration and the normalization procedure.

Theoretical calculations of the cross sections for "optically allowed" dipole transitions should be more accurate than those for the "optically forbidden" quadrupole transitions, because the results for dipole transitions are less sensitive to the wave functions. The renormalized cross sections show relative dependence on ion energy, i.e., the shape is essentially the same for the various experimenters, although the absolute magnitudes vary strongly.

The Born approximation calculations[52,68] do not give good agreement in the shape of the cross-section curve. Because the reported data lie on both sides of the Born approximation curve, it is inappropriate to speculate on the agreement in absolute magnitude at the high energy end of the range. It is clear that the peak in the Born approximation is located at too low an energy and that the magnitude is too high at the low energies.

The examples discussed cover only a small fraction of the optical studies of the excitation of helium in proton and heavy-ion collisions. The reader is referred to Ref. 10 for a more complete listing of the optical experiments involving helium.

6. Differential Cross-Section Measurements

The energy-loss spectrometry technique can also be used to measure cross sections that are differential in ion scattering angle. The technique previously discussed (Section 2) is used to obtain spectra at various ion scattering angles. Differential cross sections are obtained from these data. Results have been published for the excitation of helium by helium[69] and lithium ions.[70]

Several changes have been incorporated into the apparatus to make accurate angular measurements possible.[69] Ions produced in a low-voltage discharge source are focused and mass analyzed by a Wien filter. Mass-selected ions are then accelerated and steered through a variable-angle collimator into a target chamber containing the gas under study (Fig. 19).

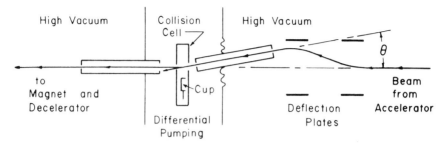

Fig. 19. Schematic drawing of variable-angle collision chamber.

After traversing the scattering chamber, the ions pass through an exit collimator, and the transmitted beam is magnetically analyzed to remove any products of charge-changing collisions. A set of movable slits beyond the magnet may be positioned accurately in both the vertical and horizontal planes.

When measurements differential in scattering angle are made, a series of runs with different incident-ion beam angles is made. The angle at which the beam enters the target chamber can be varied by changing the entrance collimator angle θ and adjusting the voltage on two pairs of vertical deflection plates that precede the collimator.

The systematic errors involved in making the angular measurements have been discussed in detail for various geometries and different angular ranges.[71] With the experimental arrangement used in these experiments, variations in the length of the interaction region as a function of scattering angle can be accurately calculated. Also, because of the small angles, the change in the acceptance angle of the detector as a function of scattering angle is less than 1%. Very small angles, however, maximize the errors in angular dependence, because an uncertainty is introduced into the scattering angles by the finite sizes of the apertures.

The total angular resolution is a convolution of the acceptance angle of the detector and the divergence of the incident beam. The convolution is obtained by plotting $[I_{10}(\theta)]_{n=0}$, the unscattered current detected in the analyzer, as a function of incident beam angle when there is no gas in the collision cell. This measurement is consistent with the values of the angular acceptance and beam divergence obtained from separate experiments and is used to determine the angular resolution at each energy.

A set of energy-loss spectra[69] for He^+ incident on He is shown in Fig. 20. The first peak at 0-eV energy loss corresponds to the initial beam or to ions which have been elastically scattered. The double peak between 19 and 22 eV is due to particles which have undergone discrete energy losses upon excitation of the $He(n=2)$ states of the target atom at 19.815, 20.611, 20.959, and 21.213 eV. The peak at 40.8 eV is due to the excitation of the

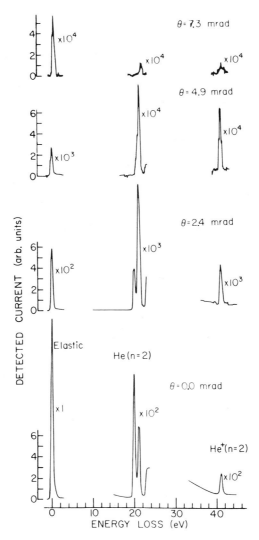

Fig. 20. Energy-loss spectra for 50-keV helium ions incident on a helium target taken at various scattering angles.

$He^+(n=2)$ states of the projectile ion. Structure resulting from the $He(n>2)$ states is observed as an unresolved peak just before the ionization continuum.

Many of the qualitative features of $He^+ + He$ angular scattering at 50 keV are apparent from Fig. 20. The elastic peak, which accounts for more than 96% of the total detected current at this target density, decreases by more than two orders of magnitude within 2.4×10^{-3} rad (center of mass).

Note that because the angular resolution at this energy is 2×10^{-3} rad (center of mass) the spectrum obtained at 2.4×10^{-3} rad includes contributions from particles that have been scattered by as little as 1.5×10^{-3} rad (center of mass). At larger angles, the elastic peak is still the predominant feature, but it is of the same order of magnitude as the inelastic peaks. The shift in position with angle of all the peaks agrees well with the expected energy loss resulting from pure elastic scattering.

The most striking feature of this set of spectra is the change in relative intensities of the He($n=2$) excitations. At $\theta = 0$, the He($2\,^3S$) state accounts for more than 50% of the total He($n=2$) peak, whereas at $\theta > 4.8\times 10^{-3}$ rad, it makes a very small contribution. This is similar to energy-loss spectra obtained at 0.6 keV by Lorents et al.,[72] who observed that initially the $2\,^3S$ contribution decreases with increasing angle. However, they made their measurements at much larger angles ($>2\times 10^{-2}$ rad, center of mass), and it is clear from their data that the contribution of the $2\,^3S$ does not remain low but appears to oscillate as a function of angle. This feature might still hold true at higher energies, but the total signal decreases so rapidly with increasing angle that measurements at larger angles cannot be made.

The He$^+$($n=2$) excitation peak is also interesting because of its relatively broad angular dependence. In this particular set of data, it has decreased by a factor of only 2×10^2 over the same angular range in which the He($n=2$) peak has decreased by a factor of 2×10^3 and the elastic peak has decreased by more than 2×10^4. This indicates that at least in He$^+$–He collisions the angular dependence of the cross section cannot be estimated from total scattering measurements that do not distinguish between the scattered ions.

The expression used to extract apparent differential cross sections from energy-loss data is

$$\frac{\Delta\sigma_p}{\Delta\Omega} = \left(\frac{1}{nL\,\Delta\Omega}\right)\frac{[I_{1p}(\theta)]_f}{(I_{10})_f} \qquad (6.1)$$

in which $[I_{1p}(\theta)]_f$ is the current under the appropriate peak in the energy loss spectrum, which results from singly charged particles that have lost energy in the interaction, p, and have been scattered into the solid angle $\Delta\Omega$ at θ; $(I_{10})_f$ is final current resulting from the elastic beam integrated over all angles to obtain the total beam current; L is the length of the interaction region; n is the average target density; and $\Delta\Omega$ is the total solid angle subtended by the detector. Equation (6.1) automatically incorporates effects of charge exchange and other inelastic processes. It is subject to the same limitations discussed in Section 2 on energy-loss spectrometry.

A further limitation in the case of differential measurements is the possibility of multiple collisions involving one energy-loss process and one or more elastic collisions. As a practical criterion, $\sigma_e(\theta > \frac{1}{2}\Delta\theta)$, the cross section for elastic scattering through angles greater than $\frac{1}{2}\Delta\theta$, was measured. In this definition for $\sigma_e(\theta > \frac{1}{2}\Delta\theta)$, $\Delta\theta$ is the angular resolution of the apparatus in the vertical plane. With this definition for σ_e, the approximation $\exp(-\sigma_e nL) \approx 1 - \sigma_e nL$ is correct to within 1% at all target densities, indicating that for practical purposes the single-collision criterion is met.

From sets of spectra taken at various angles, data were derived as pairs of current (integrated over the appropriate energy loss) versus angle for each process. Because of the rectangular nature of the beam collimators and analyzing apertures, each value, $[I_{1p}(\theta)]_f$, represents current scattered into a rectangular solid angle centered at θ. In order to calculate the total current, the detection region was divided into concentric rings of width $\Delta\theta$ which had a central circle of diameter $\Delta\theta$. Then from the data, the average current detected per solid angle, $J_p(\theta_i)$, at each ring was calculated. In terms of these current densities, Eq. (6.1) becomes

$$\left(\frac{d\sigma_p}{d\Omega}\right)_{\text{av}} = \left(\frac{1}{nL}\right)\frac{J_p(\theta_i)}{(I_{10})_f} \tag{6.2}$$

and can be interpreted to give the differential cross section $d\sigma_p(\theta)/d\Omega$ averaged over the interval $\theta_i - \frac{1}{2}\Delta\theta$ to $\theta_i + \frac{1}{2}\Delta\theta$. The current scattered into each ring can be calculated by multiplying the current density of the ring times the solid angle subtended by the ring, and the total current can be found by adding up the currents in all the rings.

The differential cross-sectional data can often be displayed in reduced plots, which demonstrate the important features of the data while also permitting data covering a large ion-impact energy range to be displayed on a single plot.[73] Plots of ρ versus τ, in which $\rho = \theta (\sin\theta)(d\sigma/d\Omega)$ and $\tau = E\theta$, make it possible to compare data obtained at different impact energies. Figure 21 shows such plots for the He($n = 2$) excitation data. In spite of the large energy range, common curves can be drawn within the error bars (not shown) of most of the points. A common differential excitation cross-section curve plotted as ρ versus τ is to be expected for an excitation mechanism involving the crossing of potential energy curves at a definite impact parameter. Here, τ is inversely related to the impact parameter, and changes in slope would be expected at values of τ corresponding to the crossing of the potential energy curves of the ground and excited states of the colliding ion–atom system.

A curve-fitting technique was used to estimate the contribution of each state to the total He($n = 2$) excitation.[74] This involved superimposing the shape of the elastic peak at the expected location of each state in the energy-loss spectrum and, by a least-squares fit, finding the height of the peak at each location such that the sum of the four peaks best reproduced the

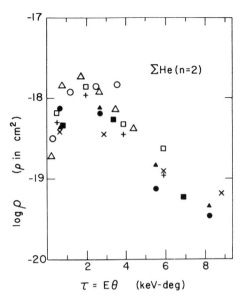

Fig. 21. Plot of ρ versus τ for the He($n = 2$) excitation peak where $\rho = \theta(\sin\theta)(d\sigma/d\Omega)$ and $\tau = E\theta$. Data points (values in keV): ○, 25; △, 30; □, 40; +, 50; ●, 70; ▲, 80; ■, 100; ×, 120.

data. Details concerning the limitations and uncertainties involved in applying this technique are included in Ref. 74. Although this procedure gives very reliable results for the He($2\,^3S$) state, there are often large fluctuations in the results for the other three states.

Figure 22 shows smoothed-out plots of the results of the curve-fitting process for finding the contribution of each He($n = 2$) state to the total He($n = 2$) peak. The cross section for excitation to the He($n = 2$) state is shown in Fig. 21. The data shown in Fig. 22 are given as fractional contributions versus the reduced angle $E\theta$ in order to illustrate the relative behavior of each state. Although there are large uncertainties in some of the contributions, the trends illustrated in the figures are generally observed in all the data.

He($2\,^3S$). The $2\,^3S$ state definitely dominates the He($n = 2$) structure at energies from 40 to 100 keV. At all energies below 100 keV, its fractional contribution decreases sharply with angle. At 100 keV, it contributes more than 30% to the He($n = 2$) peak at all angles, whereas at 120 and 140 keV the fractional contribution of the $2\,^3S$ state actually increases with increasing angle.

He($2\,^1P$). The $2\,^1P$ state could almost be considered the complement of the $2\,^3S$ state because its contribution sharply increases with angle at every energy at which the $2\,^3S$ contribution decreases with angle.

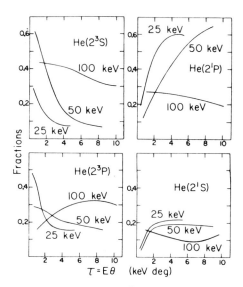

Fig. 22. Fractional contributions of the He($2\,^3S$), He($2\,^1P$), He($2\,^3P$), and He($2\,^1S$) states to the He($n = 2$) excitation peak (Fig. 21).

He($2\,^3P$). The $2\,^3P$ state exhibits similar angular dependence to that of the $2\,^3S$ state except that the change in shape occurs at lower energies. While the $2\,^3S$ contribution does not flatten out until 100 keV, the contribution of the $2\,^3P$ state is relatively uniform at 50 keV and definitely increases with angle at higher energies.

He($2\,^1S$). The $2\,^1S$ state never contributes more than 24%. There is more uncertainty in its behavior, because the fitting technique is less reliable for small contributions. Nevertheless, the shape of the 25- and 50-keV curves is typical of all observations in that energy range, i.e., the contribution is less than 8% at $\theta = 0$ and increases to ~20% at larger angles. There appears to be an abrupt change at 70 keV, above which the $2\,^1S$ contribution is 15–20% at $\theta = 0$.

In a ρ versus τ plot for the elastic scattering of lithium ions by helium,[70] the low energy data of Lorents and Conklin[75] and Francois et al.[76] seem to have slightly higher values and less slope than data taken at 15–100 keV.[70] However, if one considers the differences in energy, technique, and normalization, the similarities in the curves are more striking than the differences. All the data exhibit curves of similar shape and decrease somewhat more rapidly with large values of τ as the ion energy increases. The differences in magnitude are small in relation to the large range in energy.

The data for the 21- and 60-eV peaks corresponding to excitations of the helium target do not exhibit this behavior. The maximum in the ρ versus

τ plots for these transitions moves to lower τ with increasing energy. The possibility of a velocity dependence in R, the radius at which the proposed curve crossings occur, was noted by Lorents and Conklin.[75] This effect and the implications of the constant ratios of He($2\,^1S$) to He($2\,^1P$) excitations in the 21-eV peak have been discussed by Park et al.[77]

The calculations of Olson[78] indicate that plots of $\rho = \theta \sin \theta (d\sigma/d\Omega)$ versus $E^{3/2}\theta$ provide a reduced plot for collisions involving a rotational coupling. Figure 23 shows data for the 21-eV peak in the ion-energy spectrum which is due to the excitation of the helium target to the $n = 2$ states by incident lithium ions. The low-energy data of Lorents and Conklin[75] and that of Francois et al.[76] and the 15–100 keV data of Park et al.[70] are shown in a ρ versus $E^{3/2}\theta$ plot. As Olson[78] predicted, many of the threshold effects have been removed, and the velocity dependence has been largely accommodated. The magnitude of ρ at the peak in the curve is not expected to be exactly identical over the entire energy range even if the threshold effects are removed by the ρ versus $E^{3/2}\theta$ plot, because the branching ratio between the excitation and charge-exchange channel depends on the velocity.[79] However, a common curve could be drawn

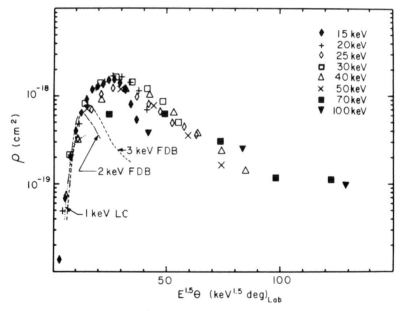

Fig. 23. Apparent reduced-cross-section measurements for excitation of the He($n = 2$) states by Li$^+$, $\rho = \theta \sin \theta (d\sigma/d\Omega)$ versus $E^{1.5}\theta$. The term ρ is expressed in center-of-mass coordinates and the identifying ion energies and $E^{1.5}\theta$ are expressed in laboratory coordinates. The curve LC is derived from the data of Lorents and Conklin.[75] The curve FDB is from the data of Francois et al.[76]

through the points, which would provide within a factor of 2 a representation of the data covering the entire energy range.

Cross sections differential in ion scattering angle could also be obtained with optical techniques and by correlating the emitted photon with the scattered ion (see Refs. 80 and 81). The emphasis in the coincidence measurements has not been placed on direct excitation measurements, and little information is available. The reader is referred to Chapter 3 of this text for a discussion of these coincidence measurements.

7. Excitation of Simple Molecules

The number of possible excitation channels in a collisional process is greatly increased when molecular targets are substituted for atomic targets. The complications involved in making the measurements are correspondingly increased. Much of the experimental effort has been concentrated on N_2, O_2, and CO_2 because of their obvious importance in understanding the aurora and other atmospheric phenomena. The extensive literature concerning proton and atomic hydrogen collisions with atmospheric constituents has been reviewed by McNeal and Birely.[82] The general literature through 1969 is reviewed by Thomas.[10]

Both optical and energy-loss techniques meet major difficulties when the collisional excitation of molecules is studied. The electronic states produce emissions from a series of vibrational states each of which has rotational structure. The energy-loss apparatus frequently has inadequate resolution to resolve all of this detailed structure. In optical experiments, it is frequently found that the rotational structure extends beyond the band pass of the spectrometer. Emissions of the atomic lines from dissociation fragments often overlap the molecular emissions in the spectra. Even if emission cross sections for molecular targets can be measured, it is usually impossible to derive level excitation cross sections because the branching ratios are not well known. In the case of the N_2^+ spectrum, measurements have been made of all the transitions that populate and depopulate the excited states, and a direct estimate of some excitation cross sections is possible.[82] This case is an exception.

The analysis of the data must take into account the fact that the target molecule will not necessarily be in its lowest vibrational and rotational state before the collision. The measured cross section is a weighted average of the cross sections over the Boltzmann distribution of the states of the molecule at the target temperature. The existence of metastable states in the molecule, notably triplet states, could result in a nonthermal distribution of states in the target as a result of excitation produced by the incident ion beam.

The energy-loss spectra of heavy ions incident on molecules can be understood in terms of the Franck–Condon principle[83,84] and the Wigner spin conservation rule.[85] The Franck–Condon principle states that if the collision occurs in a very short interval, the nuclear separation in a molecule does not change during the collision. The transition produces an energy change without a change in nuclear separation; hence on a potential energy diagram in which energies of the states are given as a function of nuclear separation the transition can be represented by a vertical line. If vertical lines are drawn on the potential energy diagram from the extremes of the lowest vibrational ground state, the energy loss corresponding to the excitation of the transition can be read from the intersection of these vertical lines with the potential energy curve representing the excited state. Figure 24 shows the energy-level diagram[86] of O_2 and an energy-loss spectrum for protons incident on molecular oxygen. The spectrum is taken from the work of Moore.[5] The structure in the energy-loss spectrum can be seen to be directly related to the potential energy curves through the Franck–Condon principle.

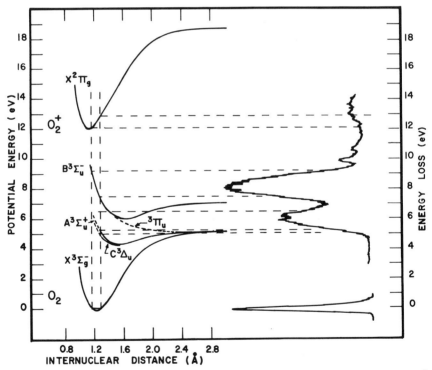

Fig. 24. Energy-loss spectrum of 3.3-keV protons scattered from O_2 adapted from Moore,[5] compared with the potential energy curves for O_2 adapted from Gilmore.[86]

The Wigner spin conservation rule,[85] which requires that the total electron-spin angular momentum remain unchanged during a collision, appears to hold quite rigorously. In typical cases, the probability of a spin-nonconservative transition is only about 0.002 that of a spin-conservative transition.[87] Moore studied inelastic $N^+ + O_2$ collisions in which each of the collision partners possesses several low-lying excited states.[87] The crossings and pseudocrossings of the potential energy surfaces associated with these low-lying states provide for spin-orbit-coupling, which could result in spin-nonconserving transitions. Moore found that such spin nonconservation was more probable in $N^+ + O_2$ collisions than in less complex collision systems. Even in this case, there is a difference of one order of magnitude in the cross sections between spin-conserving and spin-nonconserving transitions.

7.1. Excitation of Molecular Hydrogen

The ion energy-loss spectrum of molecular hydrogen shown in Fig. 25 has no resolved structure with 0.6-eV resolution.[88] Even in optical measurements, it is almost impossible to provide sufficient resolution to assure that spectral features will not interfere.[10] The spectrum shows a single peak at 12.5 eV that corresponds to the maximum in the Lyman bands.[89] The large numbers of close-lying molecular states prevents the determination of the cross section for any particular state. However, the spectrum does provide

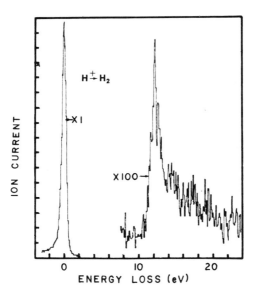

Fig. 25. Energy-loss spectrum for 50-keV protons incident on a molecular hydrogen target. (Raw data from a single run.)

upper bounds on the cross section for excitation of the molecular states. The one major peak in the spectrum corresponds to the Franck–Condon overlap region for the excitation of the $B\ ^1\Sigma_u^+$ level from the $X\ ^1\Sigma_g^+$ ground state of the hydrogen molecule. This one major peak in the spectrum cannot contribute to Lyman α radiation (1216 Å). Lyman α could result from the dissociation of excited H_2 with one of the product atoms in the $2\ ^2P$ state. This process requires 14.6 eV. The excitation of the main peak results in the emission of the Lyman bands centered at 1606 Å.

Optical studies of emissions from the collision of protons with molecular hydrogen have been centered on the emission lines of atomic hydrogen.[10,90,91] In the ultraviolet, the Lyman α emissions are the most intense. The Balmer lines of atomic hydrogen dominate the spectrum in the visible wavelengths. Both the Lyman α and Balmer lines are excited by the charge exchange $(1(00)/0^*(01))$ and by the simultaneous dissociation and excitation of the target molecule $(1(00)/1(00)^*/100^*)$. In addition to the atomic lines, the weaker but quite obvious optical spectrum of the neutral molecule is recognizable. Edwards and Thomas[91] and Dahlberg et al.[90] have studied these molecular emissions. Strong emissions from the singlet states

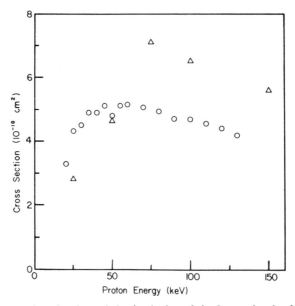

Fig. 26. Cross sections for the emission/excitation of the Lyman bands of H_2 by proton impact: ○, Data for emission of the group of bands centered at 1606 Å including the (4, 11), (5, 12), and (6, 13) vibrational bands.[90] △, Cross-section data for energy loss corresponding to the excitation of the Lyman bands. These data have been divided by 10 to aid in comparison with the emission data.[88]

were observed, but no triplet emissions were positively identified. This is to be expected for proton bombardment from considerations of spin conservation.

Dahlberg et al.[90] measured the cross section for the emission of the (4, 11), (5, 12), and (6, 3) vibrational transitions of the ultraviolet Lyman bands, $B\ ^1\Sigma_u^+ \to X\ ^1\Sigma_g^+$. In their experiment, which was confined to a study of the vacuum ultraviolet emissions, they employed a vacuum Czerny–Turner monochromator which was placed at 90° to the beam path. The photons were detected by a photomultiplier with a sodium salicylate screen. The measurements were relative and were normalized to the H^+ measurement of the Lyman α emission for H^+ impact on N_2, an experiment which has since been shown to contain errors.[10]

The sum of the absolute magnitude of the cross sections for the (4, 11), (5, 12), and (6, 3) vibrational transitions of the ultraviolet band ($B\ ^1\Sigma_u^+ \to X\ ^1\Sigma_g^+$) obtained by Dahlberg et al.[90] is shown in Fig. 26. Energy-loss data[88] for the integral of the inelastic cross section measured over the region of the spectrum corresponding to the energy loss required for the Lyman excitation of the $B\ ^1\Sigma_u^+$ vibrational levels is also shown. The energy-loss data have been divided by 10 so both curves can be displayed in the same figure. The large difference in magnitude is not significant, because different things were measured in the two experiments. The energy-loss experiment included 19 vibrational levels. In their experiments, Dahlberg et al. measured the emission in a 16-Å band-width centered about 1606 Å. The group of bands transmitted included the (4, 11), (5, 12), and (6, 13) vibrational bands.

Edwards and Thomas[91] studied the emissions from the $G\ ^1\Sigma_g^+ \to B\ ^1\Sigma_u^+(0-0)$, 4632 Å, and the $I\ ^1\Pi_g \to B\ ^1\Sigma_u^+(1-0)$, 4176 Å, transitions. (See Fig. 27.) Although they were unable to integrate over the entire rotational structure of these transitions, they estimated that at least 90% of the rotational structure was included. Because no known transitions cause a cascade population of the $G\ ^1\Sigma_g^+$ and $I\ ^1\Pi_g$ states, the emission cross sections were believed to be proportional to the cross sections for the excitation of these states; however, level-excitation cross sections could not be calculated because the branching ratios were not known.

The data on the excitation of molecular states of hydrogen are very sparse. More data are available on the emission of the hydrogen atoms resulting from collisionally induced dissociation.[10] These data are not reviewed here.

7.2. Excitation of Molecular Nitrogen

The excitation of N_2 has been much more thoroughly studied than has H_2. A high-resolution (0.15-eV FWHM) energy-loss spectrum of 2.9-keV protons incident on a nitrogen target is shown in Fig. 28. This spectrum is

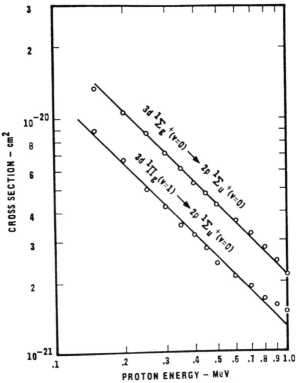

Fig. 27. Approximate cross sections for the emission of the $3d\ ^1\Sigma_g^+\ (\nu = 0) \to 2p\ ^1\Sigma_u^+\ (\nu = 0)$ $[G\ ^1\Sigma_g^+ \to B\ ^1\Sigma_u^+\ (0\text{–}0)]$ and the $3d\ ^1\Pi_g\ (\nu = 1) \to 2p\ ^1\Sigma_u^+\ (\nu = 0)\ [I\ ^1\Pi_g \to B\ ^1\Sigma_u^+\ (1\text{–}0)]$ bands of molecular hydrogen by proton impact.[91]

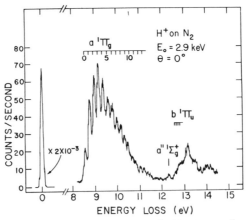

Fig. 28. Energy-loss spectrum of 2.9-keV protons incident on molecular nitrogen from Moore.[5]

from the work of Moore.[5] Schowengerdt and Park[92] have obtained energy-loss spectra for energies from 20 to 120 keV. These spectra have poorer resolution but display the same general features.

The features displayed in the energy-loss spectrum shown in Fig. 28 can be identified with known states of the N_2 molecule.[93] The dominant feature is identified as the Lyman–Birge–Hopfield (LBH) transition $X\ ^1\Sigma_g^+ \to a\ ^1\Pi_g$. The system can be identified from the location and spacing of the individual vibrational states of the band system. The relative intensity of the first six vibrational components of this transition is in good agreement with theoretical predictions based on Franck–Condon factors.[5] The structure between 13 and 14 eV is identified as a result of the excitation of the $b\ ^1\Pi_u$ state of N_2.[87]

Both Schowengerdt and Park[92] and Moore[5] report cross sections for the excitation of the LBH bands ($X\ ^1\Sigma_g^+ \to a\ ^1\Pi_g$) (Fig. 29). The cross sections obtained, however, are actually cross sections for energy-loss corresponding to the energy losses of the LBH bands. The later work of Moore[94] indicates that there is a contribution from the $X\ ^1\Sigma_g^+ \to w\ ^1\Delta_u$ transition to the observed structure in the N_2 spectra at low impact energies. It is not known if this transition makes a similar contribution at higher impact energies.

Spectral scans (see Refs. 95–98) of the optical emission from nitrogen gas under the bombardment by protons are dominated by the vibrational bands of the N_2^+ first negative system ($B\ ^2\Sigma_u^+ \to X\ ^2\Sigma_g^+$). Some bands of the

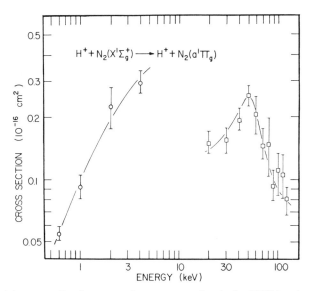

Fig. 29. Total cross section for energy-loss corresponding to the LBH bands of molecular nitrogen. The data are reported as cross sections to the $a\ ^1\Pi_g$ state; however, contributions from an underlying transition are likely (see Ref. 94): ○, Data of Moore[5]; □, data of Schowengerdt and Park.[92]

N_2 second positive system ($C\,^3\Pi_g \to B\,^3\Pi_g$) are also observed, although this signal is very weak. The atomic nitrogen lines and the Lyman α lines are the most prominent emissions of the spectrum below 2000 Å. The LBH bands ($a\,^1\Pi_g \to X\,^1\Sigma_g^+$), which dominate the energy-loss spectra, are optically forbidden, and while they appear in the spectra, the bands are significantly weaker than the atomic lines.

The only optical measurements of direct excitation of a nitrogen molecule by protons are those of the emission cross sections for the $C\,^3\Pi_u \to B\,^3\Pi_g$ transition. This excitation to the $C\,^3\Pi_u$ state from the $X\,^1\Sigma_g^+$ ground state by protons involves the violation of spin conservation. This cross section is, therefore, very small for proton bombardment. The excitation of a triplet state can proceed readily when the bombarding projectile carries an electron, which can be exchanged with the target without requiring a simultaneous state-change in the projectile. Thus, atomic hydrogen or helium ions can excite singlet–triplet transitions because they have a single electron that they can exchange without changing their principal quantum number. Because the cross section for bombardment by a hydrogen atom is quite large, great care must be exercised to ensure that the observed emission is not the result of atomic contamination of the proton beam.

The measurement of the cross section for emission from the $C\,^3\Pi_u \to B\,^3\Pi_g$ transition is shown in Fig. 30. The work of Thomas et al.,[98]

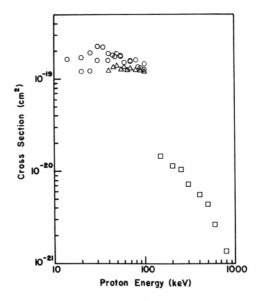

Fig. 30. Emission cross section for $C\,^3\Pi_u \to B\,^3\Pi_g$ transition for molecular nitrogen under proton impact: ○, Dahlberg et al.[95]; ▲, Hoffman et al.[96]; □, Thomas et al.[98]

Dahlberg et al.,[95] and Hoffman et al.[96] is shown. The emission cross sections are believed to be directly proportional to the cross section for formation of the $C\,^3\Pi_u$ state, because there are no known transitions which might populate the $C\,^3\Pi_u$ state by cascade.

The data of Dahlberg et al.[95] are in fairly good agreement with those of Hoffman et al.[96] if one considers the large scatter in the earlier data of Dahlberg et al. The agreement is perhaps fortuitous. The cross-sectional data of Thomas et al.[98] do not seem to fit smoothly onto the lower-energy data. The normalization procedure used by Dahlberg et al.[95] has been subject to some criticism[10]; however, the measurements of Hoffman et al.[96] and Thomas et al.[98] are both absolute. Both groups of experimenters paid careful attention to the possibility of neutral contamination of the projectile beam. The apparent discrepancy cannot be resolved without further experimental measurements.

Collisional-excitation studies of molecular nitrogen have been dominated by the measurements of the emission at 39.14 nm corresponding to the $B\,^2\Sigma_u^+ \to X\,^2\Sigma_g^+ (0\text{--}0)$ band of N_2^+. The emission cross section for this state is very large ($\sim 8 \times 10^{-17}$ cm^2 for 10-keV protons). The excitation process is through either charge exchange (1(00)/0(01)*) or direct collisional ionization (1(00)/1(01)*e). The maximum cross section for direct excitation of the entire LBH band by comparison is $\approx 3 \times 10^{-17}$ cm^2. Because there are no known states of N_2^+ that populate the $B\,^2\Sigma_u^+$ state by cascade and because the state has a short radiative lifetime, the emission cross section for this transition is essentially equal to the excitation cross section.[82] The cross sections for the excitation of the N_2^+ states have been discussed in detail in Ref. 82, and because the excitation processes are dominated in this case by charge exchange, the data are not repeated here.

7.3. Excitation of Molecular Oxygen

An energy-loss spectrum for 3.3-keV protons incident on molecular oxygen is shown in Fig. 24. This spectrum is taken from Moore.[5] The peak at 6-eV energy-loss has been attributed to the excitation of either the $A\,^3\Sigma_u^+$ state or the $C\,^3\Delta_u$ states.[5,99] Excitation of the $C\,^1\Sigma_u^-$ state would energetically make a possible contribution to this peak; however, the excitation is forbidden by the spin-conservation rule. Essentially all of this peak is above the $O(^3P)+O(^3P)$ dissociation limit.

The second inelastic peak in Fig. 24 is attributed to the excitation of the $B\,^3\Sigma_u^-$ state. This state is the upper state of the Schumann–Runge transition. The excitations, which are to high vibrational states of the $B\,^3\Sigma_u^-$ state, result in dissociation to the $O(^3P)+O(^1D)$ states. Thus, the structure below 10 eV results in dissociation of the molecule.

Park et al.[100] have obtained energy-loss spectra for proton energies between 20 and 110 keV. The differences between the spectra obtained for proton energies above 20 keV and those shown in Fig. 24 cannot be entirely attributed to the poorer resolution obtained at the higher energies. The peak at 6 eV in Fig. 24 appears to be missing from the spectra taken at proton energies of 20 keV and above. Even with a 2-eV resolution, a peak of the magnitude observed by Moore at 6-eV energy loss should have been clearly resolved. The energy-loss spectrum of protons incident on O_2 changes markedly at low energy,[4] and the relative magnitude of the 6-eV peak apparently decreases with increasing energy above 3 keV.

Fig. 31 shows the cross section for excitation of molecular oxygen. Because essentially all of these excited molecular states lead to dissociation, the curve could also represent the cross section for dissociation of O_2, $(1(00)/1(00)^*/100)$ (see Fig. 24). As already noted, at high proton-energies, the excited state is dominantly the $B\ ^3\Sigma_u^-$ state. At lower energies, there is an appreciable contribution from a lower state, probably the $C\ ^3\Delta_u$ state.[99]

The theoretical calculations of Green[101] and Breen[102] are also shown in Fig. 31. Both of these calculations are for direct excitation of the Schumann–Runge continuum by protons and hence do not include any contributions from excitation of the $C\ ^3\Delta_u$ or other states. The theoretical curve of

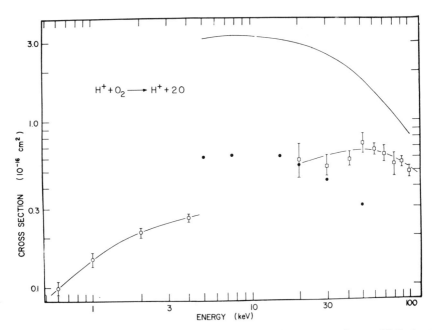

Fig. 31. Total cross sections for dissociative excitation of O_2 by proton impact: □, Park et al.[100]; ○, Moore[5]; ●, Breen[102]; ———, Green.[103]

Green[101] was determined from generalized oscillator strengths obtained by Lassettre et al.[103] from inelastic electron-scattering experiments. Breen used an impact parameter method. The agreement of the data with the theoretical curves is not good either in magnitude or in curve shape. The lack of agreement is not unexpected since the approximations employed in both types of calculations are not valid for low-energy protons.

Optical scans over the vacuum ultraviolet region of the spectrum of O_2 excited by protons indicate an absence of molecular emission from O_2. The $(b\ ^4\Sigma_g^- \to a\ ^4\Pi_u)$ first negative and $(A\ ^2\Pi_u \to X\ ^2\Sigma_g^+)$ second negative band systems of O_2^+ are the most intense spectral features.[104] This excitation results from charge-transfer excitation, $(1(00)/0(01)^*)$, and simultaneous excitation and ionization, $(1(00)/1(01)^*e)$. The atomic lines are quite strong. The observed intensity of 1304-Å ^3P–^3S resonance corresponds to a cross section of $\sim 6 \times 10^{-17}\,\mathrm{cm}^2$ for 12.5-keV protons.[104] This value is consistent with the cross sections measured in the energy-loss experiment for excitation of the $B\ ^3\Sigma_u^-$ and $C\ ^3\Delta_u$ states, which produce an oxygen atom in the ^3P state after dissociation. The available data on the emission of the atomic and molecular states resulting from charge-exchange excitation processes have been reviewed in Refs. 10 and 82 and are not repeated here.

7.4. Excitation of Singlet–Triplet Transitions

Ion energy-loss spectrometry permits the study of states which have optically forbidden decay transitions. The excitation of the $He(2\ ^3S)$ state by He^+ bombardment has already been mentioned. In the case of the excitation of neon by protons, hydrogen molecular ions, and helium ions, significant excitation of the triplet states has been observed only for the helium ions.[105] In ion–molecule collisions, the possibility of electron exchange permits the multiplicity of the target molecule to change while the electron spin angular momentum of the ion–molecule system is conserved. Singlet–triplet transitions can also be studied in electron energy-loss spectra[106]; however, the experiment is not quite so straightforward. With heavy-ion bombardments, the same apparatus can also be used with protons so that the same target can be studied without the possibility of electron-exchange excitation. This permits the identification and study of singlet–triplet transitions.

The excitation of the triplet states occurs predominantly through electron exchange. If the target atom in a singlet ground state exchanges one of its electrons for a projectile electron with opposite spin, the target is left in a triplet state with two electrons of the same spin but with different principal quantum numbers.

Essentially half of all electronic states of organic molecules are triplet states. These triplet states cannot be studied either by conventional photon-absorption studies or other conventional spectroscopic methods if the ground state of the molecule is a singlet state because they have an optically forbidden decay transition. As a result, the triplet states have not been studied with the care afforded the singlet states. A similar argument can be applied to singlet states of polyatomic molecules with triplet ground states. Molecules with singlet ground states excited to low-lying triplet states are usually metastable because of the requirement to conserve spin in photon emission. The excited metastable triplet states provide a long-lived source of energy for chemical processes which would not otherwise be energetically favored. Studies of photochemical kinetics indicate that triplet states are involved in the production or decomposition of substances in which there is energy available to populate molecular triplet states.[107] The conversion of solar radiation into chemical energy by biological systems is an obviously important example.

Figure 32 shows the energy-loss spectra of ethylene and butadiene for proton and helium ions. This figure is taken from the work of Moore.[108] The data were obtained with an energy resolution of 0.35 eV for 3-keV ions. The obvious differences between helium ion bombardment spectra and spectra under proton bombardment are due to the excitation of the triplet states by the helium ion. The most intense feature in each of these spectra is a peak at ~4 eV that corresponds to the transition from the singlet state to the lowest-lying triplet state. Moore[108] finds that the maximum of the peak, which results from the singlet–triplet excitation, occurs at 4.3 eV in the case of ethylene and is little changed in the case of the alkyl-substituted ethylene. The peak location is shifted to slightly higher energies for 1,1-difluoroethylene and to slightly lower energies for the chloro-substituted ethylenes. In butadiene the maximum of the peak resulting from the lowest triplet state is located at 3.2 eV. In addition, a peak appears at 4.9 eV, which also lies below the lowest-energy singlet state. The proton spectra are dominated by the optically allowed transitions. The first Rydberg transition and the transition to the $^1B_{1u}$ state are observed in proton bombardment.

The T-state and the V-state of ethylene have the same electronic configuration except for the difference in electron-spin multiplicity. The singlet–triplet splitting of 3.5 eV can be determined directly from the spectra.

Figure 33 shows the energy-loss spectra of pyridine for proton and helium ion bombardments.[109] Pyridine is expected to have three triplet states lying below or just above the lowest excited singlet state; however, the singlet–triplet transition is not observed in optical studies of the gas phase. The proton-impact spectrum displays a peak attributed to the lowest-energy singlet transition, $\tilde{X}\,^1A_1 \to \tilde{A}\,^1B_1$, and the $\tilde{X}\,^1A_1 \to {}^1B_2$ transition. The peak at 6.3 eV is attributed to the $\tilde{X}\,^1A_1 \to \tilde{C}\,^1A_1$ transition, and the peak at 7.3 eV to transitions to the split state ($^1B_2$, $^1A_1$). These three peaks also appear in the helium-ion-impact spectrum. However, the broad bands,

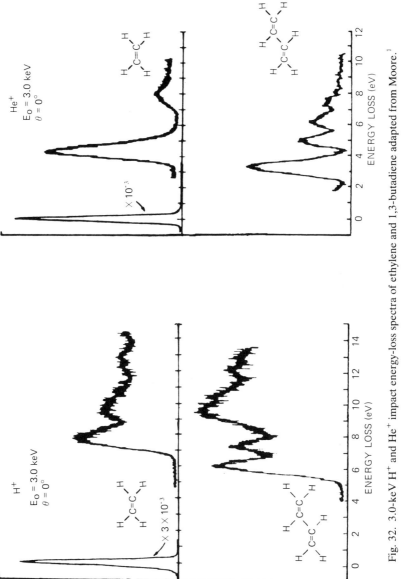

Fig. 32. 3.0-keV H$^+$ and He$^+$ impact energy-loss spectra of ethylene and 1,3-butadiene adapted from Moore.[1]

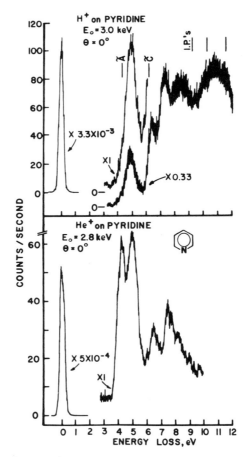

Fig. 33. 3.0-keV H^+- and He^+-impact spectra of pyridine taken from Doering and Moore.[109]

which were attributed to Rydberg transitions, are missing. The structure of interest in the helium ion spectrum is the peak at 4.1 eV, which is 0.2 eV below the lowest-energy singlet–triplet transition at 4.3 eV. Doering and Moore[109] attribute this peak to a singlet–triplet transition. They did not observe any lower-lying triplet states and concluded that in view of their experience in similar studies, lower-lying triplet states are unlikely.

8. Discussion

The general features of excitation cross sections in ion–atom intermediate energy collisions can be understood in terms of the energy-level scheme of the target. The cross section for excitation to a level, which can also be reached by a collisionally induced optically allowed transition, is generally larger than the cross section for an optically forbidden transition. The energy dependence of the cross section for an optically forbidden transition is generally more sharply peaked in impact energy than is the cross section of an optically allowed transition. However, the energy dependence of the excitation cross-section curves has a generally similar shape for proton energies between 15 and 500 keV. The cross-section curves show a single maximum with a smooth shape and monotonic decrease with increasing energy for both optically allowed and optically forbidden transitions. The optically allowed transitions fall off roughly as $(1/E)\ln E$, whereas the optically forbidden transitions fall off as $(1/E)$.[110] The location of the maximum in the curves is crudely given by the adiabatic criterion of Massey.[1] The maximum occurs at $v \approx aQ/h$, in which v is the impact velocity, a is the effective interactive distance, Q is the energy change, and h is Planck's constant. The adiabatic criterion cannot be applied too seriously, because it does not take into account whether the transition is allowed or forbidden. The difference between excitation cross sections for optically forbidden and optically allowed transitions with the same Q has already been noted.

For a term series, the cross sections appear to be roughly proportional to n^{-3}.[9,67] As can be seen from the data of Hippler and Schartner[50] on the $2\,^1P$, $3\,^1P$, and $4\,^1P$, the n^{-3} rule is only a crude guide and not a source of accurate cross sections. Similar observations have been made from energy-loss spectra data.[8]

These crude descriptions of the behavior of the cross section curves are strongly modified when a complete explanation based on quantum mechanical calculations is employed. Recent theoretical studies have become sufficiently sophisticated to provide an adequate description of the data. These studies have been confined to a few cases, primarily the low-lying states of atomic hydrogen and helium. In these cases, the theory is as fully developed as the experimental work, but much work remains to be done in both theory and experiment.

Born approximation calculations tend to overestimate the cross sections near the peak in the cross-section curves. They are completely inadequate at energies below the maximum in the curves. Distortion and second Born calculations appear to be inadequate in describing the data near the peak in the cross-section curves. The fit to the data is satisfactory to lower proton energies than is the case with the first Born approximation calculations.

The close coupling calculations in which exchange is included appear to provide an adequate description of the data over the maximum in the cross section curves. The introduction of the exchange channel makes a marked improvement in the fit to the data. The inclusion of a large number of high states in the close-coupling calculations does not appear to improve the fit to the cross-sectional data for lower-lying states. The inclusion of pseudostates to provide strong overlap with the intermediate states of helium and to simulate the molecular features at small separations[33] has the effect of improving the fit to the data at low proton energies. However, the fit to the cross section at higher ion energies is made much poorer by the addition of these pseudostates.

The second-order diagonalization technique employed by Baye and Heenen[59] seems to provide results essentially identical to the close coupling calculations on proton–helium calculations. The 20-state diagonalization method applied to proton–atomic hydrogen collisions does not, however, provide a satisfactory fit.

The second-order potential technique appears to provide a satisfactory result at proton energies well above the peak in the cross section curve but is not suitable for applications at lower proton energies.

The Glauber and distorted eikonal calculations can be applied for structureless projectiles. These approaches appear to give satisfactory results, although the theoretical cross section curves tend to be less peaked than the data. The theoretical curves tend to underestimate the cross section at the peak in the cross-section curves.

It thus appears that several viable theoretical approaches are available. However, it must be noted that neither experiment nor theory can claim better than 10% accuracy and that in many cases the experimental data which are available are inadequate to test theoretical predictions.

The discussion in this chapter has touched on only a small sample of the available emission and excitation measurements which have been reported in the literature. Even with the bulk of these data, it is obvious that few of the many possible transitions have been studied to reasonable accuracy. The most thorough studies have been made on the excitation of atmospheric constituents because of their obvious importance. Even for atmospheric gases, only in the cases of nitrogen and oxygen are the data sufficient to develop effective auroral cross sections for any emission feature of the target gas.

The lack of data on molecular excitation is most significant. The molecular emissions do not always dominate the optical emission spectra. Molecular excitations appear, however, to dominate the energy-loss spectra. In all of the cases in which ion energy-loss spectra are available for molecular targets, the most prominent feature of the spectra has been the result of the excitation of a molecular state. Further, these excitations represent a large fraction of the total inelastic cross section.

A great deal of work needs to be done simply to provide needed data. Many of the early measurements need to be remeasured. As can be seen from Figs. 15 and 17, there is a great deal of scatter in the available data. In many cases, it is not possible to choose between widely differing cross section values that are obtained by equally qualified researchers. The only solution is to remeasure with modern equipment and to use great care to avoid the many pitfalls inherent in excitation cross-section measurements.

The task of identifying molecular energy levels with heavy-ion energy-loss spectrometry has been performed at relatively low impact-energies. Because the heavy ions can excite states that are optically forbidden to radiate to the ground state as in singlet–triplet processes and because the spectral features which appear can be changed by changing the incident ion, energy-loss spectrometry provides a valuable diagnostic tool.

Heavy ion energy-loss spectrometry is in its infancy. Hopefully, it will be possible to improve the resolution in the energy-loss spectra sufficiently to permit measurement of cross sections for close-lying states. The energy-loss spectra only provide information on direct excitation processes. This limitation has the advantage of avoiding the confusion in the identification of the source of the observed excitation.

The optical emission spectra include the emission from all the atomic and molecular lines that can be excited by direct or indirect processes. The complexity of the resulting structure provides a wealth of information on all the processes involved in a collision. The same complexity makes interpretations difficult.

The optical and energy-loss techniques seem to provide very good agreement when the process being studied can be clearly shown to be the same in both. The excitation cross sections for atomic hydrogen that are measured by using the two techniques give satisfactory agreement in the energy region where measurements are available from both. Similar agreement is indicated in the excitation of the $2\,^1P$ state of helium. There are surprisingly few cases in which a real comparison can be made. The energy-loss technique has been applied to low-lying states with large energy separation. Radiation from these states is in the vacuum ultraviolet and is not easily studied by optical techniques. On the other hand, the energy-loss technique does not have the high resolution required to study the states which provide optical emissions in the visible region. The two techniques are complementary and permit unambiguous interpretation of the observations. Both techniques are needed to provide accurate cross sections.

ACKNOWLEDGMENT

This work was supported by the National Science Foundation.

References

1. N. F. Mott and H. S. W. Massey, *The Theory of Atomic Collisions*, Oxford Univ. Press, London and New York, 1949.
2. J. B. Hasted, *Physics of Atomic Collisions*, Butterworth, London, 1964.

3. R. G. Newton, *Scattering Theory of Waves and Particles*, McGraw-Hill, New York, 1966.
4. J. H. Moore, Jr., *J. Chem. Phys.* **55**, 2760 (1971).
5. J. H. Moore, Jr., *J. Geophys. Res.* **77**, 5567 (1972).
6. C. E. Kuyatt and J. A. Simpson, *Rev. Sci. Instr.* **38**, 103 (1967).
7. J. T. Park and F. D. Schowengerdt, *Rev. Sci. Instr.* **40**, 753 (1969).
8. D. R. Schoonover and J. T. Park, *Phys. Rev. A* **3**, 228 (1971).
9. F. J. de Heer, in *Advances in Atomic and Molecular Physics* (D. R. Bates, ed.), Academic Press, New York, 1966, pp. 327–384.
10. E. W. Thomas, *Excitation in Heavy Particle Collisions*, Wiley-Interscience, New York, 1972.
11. E. W. Thomas and G. D. Bent, *Phys. Rev.* **164**, 143 (1967).
12. D. H. Crandall, G. York, V. Pol, and J. T. Park, *Phys. Rev. Lett.* **28**, 397 (1972).
13. P. M. Stier, C. F. Barnett, and G. E. Evans, *Phys. Rev.* **96**, 973 (1954).
14. S. C. Liang, *J. Appl. Phys.* **22**, 148 (1951).
15. H. Ishii and K. Nakayama, *1961 Vacuum Symp. Trans.*, 1962, p. 519.
16. N. G. Utterback and T. Griffith, *Rev. Sci. Instr.* **37**, 866 (1966).
17. J. C. de Vos, *Physica* **20**, 690 (1954).
18. R. D. Larrabee, *J. Opt. Soc. Am.* **49**, 619 (1959).
19. D. Hasselkamp, R. Hippler, A. Scharmann, and K. H. Schartner, *Z. Phys.* **248**, 254 (1971).
20. J. T. Morgan, J. Geddes, and H. B. Gilbody, *J. Phys. B* **6**, 2118 (1973).
21. J. H. Birely and R. J. McNeal, *Phys. Rev. A* **5**, 692 (1972).
22. E. P. Andreev, V. A. Ankudinov, and S. V. Bobashev, *Zh. Eksp. Teor. Fiz.* **50**, 565 (1966); *Soc. Phys. JETP* **23**, 375 (1966).
23. W. E. Kauppila, P. J. O. Teubner, W. L. Fite, and R. J. Girnius, *Phys. Rev. A* **2**, 1759 (1970).
24. D. H. Crandall and D. H. Jaecks, *Phys. Rev. A* **4**, 2271 (1971).
25. T. Kondow, R. J. Girnius, Y. P. Chong, and W. L. Fite, *Phys. Rev. A* **10**, 1167 (1974).
26. R. F. Stebbings, R. A. Young, C. L. Oxley, and H. Ehrhardt, *Phys. Rev. A* **138**, A1312 (1965).
27. W. L. Fite and R. T. Brackman, *Phys. Rev.* **112**, 1151 (1958).
28. R. A. Young, R. F. Stebbings, and J. W. McGowan, *Phys. Rev.* **171**, 85 (1968).
29. D. R. Bates and G. Griffing, *Proc. Phys. Soc. (London)* **66**, 64 (1953).
30. D. R. Bates, *Proc. Phys. Soc. (London)* **77**, 59 (1961).
31. M. R. Flannery, *J. Phys. B* **2**, 1044 (1969).
32. D. Rapp and D. Dinwiddie, *J. Chem. Phys.* **57**, 4919 (1972).
33. I. M. Cheshire, D. F. Gallaher, and A. J. Taylor, *J. Phys. B* **3**, 813 (1970).
34. V. Franco and B. K. Thomas, *Phys. Rev. A* **4**, 945 (1971).
35. J. T. Park, J. E. Aldag, and J. M. George, *Phys. Rev. Lett.* **34**, 1253 (1975).
36. A. E. Kingston, B. L. Moiseiwitsch, and B. G. Skinner, *Proc. R. Soc. A* **258**, 273 (1960).
37. B. L. Moiseiwitsch and R. Perrin, *Proc. Phys. Soc.* **85**, 51 (1965).
38. A. R. Holt and B. L. Moiseiwitsch, *J. Phys. B.* **1**, 36 (1968).
39. D. S. F. Crothers and A. R. Holt, *Proc. Phys. Soc.* **88**, 75 (1966).
40. J. Van den Bos and F. J. DeHeer, *Physica* **34**, 333 (1967).
41. D. R. Bates, *Proc. Phys. Soc. (London)* **73**, 227 (1959).
42. L. Wilets and D. F. Gallaher, *Phys. Rev.* **147**, 13 (1966).
43. B. H. Bransden and J. P. Coleman, *J. Phys. B* **5**, 537 (1972).
44. B. H. Bransden, J. P. Coleman, and J. Sullivan, *J. Phys. B* **5**, 546 (1972).
45. J. Sullivan, J. P. Coleman, and B. H. Bransden, *J. Phys. B* **5**, 2061 (1972).
46. D. F. Gallaher and L. Wilets, *Phys. Rev.* **169**, 139 (1968).
47. D. Baye and P. H. Heenen, *J. Phys. B* **6**, 105 (1973).
48. C. J. Joachain and R. Vanderpoorten, *J. Phys. B* **6**, 662 (1973).

49. J. T. Park and F. D. Schowengerdt, *Phys. Rev.* **185**, 152 (1969).
50. R. Hippler and K. H. Schartner, *J. Phys. B.* **7**, 618 (1974).
51. P. N. Clout and D. W. O. Heddle, *J. Opt. Soc. Am.* **59**, 715 (1969).
52. K. L. Bell, D. J. Kennedy, and A. E. Kingston, *J. Phys. B* **1**, 1037, 218 (1968).
53. Y. K. Kim and M. Inokuti, *Phys. Rev.* **175**, 176 (1968).
54. R. J. Bell, *Proc. Phys. Soc.* **78**, 903 (1961).
55. A. R. Holt, J. Hunt, and B. L. Moiseiwitsch, *J. Phys. B* **4**, 1318 (1971).
56. M. R. Flannery, *J. Phys. B* **3**, 306 (1970).
57. J. Van den Bos, *Phys. Rev.* **181**, 191 (1969).
58. C. J. Joachain and R. Vanderpoorten, *J. Phys. B* **7**, 817 (1974).
59. D. Baye and P. H. Heenen, *J. Phys. B* **6**, 1255 (1973).
60. S. Begum, B. H. Bransden, and J. Coleman, *J. Phys. B* **6**, 837 (1973).
61. A. H. Gabriel and D. W. O. Heddle, *Proc. Phys. Soc. (London) A* **258**, 124 (1960).
62. A. Denis, M. Dufay, and M. Gaillard, *Compt. Rend.* **264**, B440 (1967).
63. J. G. Dodd and R. H. Hughes, *Phys. Rev.* **135**, A618 (1964).
64. J. Van den Bos, G. J. Winter, and F. J. DeHeer, *Physica* **40**, 357 (1968).
65. B. H. Bransden and M. R. Issa, *J. Phys. B* **8**, 1088 (1975).
66. J. M. Robinson and H. B. Gilbody, *Proc. Phys. Soc. (London)* **92**, 589 (1967).
67. V. I. Ochkur and A. M. Petrunkin, *Opt. Spektrosk.* **14**, 457 (1963); *Opt. Spectrosc.* **14**, 245 (1963).
68. J. Van den Bos, *Physica* **42**, 245 (1969); **41**, 213 (1969).
69. V. Pol, W. Kauppila, and J. T. Park, *Phys. Rev. A* **8**, 2990 (1973).
70. J. T. Park, V. Pol, J. Lawler, J. George, J. Aldag, J. Parker, and J. L. Peacher, *Phys. Rev. A* **11**, 857 (1975).
71. E. A. Silverstein, *Nucl. Instr. Methods* **4**, 53 (1959).
72. D. C. Lorents, W. Aberth, and V. W. Hesterman, *Phys. Rev. Lett.* **17**, 849 (1966).
73. F. T. Smith, R. P. Marchi, H. Aberth, D. C. Lorents, and O. Heinz, *Phys. Rev.* **161**, 31 (1967).
74. G. W. York, J. T. Park, V. Pol, and D. H. Crandall, *Phys. Rev. A* **6**, 1497 (1972).
75. D. C. Lorents and G. M. Conklin, *J. Phys. B* **5**, 950 (1972).
76. R. Francois, D. Dhuicq, and M. Barat, *J. Phys. B* **5**, 963 (1972).
77. J. T. Park, V. Pol, J. Lawler, and J. George, *Phys. Rev. Lett.* **30**, 1013 (1973).
78. R. Olson, Stanford Research Institute, private communication.
79. C. Lesech, R. McCarroll, and J. Baudon, *J. Phys. B* **6**, L11 (1973).
80. D. H. Jaecks, in *The Physics of Electronic and Atomic Collisions; Invited Lectures and Progress Reports VIII ICPEAC* (B. C. Cobic and M. V. Kurepa, eds.), Institute of Physics, Beograd, Yugoslavia, 1973.
81. J. Macek and D. H. Jaecks, *Phys. Rev. A* **4**, 2288 (1971).
82. R. J. McNeal and J. H. Birely, *Rev. Geophys. Space Phys.* **11**, 633 (1973).
83. J. Franck, *Trans. Faraday Soc.* **21**, 536 (1925).
84. E. U. Condon, *Phys. Rev.* **32**, 858 (1928).
85. E. Wigner, *Gott. Nachr. IIa*, 375 (1927).
86. F. R. Gilmore, *J. Quant. Spectrosc. Radiat. Trans.* **5**, 369 (1965).
87. J. H. Moore, Jr., *Phys. Rev. A* **8**, 2359 (1973).
88. V. Pol, J. George, and J. T. Park, *Bull. Am. Phys. Soc.* **18**, 1516 (1973).
89. G. Heizberg and L. L. Howe, *Can. J. Phys.* **37**, 636 (1959).
90. D. A. Dahlberg, D. K. Anderson, and I. E. Dayton, *Phys. Rev.* **170**, 127 (1968).
91. J. L. Edwards and E. W. Thomas, *Phys. Rev.* **165**, 16 (1968).
92. F. D. Schowengerdt and J. T. Park, *Phys. Rev. A* **1**, 848 (1970).
93. G. Herzberg, *Spectra of Diatomic Molecules*, Van Nostrand–Reinhold, Princeton, New Jersey, 1950.
94. J. H. Moore, Jr., *Phys. Rev. A* **9**, 2043 (1974).

95. D. A. Dahlberg, D. K. Anderson, and I. E. Dayton, *Phys. Rev.* **164**, 20 (1967).
96. J. M. Hoffman, G. J. Lockwood, and G. H. Miller, *Phys. Rev. A* **7**, 118 (1973).
97. J. H. Birely, *Phys. Rev. A* **10**, 550 (1974).
98. E. W. Thomas, G. D. Bent, and J. L. Edwards, *Phys. Rev.* **165**, 32 (1968).
99. J. Durup, *Chem. Phys.* **2**, 226 (1973).
100. J. T. Park, F. D. Schowengerdt, and D. R. Schoonover, *Phys. Rev. A* **3**, 679 (1971).
101. T. A. Green, *Phys. Rev.* **157**, 103 (1967).
102. R. G. Breen, *J. Chem. Phys.* **45**, 3876 (1966).
103. E. N. Lassettre, S. M. Silverman, and W. E. Krosnow, *J. Chem. Phys.* **40**, 1261 (1964).
104. J. H. Birely, *Phys. Rev. A* **11**, 79 (1975).
105. G. W. York, Jr., J. T. Park, V. Pol, and D. H. Crandall, *Phys. Rev. A* **6**, 1497 (1972).
106. E. N. Lassettre, in *Chemical Spectroscopy and Photochemistry in the Vacuum-Ultraviolet* (C. Sandorfy, P. J. Ausloos, and M. B. Roben, eds.), D. Reidel Publ., Dordrecht, Holland, 1974.
107. J. G. Calvert and J. N. Pitts, Jr., *Photochemistry*, Wiley, New York, 1966.
108. J. H. Moore, Jr., *J. Phys. Chem.* **76**, 1130 (1972).
109. J. P. Doering and J. H. Moore, Jr., *J. Chem. Phys.* **56**, 2176 (1972).
110. H. S. W. Massey, *Rept. Progr. Phys.* **12**, 248 (1949).

2

Charge Transfer in Atomic Systems

R. E. Johnson and J. W. Boring

1. Introduction

In considering the topic of collisions between atomic and molecular particles, one must pay particular attention to the processes that involve the transfer of an electron from one of the colliding partners to the other, since frequently this is one of the most probable processes. This process plays a predominant role in a considerable body of physical phenomena. For instance, in the charge balance in the ionosphere,[1] as well as in laboratory plasmas, the charge-transfer process is an important and often dominant intermediate step. The process has also been experimentally useful for obtaining neutral beams with known distributions of states which can be used then to study other phenomena.[2] Here we are also interested in charge transfer both as a means for obtaining certain atomic spectra and for understanding interatomic forces. We are concerned in this chapter with events in which the participants in the collisions are atomic systems, i.e., atoms and ions, and in which a single electron is transferred. The following equations indicate the events to be considered:

$$A^+ + B \to A + B^+$$
$$A^{2+} + B \to A^+ + B^+ \tag{1.1}$$

R. E. Johnson and J. W. Boring • Department of Nuclear Engineering and Engineering Physics, University of Virginia, Charlottesville, Virginia 22901.

It is seen that we have limited ourselves to a consideration of only singly and doubly charged ions. Although the subject of charge transfer for highly charged ions is interesting, there is relatively little information available for these processes. We have also limited ourselves to considerations of events in which the target particle [B in Eq. (1.1)] is a neutral atom, but even with these limitations we are left with a large range of possible processes, particularly when one considers that each of the reactants shown in Eq. (1.1) may be in any one of a number of internal electronic states. Although for simplicity in notation we have not specifically indicated by symbols in Eq. (1.1) that the colliding partners may be in electronically excited states, we certainly intend to consider these possibilities, since some of the more interesting features of charge transfer come about because of the existence of these states.

In discussing charge transfer it is important to consider the electronic states of the reactants before and after the collision takes place. In a general consideration of charge-transfer processes an important parameter is the total change in internal energy that takes place during the collision, Q (sometimes called the energy defect in the collision). The energy Q can either be positive (a decrease in internal energy during the collision, with a corresponding increase in kinetic energy) or negative (an increase in internal energy and a decrease in kinetic energy). The former is called an exothermic reaction and the latter an endothermic one. The energy Q can also be zero as is seen in the discussion that follows.

For the processes indicated by Eq. (1.1) there are some special cases that are worthy of note. If A and B are of the same species, then electron transfer for the singly charged incident ion becomes

$$A^+ + A \rightarrow A + A^+ \tag{1.2}$$

This process is called symmetrical charge transfer, and if the reactants are all in their ground electronic states such that the total internal energy before the collision is the same as that after the collision ($Q = 0$), the event is also called resonant charge transfer because of the generally high probability of these processes (especially at low kinetic energies). It is also possible for the asymmetrical case to be resonant or nearly resonant if there are appropriate electronic states available so that Q is nearly zero, and these processes are called acccidentally resonant.

Having discussed the processes to be considered here, let us proceed to ask what aspects of these collisions are of interest to us. In collisions between heavy particles it is generally meaningful to speak in terms of the classical trajectories of the nuclei involved, although we must usually resort to quantum mechanics to describe the motion of the electrons during the collision. This is referred to as a semiclassical treatment. Thus, we may represent a typical collision as shown in Fig. 1. In this figure we have chosen

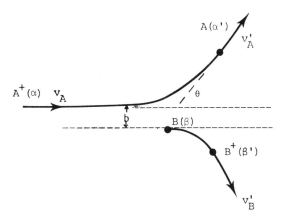

Fig. 1. Charge-transfer collision with change in internal states of the collision partners. The scattering angle of the projectile (θ) and the impact parameter (b) are indicated.

the reference frame in which the target particle B is initially at rest (closely approximating many laboratory conditions when the velocity v_A of the incident ion is reasonably large and B is a thermal gas). The incident particle has a velocity v'_A after the collision and the target particle has a velocity v'_B. The classical impact parameter is b and the laboratory scattering angle of the incident particle is θ. The state of A^+ before the collision is indicated by α and that of A after the collision by α'; correspondingly, the internal state of B before the collision is β and that of B^+ after the collision is denoted by β'.

Ideally, one would like to know for any desired species and initial states, A^+ (α) and B (β), what the probability is of the collision resulting in any possible combination of final states, as a function of v_A and b. It is also desired that the relationship between b and θ be known. This information is far from being realized at the present time. In looking at the problem from the theoretical point of view, one must consider the interaction potentials that represent the transient molecular ion states formed during the collision and possible electronic transitions between these states in order to be able to obtain the desired probabilities. Except in special cases such calculations are severely hampered by a lack of information about the appropriate potentials as well as the quantum mechanical difficulties in calculating accurate transition probabilities between the potential curves. From the experimental point of view, it is difficult to obtain the kind of detailed information one would like to have for comparison with available calculations. Most experiments have involved measurements that either sum over the final states or effectively integrate over θ (i.e., b), or do both (total cross sections). It is clear that it is desirable to have experiments performed that give as much detail as possible, but this can, of course, only be achieved at the expense of greater experimental complexity.

We will discuss here general topics having to do with the experimental study of charge-transfer events such as those indicated by Eq. (1.1). We will consider the most general requirements for such measurements and the major difficulties we expect to encounter in making them. A somewhat more detailed discussion of experimental techniques is given in Section 3.

Charge-transfer measurements are normally performed by forming a beam of the appropriate ions and allowing these ions to collide with the desired target atoms. When the source of target atoms is simply a gas at room temperature, then there is no question that the great majority of the target atoms are in their ground electronic states, moving with thermal kinetic energies. The electronic state of the incident ions is a somewhat more difficult question since it has been shown[3] that with conventional ion sources some care is required to be assured that the beam does not have an unknown mixture of ions in various long-lived excited states. If one desires to study collisions in which the incident ions are in known excited electronic states, rather than in their ground states, then the experimental difficulties of establishing that the beam has the desired composition are greatly increased. If one wants to study collisions in which the incident ion is multiply charged, there is generally no particular difficulty in producing doubly charged ions in the conventional ion sources (although the intensity obtained is usually much less than that for singly charged ions), but creating ions with charges greater than 2 in sufficient quantity is generally quite difficult, requiring that special ionization techniques be used. Most charge-transfer measurements have been performed at incident-ion energies greater than around 10 eV, since in the use of the usual techniques at lower energies one encounters serious experimental difficulties due to space charge spreading of the ion beam and the effects produced by surface charges within the apparatus.

In considering the general difficulties that arise when attempting to detect the results of collisions in which charge transfer has taken place, a great deal depends on the amount of information one seeks concerning the details of the collisions. If total cross sections are to be measured by collecting the positive ions produced in a gas when a beam of ions passes through the gas, then the principal task is to be assured that the ions collected result from the desired events rather than other possible processes such as ionization. If one wishes to obtain the greater detail given by differential cross sections, then the angular distribution of the incident particles after receiving an electron from the target atoms is to be measured. The principal problems associated with this type of measurement occur when the incident ions are singly charged, resulting in neutral particles whose number must be detected in an absolute manner. Neutral atoms in their ground state can be successfully detected by electron-multiplier detectors when their kinetic energy exceeds 1 keV, but even at these energies it is not a simple task to determine the absolute efficiency of the detector, which is required in order to obtain absolute differential cross sections. When the kinetic energy of the neutral products of a charge transfer process is less than 1 keV the difficulty is compounded because of the lack of sensitive detectors for low-energy neutral particles. When the charge transfer process to be studied involves the use of doubly charged incident ions which receive a single electron from the target atom, then the detection problems are considerably simplified since it is much easier to detect the singly charged ions that result than it is to detect neutral atoms.

For information concerning the final internal states of the collision products, some study of the energetics of the interaction must be made. When doubly charged incident ions are used the single charge on the ion that results after charge transfer allows one to measure its kinetic energy by the usual techniques, such as electrostatic analysis. For neutral product particles it is usually necessary to use pulsed beam and time-of-flight techniques to obtain their kinetic energy. Another possibility for studying the energetics of charge-transfer processes is to observe the photons that result from transitions between electronic states either during the collision or shortly after. One significant advantage of the observation of the light emitted is that much greater energy resolution can be obtained than is usually possible for kinetic energy measurements. The greatest disadvantage to studies of the light emitted is the necessity of being assured that the light observed is coming from the events in which one is interested. This difficulty can be greatly relieved by studying coincidences between photons emitted and particles scattered through a known angle. This latter type of experiment allows the possibility of relating scattering angle to impact parameter, while at the same time retaining the good energy resolution of optical spectroscopy. These advantages, of course, come at the expense of much more complex apparatus compared to that required for the measurement of the total cross sections and are limited to the study of optically allowed transitions. In all experimental studies of absolute charge-transfer cross sections one of the most difficult quantities to measure accurately is the number density of the target particles. When the target particles are gas atoms within a gas cell this amounts to requiring an accurate determination of the pressure within the cell, a procedure that in many cases requires a great deal of care in order to obtain cross sections that are accurate to a few percent. If the charge-transfer events are studied in a crossed-beam configuration, then the determination of the target number density is even more difficult than it is for the gas cell, and more indirect methods must be employed to make this measurement.

In considering theoretical studies of charge transfer one is concerned with transitions between the electronic states before the collision to some final states. In calculating differential cross sections one requires a knowledge of interaction potentials corresponding to the states that participate in the collision, as well as the probabilities of transitions between these states. For most systems it is difficult to obtain both the interatomic forces and the transition probabilities, making the theoretical difficulties in obtaining accurate cross sections comparable to those encountered in attempting experimental studies of reasonably complex systems. In essence one is reduced to making approximations except for the simplest processes. In an approximate calculation done for comparison with experiment, numerical accuracy is often much less important than the physical insight which is brought to bear on the collision process, since those collision partners that are easily studied in the laboratory are often *not* the most important species involved in physical phenomena for which atomic cross sections play a role.

If, instead of desiring to compute differential cross sections, one is interested in cross sections integrated over angle (or impact parameter b), the difficulties are reduced considerably. One often obtains sufficient information from knowledge of the maximum interatomic separation for which a given reaction will take place and an average transition probability within the reaction region. A considerable body of literature exists describing such methods in the fields of nuclear collisions,[4] electron–atom collisions,[5] as well as ion–atom[6] and molecular collisions.[7] For the latter cases one of the most widely used methods is the so-called Landau–Zener–Stueckelberg[6,7] approximation which we will consider later.

In the following sections we will describe in more detail some of the theoretical and experimental methods that we feel are most useful for these charge-transfer cross sections. The state of the art of the theory of atomic collisions is considerably advanced as compared, for instance, to that for molecular collisions, and we spend in the next section considerable time describing some of the more important aspects of ion–atom collisions in the energy range of interest.

2. Theory

2.1. Introduction

As stated earlier, in the application of quantum mechanics to charge transfer collisions severe approximations need to be made to obtain results for all but the simplest systems. A complete discussion of many of the most important methods is contained in a recent text by Mapleton on charge transfer[8] as well as other texts on the general field of atomic collisions.[6] It is seen that the appropriate method for a particular collision depends primarily on the relative velocity of motion of the colliding particles. In this section, we restrict ourselves to a brief discussion of some of the concepts involved in charge transfer problems, paying particularly close attention to the semiclassical method since this description provides a useful conceptual interpretation of experimental observations, as well as a good starting point for a reasonably accurate calculation.

The primary quantities of interest are cross sections differential in angle and energy for a given collision. These provide the most direct connection between experiment and theory, and the most rigorous test of approximate methods and interpretations. Closely related to this quantity is the energy-loss spectrum as a function of angle (or impact parameter) which shows the relative importance of the various channels at each scattering angle. The probability of observing an energy loss, associated with a given reaction and angle of scattering, is proportional to the differential cross section for that reaction. Also of interest, although they often allow a variety of interpretations, are such summed quantities as the cross section integrated over angle and reaction rates, or total differential cross sections, i.e., summed over energy loss or final states.

For incident-ion energies such that the relative velocity of the colliding particles is of the same order of magnitude or smaller than the velocity of the outer-shell electrons (generally around one atomic unit,* 1 a.u., of velocity for a proton ~25 keV), reactions can best be viewed as transitions between a set of quasi-molecular electronic states formed by the collision partners (cf. Chapter 3, Section 2). Associated with each state is an interaction potential which depends on the internuclear separation R, and which is represented as a potential curve that is a sum of the nuclear–nuclear repulsion and the average electronic binding energy. These potential curves determine the nuclear motion, which is often treated classically because of the large mass, hence small de Broglie wavelength, associated with the nuclei. For charge-transfer reactions, transitions between the quasi-molecular states often occur at internuclear separations on the order of 2–5 a.u. and this is generally the region where envisioning the system as composed of separate atomic and ionic species breaks down. For smaller R the outer electrons, at these collision velocities, must be treated as moving in the combined field of the two charge centers. Therefore, in a collision in which the colliding particles penetrate to smaller R, their outer electrons will have a probability of following either nucleus out subject to constraints on changes in energy, momentum, and angular momentum.

At considerably higher velocities of relative motion the charge-transfer reaction can be pictured as a two-step process and is best described by the second Born approximation.[8] This involves an energy transfer from the incident ion to the target electron followed by a "second collision" involving

---

*In atomic units $\hbar = 1$, $m_e = 1$, and $e = 1$. Therefore, 1 a.u. of energy is twice the binding energy of the hydrogen atom, 27.2 eV, and 1 a.u. of length is the radius of the Bohr orbit for ground state hydrogen, etc.

the target nucleus which allows the electron to have the correct momentum to be carried off by the incident ion. This is clearly a complementary view to the previous one and generally involves considerably smaller internuclear separations for the transition.

In the following we will discuss the charge-transfer process from the first point of view, i.e., as a transition between quasi-molecular electronic states, since our primary interest is in collision velocities where this approximation is valid, and even at higher velocities the concept of such states can often prove useful. In implementing this scheme one first chooses a set of electronic wave functions to describe the quasi-molecular states associated with each set of atomic states $\alpha\beta$ at $R \to \infty$, as described earlier. A possible and unique choice for this set are the adiabatic wave functions, solutions to the electronic Hamiltonian with fixed nuclei at each value of R. One uses this set at very low velocities and transitions are induced by the nuclear motion (i.e., the Born–Oppenheimer approximation[6] breaks down). For velocities of interest here other sets, called diabatic wave functions, are more appropriate for describing a collision[9] (cf. Chapter 3, Section 2). One such set of states is constructed of molecular orbitals for which the electron–electron interactions are averaged.[10,11] Because the diabatic states are not exact solutions of the electronic Hamiltonian, the energy levels or potential curves associated with states of the same symmetry may cross (i.e., become degenerate for some R) and transitions are attributed to an interaction potential coupling any two states. For the charge-transfer reaction the coupling is related to the exchange interaction, the sharing of the electron by the two centers, which has an exponential dependence on R.[11] If a crossing occurs at a separation R_x, as in Fig. 2, transitions generally occur with high likelihood in the vicinity of R_x depending on the strength of the coupling.

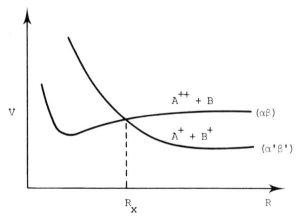

Fig. 2. Crossing of diabatic potential energy curves at internuclear separation R_x.

Quite often the general behavior of the diabatic potential curves and the approximate crossing points can be estimated from simple considerations of the molecular orbitals without solving for the wave function,[10,11] and hence the usefulness of this procedure.

Using a set of wave functions one can solve for the nuclear motion quantum mechanically or classically. In the latter case, elastic scattering trajectories and hence cross sections are easily obtained for each potential associated with a given state. However, when transitions occur the classical trajectory considerations become ambiguous. This is not too important for integrated cross sections, but for angular differential results it is very important. The method described in the following section attempts to deal with this.

2.2. Semiclassical Cross Sections Differential in Angle and Energy

In this section we will discuss a semiclassical method for describing cross sections which are differential in angle and energy of the scattered particle, which is that, where the final state of the scattered particle is analyzed. We describe the result in one form that is intuitively appealing and useful for determining the principal features of the differential cross sections. In wave mechanics we imagine that the scattering intensity at a given angle is the sum of contributing partial waves (wavelets) originating at all parts of the incident wave front which we often parametrize with an impact parameter (the perpendicular component of the distance from the point in the wave front to the scattering center). In this sum constructive interference (a region of stationary phase) for each scattering angle is found to occur for a few regions of impact parameter.[12] Each of these impact parameters can be associated with a classical trajectory, leading to scattering at the angle of interest. For a given transition, $\alpha\beta \rightarrow \alpha'\beta'$, the contributing classical trajectories can be deduced from the diagram of potential energy curves if switching between curves (states) is considered to occur for well-defined regions of R where the coupling is optimum[13] (e.g., at a curve crossing). Associated with each trajectory (j) and corresponding impact parameter $b^{(j)}$ is a classical action $A^{(j)} = [\int \mathbf{p}^{(j)} \cdot d\mathbf{R} - \int \mathbf{p} \cdot d\mathbf{R}]$ (where $\mathbf{p}^{(j)}$ is the center-of-mass momentum along the trajectory and $\mathbf{p}$ is the incident momentum) and a classical, elastic scattering cross section $\sigma^{(j)}(\theta)$ (where θ is the center-of-mass scattering angle) determined from the potential energy curves.[6] The differential cross section at any angle is a sum of interfering amplitudes associated with each classical trajectory:

$$\sigma_{\alpha\beta \rightarrow \alpha'\beta'}(\theta) \approx \left| \sum_j [P^{(j)}_{\alpha\beta,\alpha'\beta'} \cdot \sigma^{(j)}(\theta)]^{1/2} \exp\left[i\frac{A^{(j)}}{\hbar} - i\gamma^{(j)}\right] \right|^2 \quad (2.1)$$

In this expression $P^{(j)}_{\alpha\beta,\alpha'\beta'}$ is the total transition probability for $\alpha\beta \rightarrow \alpha'\beta'$ on the trajectory (j) (a number of transitions between intermediate states may occur), $\gamma^{(j)}$ is an additional constant phase factor, generally $\pm\pi/4$,[12,13] and only *differences* in the classical action between any two trajectories are evaluated. For other interference effects, such as rainbow scattering, where for a *single* potential more than one impact parameter may contribute to a given scattering angle, e.g., a potential which is attractive for large R and repulsive at small R, simple modifications to Eq. (2.1) can be made.[12] Beyond this the validity of Eq. (2.1) depends on the transition regions being well localized in R. The transformation to laboratory angles and differential cross section when the internal energy is changed by Q is straightforward, involving a multiplicative kinematic factor.

To help elucidate the trajectory consideration in Eq. (2.1), we consider two familiar examples involving only two states throughout the collision. First, for curve crossing (Fig. 2), these two trajectories will result in a transition described by the potential functions given in Fig. 3 if the distance of closest approach during the collision is smaller than R_x. The differential cross section for a transition then becomes, using Eq. (2.1),

$$\sigma_{\alpha\beta \to \alpha'\beta'}(\theta) \approx P_{\alpha\beta,\alpha'\beta'} \left\{ \sigma^{(1)}(\theta) + \sigma^{(2)}(\theta) \right.$$
$$\left. + 2[\sigma^{(1)}(\theta)\sigma^{(2)}(\theta)]^{1/2} \cos\left(\frac{A^{(1)} - A^{(2)}}{\hbar} - \gamma^{(1)} + \gamma^{(2)}\right) \right\} \quad (2.2)$$

where we have assumed that the transition probability $P^{(1)}_{\alpha\beta,\alpha'\beta'} \approx P^{(2)}_{\alpha\beta,\alpha'\beta'}$. This reduces to the standard curve crossing expression when the cross sections for the two trajectories are about equal.[6] A closely related example is symmetric resonance-charge transfer:

$$A^{2+} + A \to A + A^{2+}, \quad A^{+} + A \to A + A^{+}$$

Now the two states involved are the gerade (g), symmetric with respect to inversion, and ungerade (u), unsymmetric, states of the molecule which are

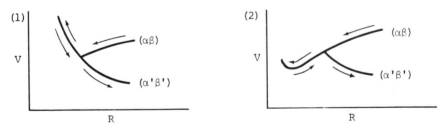

Fig. 3. Potential curves from Fig. 2 associated with the two trajectories leading to a transition $\alpha\beta \to \alpha'\beta'$. Arrows indicate relative motion of the colliding particles during the collision.

degenerate at large R. There is a probability of $\frac{1}{2}$ for being in either state. On exiting these states can lead to either elastic scattering or charge transfer[15]:

$$\sigma_{\pm}(\theta) \approx \tfrac{1}{4}\left\{\sigma^{(g)}(\theta) + \sigma^{(u)}(\theta) \pm [2\sigma^{(g)}(\theta) \cdot \sigma^{(u)}(\theta)]^{1/2} \cos\left(\frac{A^{(g)} - A^{(u)}}{\hbar}\right)\right\} \quad (2.3)$$

where the + implies elastic scattering and − charge transfer. At very low energies, because the nuclei are identical, contributions from angles $(\pi - \theta)$ must be included.[16] If in this expression $\sigma^{(g)}(\theta) \approx \sigma^{(u)}(\theta)$, then one obtains the well-known expression for the charge transfer probability:

$$P \approx \sin^{2}\left(\frac{A^{(g)} - A^{(u)}}{2\hbar}\right) \quad (2.4)$$

In both of these examples the characteristic oscillatory structure of differential cross sections is seen to be due to the differences in action for the two trajectories.

2.3. Transition Probabilities and Energy-Loss Spectra

At those scattering angles and energies* for which the deflections associated with the different trajectories are all roughly the same, a single average trajectory and elastic scattering cross section $\bar{\sigma}(\theta)$ may be used. This occurs, for instance, when the deflections are primarily determined by the repulsive core. For these cases the internuclear separation may be treated as a simple function of time and the motion of the nuclei can be thought of as creating a time-dependent field for the electrons. This results in the impact parameter equations for the transition probabilities. The differential cross section becomes

$$\sigma_{\alpha\beta \to \alpha'\beta'}(\theta) \approx P_{\alpha\beta,\alpha'\beta'}\bar{\sigma}(\theta) \quad (2.5)$$

with $P_{\alpha\beta,\alpha'\beta'}$ a sum of transition probabilities $P^{(j)}_{\alpha\beta,\alpha'\beta'}$ and phase factors in Eq. (2.1). In this method the action becomes $\int_{-\infty}^{\infty} V^{(j)}\, dt$, where $V^{(j)}$ is the interaction energy along a trajectory. One solves the time-dependent Schrödinger equation to obtain the transition probabilities

$$\left[H_e(\vec{r}|R) - i\hbar\frac{\partial}{\partial t_{\vec{r}}}\right]\phi(t) = 0 \quad (2.6)$$

where H_e is the *electronic* Hamiltonian and R is a function of time.[8] At high energies it is important to include the momentum change of the active electron.[8]

*Expressed conveniently in terms of the reduced scattering angle $\tau = E\theta$, which for small-angle scattering depends only on the impact parameter.[17]

As the sum of the transition probabilities $P_{\alpha\beta,\alpha'\beta'}$ over all final states is unity for any set of initial states $\alpha\beta$ the quantity $\bar{\sigma}(\theta)$ is the total differential cross section, i.e., the angular differential cross section summed over all final states or energy transfers. Therefore, the quantity $P_{\alpha\beta,\alpha'\beta'}$ can also be described as the transition spectrum at angle θ or the energy-loss spectrum at that scattering angle if the state labels are translated into the corresponding change in internal energy of the colliding system. That is, $P_{\alpha\beta,\alpha'\beta'}$ is the probability of an energy loss (or gain) associated with the difference in internal energy between the $\alpha\beta$ and $\alpha'\beta'$ states. More precisely the probability as described here should be multiplied by the density of states. This description is only valid, of course, at those τ values at which the impact parameter method is valid. The converse of this is that for many regions of τ energy-loss spectra provide a direct measure of the impact parameter probabilities provided the relationship between b and θ can be established.

At other values of τ where the different trajectories have signficantly different θ, b relationships, the *two-state* impact parameter equation can be solved locally to obtain $P^{(j)}_{\alpha\beta,\alpha'\beta'}$ *using an average trajectory in the transition region* for the time dependence in Eq. (2.6).[18] Here we consider some approximations to that solution for three classes of coupling: weak coupling and strong coupling, with or without a curve crossing.[13] For weak coupling, with coupling matrix element $V_{0f}(R)$, one has

$$P^{(j)}_{0,f} \approx \left| \frac{1}{\hbar} \int_{-\infty}^{\infty} V_{0f} \cdot \exp\left[\frac{i}{\hbar} \int_{-\infty}^{t} (V_f - V_0) \, dt' \right] dt \right|^2 \qquad (2.7)$$

where to simplify notation the subscripts 0 and f are used for the initial and final state labels, respectively, and V_0, V_f are the corresponding interaction potentials. In the calculation of $A^{(j)}$ the transition can be treated as occurring at the distance of closest approach, i.e., follow V_0 on the way in and V_f on exiting.[13] This is the impact parameter version of the widely applicable distorted wave approximation.[6] The distorted wave approximation has been shown to describe the threshold dependence at a curve crossing reasonably accurately.[18,19] In Eq. (2.7) it has become customary for simplicity to treat the time dependence by a straight-line trajectory, $R^2 \approx b^2 + v^2 t^2$, in which case analytic expressions for $P^{(j)}_{0,f}$ can often be obtained.[6]

When the distance of closest approach in a collision is such that a curve crossing has been passed, the LZS (Landau–Zener–Stueckleberg[20]) expression for transition probabilities has proven to be a useful approximate solution to the local two-state impact parameter equations.[6,18-20] In ion–atom collisions one often finds a series of crossings, as when the initial diabatic state correlates with a doubly excited state of the united atom

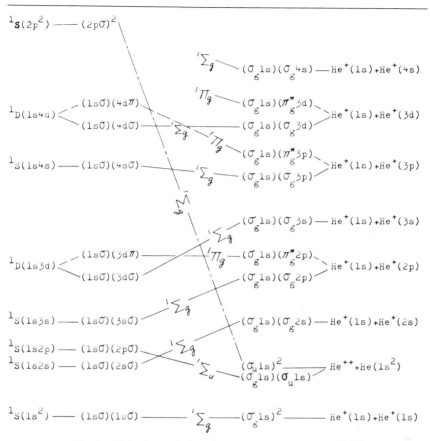

Fig. 4. Diabatic correlation of singlet gerade states in He^{2+}.[77]

($R = 0$); see Fig. 4 for $\text{He}^{2+} + \text{He}$. Each successive crossing can be treated separately if the transition regions, which have extent

$$\delta R_x \simeq \left| V_{0f} \bigg/ \left[\frac{\partial}{\partial R}(V_f - V_0) \right] \right|_{R=R_x} \quad (2.8)$$

do not overlap. The total transition probability is then the product of the individual ones. Each time a crossing point R_x is passed on a trajectory (j), the probability of staying on the same diabatic curve has the form $P_x^{(j)} = \exp(-\omega^{(j)})$, where

$$\omega^{(j)} \simeq \frac{2\pi}{\hbar} \left| V_{0f}\left(\frac{\mu \cdot \delta R_x}{p_R^{(j)}} \right) \right|_{R=R_x} \quad (2.9)$$

with μ the reduced mass, and $p_R^{(j)}$ the radial component of momentum.[11] The term in parentheses is essentially the time it takes to pass through the transition region, δR_x. From Fig. 3, for a collision involving a single crossing region the transition probability becomes $2P_x^{(j)}(1-P_x^{(j)})$; however, more complicated cases may be considered. For instance, if two crossings are encountered on a trajectory, as in Fig. 5, the total transition probability for that trajectory would be written

$$P_{0,f}^{(j)} \approx P_1^{(j)}(P_2^{(j)})^2(1-P_1^{(j)}) \qquad (2.10)$$

The LZS approximation breaks down in the threshold region where the distance of closest approach is approximately equal to R_x. Here one might use Eq. (2.7).[21] Bates[22] and Delos et al.[18] have discussed other limitations on Eq. (2.9).

If the initial state is close in energy to a continuum of states, as in the $He^{2+}+Ar$ collision, "crossings" may occur in the continuum since there are often well-defined states of the quasi molecule which correlate at large R to autoionizing states of the separated ion. For this case the dominant effect is caused by transitions between the initial state and these well-defined molecular states at the crossing, with the remainder of the continuum treated as a perturbation.[23]

For cases in which the curves do not cross, localized regions of strong coupling may still occur especially at large separations. This is particularly important for charge transfer, as in the $H^+ + Ar \rightarrow Ar^+ + H$ reaction.[24] At large R the energy defect is small. As the exchange interaction V_{0f} increases with decreasing R and the adiabatic potentials therefore diverge, a point R_x is reached at which the coupling is optimum. This can be shown by a numerical integration of the two-state impact parameter equations (Fig. 6). Here the probability of being in the excited $^1\Sigma^+$ state of ArH^+ that leads to charge transfer changes essentially once on the way in and again on the way out at well-defined internuclear separations. If V_0, V_f are the diabatic

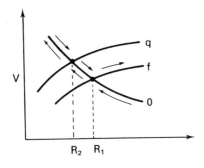

Fig. 5. A trajectory involving two curve crossings.

Fig. 6. The probability of being in the excited $^1\Sigma$ molecular ion state of ArH$^+$ during the course of a collision with impact parameter $b = 2$ a.u. Results were obtained from integrating the two-state impact parameter equations for the H$^+$ + Ar collision. R_x indicates region of strong coupling.[24]

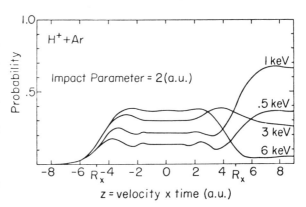

potentials of the initial and final states, then R_x is roughly defined by the relation

$$V_{0f}(R_x) \approx \tfrac{1}{2}|V_f - V_0|_{R=R_x} \qquad (2.11)$$

i.e., when the coupling is comparable to the splitting. This is the region discussed earlier where the outer electron orbitals of the ArH$^+$ molecular ion change from a system of separated atomic orbitals to molecular orbitals. A useful estimate of the transition probabilities can be deduced in this case from the Rosen–Zener model solution to the two-state equations.[24,25] For distances of closest approach much smaller than R_x the deflections do not depend significantly on contributions from $R > R_x$. Therefore, the adiabatic potentials, which include the effect of the interaction potential between the states, are used to obtain $\sigma^{(i)}(\theta)$ and $A^{(i)}$ in Eq. (2.1). Then, by analogy with the LZS method, the transition probabilities $P_x^{(i)}$ for each crossing of the transition region R_x can be obtained from [24,25]

$$4P_x^{(i)}(1-P_x^{(i)}) \approx \frac{|\int_{-\infty}^{\infty} V_{0f} \cdot \exp[(i/\hbar)\int_{-\infty}^{t} (V_f - V_0)\, dt']\, dt|^2}{|\int_{-\infty}^{\infty} V_{0f}\, dt|^2} \qquad (2.12)$$

For exponential coupling, $V_{0f} = Ae^{-\alpha R}$, if the phase in Eq. (2.12) is approximated by $Qt/\hbar$, and a straight-line trajectory is used for the integration, one obtains

$$4P_x^{(i)}(1-P_x^{(i)}) \approx \left|\frac{K_1(b^{(i)}\alpha')}{K_1(b^{(i)}\alpha)}\right|^2 \cdot \left(\frac{\alpha}{\alpha'}\right)^2 \qquad (2.13)$$

where $\alpha' = [\alpha^2 + (Q/v\hbar)^2]^{1/2}$ and the K_1 are modified Bessel functions. A similar result can obviously be obtained for the weak coupling case Eq. (2.7). These expressions provide useful approximations to the transition probabilities (or energy-loss spectra in the appropriate τ regions) and they are written explicitly in terms of the energy change Q. It should be emphasized that in applying Eq. (2.13) to the type of collisions discussed here, for Q one should use the energy difference between the states in the transition region. For charge transfer transitions occurring at *large R* with incident doubly charged ions exothermic processes will therefore be favored because of the $1/R$ potential in the final state. For transitions occurring at small R the effect of the $1/R$ potential is negligible. Equation (2.13) is intended to be used when only two states are involved in a transition. When many final states are involved a corresponding integration of Eq. (2.7) is useful. In Eq. (2.13) it can be seen that, unlike the curve-crossing case, as $v \to \infty$, $P_x^{(j)} \to \frac{1}{2}$. That is, the states behave as if they are effectively degenerate (resonant) at large R. A case of particular interest in charge-transfer collisions is where the atomic or ionic spin–orbit splittings are large[26,27] as in

$$H^+ + Xe \to H(1s) + Xe^+(^2P_j) \quad \text{or} \quad Xe^+(^2P_j) + Xe \to Xe + Xe^+(^2P_{j'})$$

collisions. At some value of R the exchange interaction becomes equivalent to the spin–orbit coupling, and in this region transitions among the spin states may occur with high likelihood depending on the relative velocity. One can also exploit this to create, for instance, a neutral beam of polarized atoms at the appropriate scattering angle.[26] For the systems under consideration in this chapter one may have both strong coupling regions and curve crossings, as in Fig. 7.

The approximate methods discussed here have proved useful in estimating and understanding the behavior of measured angular differential and integrated cross sections as will be discussed in Section 4. Needless to say there are many additional approximate methods which we have not discussed. The reader is referred to standard texts for a review of these other methods as well as accurate numerical methods for obtaining cross sections.[6]

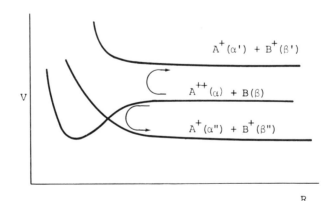

Fig. 7. Examples of transitions occurring at both a curve-crossing and non-curve-crossing region.

2.4. General Considerations

It is seen from the expressions for approximate transition probabilities [e.g., Eq. (2.13)] that the probability of a transition depends on the separation in energy of the quasi-molecular states and the rate of change of the coupling matrix. Therefore, at low collision energies, calculations of transition probabilities indicate that charge exchange occurs preferably between states of small energy defect, if the potentials do not diverge. Except at the very lowest velocities, where orbiting occurs, there is no general preference between exothermic and endothermic transitions; one has to consider the behavior of the potential curves in each case. The impact parameter formalism for transition probabilities essentially considers as coupling the two largest corrections to the Born–Oppenheimer approximation.[6,8] These two corrections have separate selection rules on the change of electronic angular momentum along the internuclear axis, M_L, when spin–orbit coupling is weak. Radial coupling, due to the rate of change in the radial coordinates, which in diabatic expressions is replaced by a potential, obeys the selection rule $|\Delta M_L| = 0$, e.g., $\Sigma \to \Sigma$, $\Pi \to \Pi$. Rotational coupling, due to the angular rate of change of the internuclear axis, occurs with greatest probability near the distance of closest approach in the collision, where the rate of rotation of the axis is greatest, and obeys the selection rule $|\Delta M_L| = 1$, e.g., $\Sigma \to \Pi$. The rotational coupling operator, which we have not treated, comes from the angular contribution to the time derivative in Eq. (2.6).[6] It also plays an important role at large R where spin–orbit coupling is important.[27]

Additional rules apply: $g \not\to u$ and $\Sigma^+ \not\to \Sigma^-$. In the impact parameter formalism it is assumed that any change in total electronic angular momentum of the separated ion–atom system can be compensated for by a corresponding change in the angular momentum of the nuclear motion, conserving angular momentum. When the impact parameters are large and coupling between the nuclear and electronic motion is weak, the total electronic angular momentum may be conserved separately. It is clear that these selection rules are very different from those for optical transitions encountered in spectroscopy. Therefore in collision spectroscopy one may observe, depending on impact parameter, transitions between states of the colliding atoms and ions which may be difficult to observe by optical spectroscopy. It is also clear from Eq. (2.1) that the differential cross sections will have considerable structure due to the interference between the contributions from different classical trajectories. For this reason, the observation, for instance, of energy loss at a single angle may lead to an erroneous interpretation of the overall relative likelihood of a given process. It also means that the differential cross sections do not frequently lend themselves to simple interpretations and care must be taken in analyzing data.

Our discussion of transitions has involved an exchange of energy between electronic and nuclear motions. Because the lifetimes for spontaneous emissions of photons are generally large compared to the collision times, optical transitions are not usually observed *during* the collision.[28] However, in very fast collisions which affect inner-shell electrons, transitions between quasi-molecular states due to induced emission have been seen recently.[28] For the collisions of interest here it may be possible to observe transitions involving autoionization during the collision since the lifetimes for this process are often comparable to the collision times involved in the kiloelectron volt region.

In the collision of a doubly charged ion with a neutral atom, double-charge transfer (see also Chapter 4) is the process which competes most efficiently with single-charge transfer at low collision energies. Therefore, a discussion of this process must be included in any complete description of collisions of the type $A^{2+} + B \rightarrow A^+ + B^+$. In fact, as can be seen from Fig. 4, single-charge transfer into excited states of He^+, in the collision of He^{2+} with He, will most likely occur via crossings with one of the "elastic" scattering potentials, which in this case can lead to either elastic scattering or resonant double-charge transfer. This is discussed further in Section 4. At large collision energies the probability for double-charge transfer can be considered as roughly the square of the probability for capture of one electron and it is therefore a second-order process having a small cross section.[10,29] In the low energy region where the quasi-molecular potential curves describe the collision, Komarov and Yanev[29] have obtained an expression for the two-electron exchange splitting for symmetric systems. In general, the exponential screening constant at large R is the sum of the screening constants for each electron determined by the overlap of the electron orbitals on one center with those for the electron on the other center of charge. Since the molecular states leading to double-charge transfer have no long-range repulsive force associated with them, they are often crossed by the single-charge-transfer states, making transitions likely. For symmetric systems such as $He^{2+} + He$, resonant double-charge transfer to the ground state dominates. However, in very nonsymmetric systems double charge exchange leaving either the projectile or target in an excited state may also be a very important process. In general, this process requires a change in two molecular orbitals. However, single-charge transfer leaving the target in an excited state is also a two-orbital process and therefore the cross sections are comparable. Further, at low energies the energy defect is as important as the coupling matrix elements in determining which states have the largest cross sections. At long range the primary difference between these processes for *nonsymmetric* systems is that $A^{2+} + B \rightarrow A^+ + B^+$ goes by curve crossing to exothermic states and $A^{2+} + B \rightarrow A + B^{2+}$ will go preferentially to near-resonant states.

3. Experimental Methods

3.1. Introduction

The purpose of this section is to provide a general survey of the several experimental arrangements for studying charge transfer events. We do not intend to discuss the details of these techniques since these have been summarized elsewhere.[30-32]

The discussion considers both measurement of total cross sections, where no information is gained concerning the angular distribution of the product particles, and measurement of cross sections that are differential in the angle of at least one of the product particles. Within each of these two main categories, there are subdivisions according to which characteristics of the collision are studied, e.g., the kinetic energies of the products or their internal states.

Before proceeding to a discussion of the methods used for making charge-transfer measurements, we will discuss briefly some general considerations that apply to all studies of the results of atomic collisions. Although the principles for performing these experiments are quite simple, a good deal of effort has been expended in the last few decades in developing reliable methods for measuring the quantities necessary for the determination of collision cross sections.

In general the experimental study of charge transfer involves the production of an ion beam which is allowed either to pass through a gas cell or in the cross-beam arrangement to intersect a second beam of neutral particles. Probably the simplest measurement to make is the total charge-transfer cross section, performed by observing the number of ions that have picked up an electron or by collecting all the slow ions produced. This of course provides one with the least detailed information. Another possibility is to observe the fast neutral particles as a function of their angular deflection. This gives a greater amount of information since it may be possible to relate the angular deflection to the corresponding impact parameter for the collision process. If one makes measurements of both the angular deflection and kinetic energy of the fast particle, then one can use energy- and momentum-conservation principles to infer the total change in internal energy of the system. This latter type of measurement is made difficult by the fact that there is no easy way to determine the kinetic energy of neutral particles, unless one resorts to time-of-flight techniques. We see from the foregoing that it is possible to learn something about the internal states of the reactants by measuring the kinetic energy of either of the final products. This method has its limitations in that energy resolution would be limited and one still has only information concerning the total change in internal energy of the states. Additional information could be obtained by observing whatever light may be produced in the collision process. This would add complexity to the experimental problem but would yield greater energy resolution and add some information about the internal states of the final products.

Our attention will also be given to some of the most salient points to be considered in measuring the density of target particles and to the devices used for detecting the collision products, since these are usually the critical factors in determining the amount of information that can be obtained and the accuracy of the results. A few brief remarks will also be made concerning the production of a beam of the desired characteristics.

It should be noted that over the years a large number of measurements of charge-transfer cross sections have been made, some of which disagree markedly from others for what is supposedly the same cross section. It has been found that many of these discrepancies can be explained on the basis of an unknown excited state population in the beam (and sometimes even an unknown species), unknown contaminants in the target gas, and errors in the measurement of the pressure of the target gas. Since in many experiments the measurement of the pressure of the target gas is the crucial factor in determining the target gas density, we will give in the following section a brief discussion of the methods used and the difficulties encountered in such studies.

3.2. Pressure Measurement

In making absolute cross-section measurements where the target particles are contained in a gas cell, it is essential that one be able to measure absolute gas pressures in the range 10^{-2}–10^{-5} torr accurately. For many years the standard method of measuring pressure in this range has been by the use of the McLeod gauge, and considerable attention has been given to the contruction and operation of these gauges.[33,34] In normal use in collision experiments, a cold trap is placed between the McLeod gauge and the target gas cell to prevent mercury vapor from contaminating the target gas. The streaming of mercury vapor from the gauge to the trap causes a difference between the desired and the measured pressure.[35] This fact seems to have been ignored for many years until it was pointed out in 1961 by Ishii and Nakayama,[36] and has undoubtedly led to many of the inconsistencies in cross-section measurements.

One of the more promising alternatives to the McLeod gauge for the absolute measurement of pressures is the capacitance manometer. Since the gas to be measured does not come into contact with a liquid, this device creates much less of a contamination hazard than instruments based on liquid manometer principles. Capacitance manometers have been shown to agree with McLeod gauges to within 2% in the pressure range 10^{-5}–10 torr under conditions in which the mercury-streaming error is expected to be negligible.[37]

3.3. Detectors

In performing charge-transfer measurements, it is necessary to obtain information concerning the nature of the collisions that result in charge transfer, thereby requiring the use of detectors that will enable one to measure the particular aspects of the collisions that are of interest. The ideal detector of particles would be one that would be able to detect single particles, and determine their mass, velocity, charge, and internal energy state with no limitations on the ranges of these quantities that can be studied. Real detectors fall far short of this ideal and the amount of information one can obtain about collision processes depends primarily on one's ability to develop a suitable system of detectors. Most detectors can be categorized according to the basic processes that underlie their ability to detect particles. We will briefly discuss detectors that use the following four basic processes:

1. Collection of the charge of the particles.
2. Detection of the temperature rise of a solid due to the absorption of the kinetic energy of the particles (thermal).
3. The emission of secondary electrons from the surface of a solid.
4. The production of free electrons in gases and semiconductors.

For details concerning the use of specific detectors, the reader is referred to texts on experimental methods.[38]

3.3.1. Charge Collection

Obviously, this method of particle detection is useful only when the particles to be detected have a net electrical charge. The most troublesome aspects of this type of detector are its relatively low sensitivity and the necessity to be assured that the reflection of the particles to be measured and the effects of secondary electron emission produce negligible errors in the measurement of particle fluxes. A widely used version of this type of collector is the Faraday cup.[39]

3.3.2. Thermal Detectors

In these detectors, the particles deposit nearly all (it is assumed) of their kinetic energy in a solid, creating a temperature rise that can be detected by arrangements that utilize thermocouple junctions,[40] thermistors,[41] or superconducting films.[42] These detectors have the advantage of being sensitive to both neutral and charged particles, but as a group they have a relatively low sensitivity compared to detectors such as electron multipliers. It should be pointed out, though, that superconducting films have been found[42] to be capable of detecting single α particles provided enough energy is deposited in the film, and that this type of detector has also been shown to be capable of detecting single He and Ar neutral atoms with energies above 150 eV.[43]

3.3.3. Secondary Electron Emission

In this detector the particle to be detected strikes a solid surface and emits one or more secondary electrons. These secondary electrons can be detected directly or their number increased by having the electrons strike other surfaces, thereby emitting more electrons. This latter technique forms the basis for the operation of electron multipliers.[44] Electron multipliers can be operated in the dc mode where one measures the amplified electron current, or in the counting mode where a single incident particle produces a pulse of electrons at the output. The latter method offers a sensitive means of detecting low fluxes of particles. Electron multipliers are limited by being unable to tolerate very large particle fluxes (but for these cases some of the previously mentioned detectors may be suitable), and by being unable to detect with good efficiency neutral particles whose kinetic energy is less than 500 eV (unless they happen to be in a metastable electronic state).

Scintillators are widely used as detectors in nuclear physics, but are generally of little use as detectors of heavy particles whose energy is less than ~10 keV. However, it is possible to make use of a combination of secondary electron emission and a scintillator to produce a sensitive detector,[45] whereby the particle to be detected ejects secondary electrons from a surface and the electrons are accelerated to an energy of 10–50 keV and then allowed to strike a scintillator to produce a light pulse which is detected by a photomultiplier.

A detector which is similar to the ones based on the emission of secondary electrons is the surface ionization detector. In this device the particle to be detected (usually an alkali metal atom) is ionized upon its approach to a hot metal surface and the resultant positive ions are detected. This type of detector is quite useful for detecting particles of low (thermal) energy, but the fact that it is limited to only a few species of particles is a serious disadvantage.

3.3.4. Ionization in Gases and Solids

Detectors that utilize the ionization produced by the particle to be detected in gases[46] and semiconductors[47] have frequently been used for the detection of both ions and neutral particles. When the medium of the detector is a gas, one generally achieves the amplification desired by operating the detector as a proportional counter. When the medium is a semiconducting solid, the most common detectors are silicon-surface barrier and lithium-drifted germanium detectors. The advantages of these detectors are high gain, low noise, and under some circumstances the ability to measure the energy of the detected particle with good resolution. The main disadvantage of these detectors is that they are limited to the detection of particles that have a kinetic energy in excess of 1 keV for light particles such as protons, and energies greater than ~10 keV for particles as heavy as Ar. Even when these detectors are capable of detecting particles at the low energy end of their range, statistical fluctuations in the number of primary ions formed reduce the usefulness of the detector for energy analysis.

3.4. Beam Production

In general the production of an appropriate beam for performing collision studies is a relatively simple matter. An ion source is used which utilizes an arc or electron bombardment to produce positive ions which are then extracted from the ion source by an electric field and accelerated to the desired energy. As is frequently the case, there may be a number of different species of ions coming from the ion source, so one usually uses magnetic analysis to select the desired species.

One of the difficulties often encountered in producing ion beams is the possibility that there will be an unknown mixture of electronically excited states,[48,49] and in many cases considerable care must be exercised to ensure that only a single electronic state is present in the beam. This difficulty has led to many of the discrepancies observed in the results of charge-transfer measurements.

In considering measurements with ion beams at low energies, the maximum beam flux available may be limited by the space charge spreading of the beam due to the mutual repulsion of the positive ions. It is for this reason that it is difficult to make measurements at energies lower than ~10 eV, unless the cross section of interest is large enough that a low-intensity beam can be tolerated. In recent years this difficulty has been somewhat avoided by going to what are called merged-beam techniques in which both of the reactant particles are a part of a beam, with these two beam components traveling with very nearly the same velocity, such that the relative velocity and the relative kinetic energy between the two types of particles are rather small. These techniques can be used to perform collision studies down to several tenths of an electron volt.[50]

3.5. Cross-Section Measurements

3.5.1. Total Cross Sections

Experiments that measure total cross sections[51–53] are in one way or another integrated over the scattering angles of the products. One can detect either the relatively fast incident particles after they have picked up an electron from the target particle or the slow ions produced from the target particles. We discuss these separately below.

3.5.1.1. Detecting Fast Particles. In measuring total cross sections by detecting the fast products one assumes that the great majority of charge-transfer events result in only a small deflection of the incident particle during the collision, so that measuring the number of these products at angles near 0° (the incident beam direction) will effectively include nearly all the events that occur. This assumption would in general be expected to be valid only for rather large incident energies (say, >10 keV), and so most such experiments are performed at these higher energies.

(a) *Growth Curve Method.* If one passes a beam of particles through a chamber or cell that initially contains no gas, then the beam is unchanged in its passage through the cell. If gas is then introduced into the cell, it is found that charge components other than that of the primary beam leave the cell, and these can be separated by the application of appropriate electric or magnetic fields transverse to their direction of motion. If the intensity of each of these components is found to depend linearly on the gas pressure in the cell, then it is implied that the charge changes that have occurred are the result of single collisions within the cell and one can therefore deduce cross sections directly from such data. If the dependence of the charge components on pressure is not linear, then it is concluded that multiple collisions have occurred and the extraction of cross sections from the results is made more complex, especially for systems in which a number of different charge states are produced. This method has been used extensively to determine cross sections for charge-changing collisions both for single-collision conditions and for multiple collisions.[51]

(b) *Beam Attenuation in Transverse Fields.* A second method for studying charge transfer is similar to the growth curve method, but it involves placing a transverse electric or magnetic field within the scattering region in the gas cell. Any collision events that result in a change of charge for the beam particles will result in their departure from the path followed by those particles that do not change their charge state, with a corresponding attenuation of the number of particles that leave a slit in the gas cell positioned so as to permit only the particles that did not undergo a change of charge to pass through.[52]

(c) *Beam-Equilibrium Method.* A third method involves the use of higher gas pressures within the collision region, resulting in multiple collisions before the particles leave the cell. In this technique, one observes the fraction F_i of the total number of particles leaving a gas cell that are in a particular charge state i, and then observes how these fractions vary as the gas density in the cell is increased. In particular, when the fraction no longer changes with increased gas density, the beam is said to have come to an equilibrium charge state, with components $F_{i\infty}$. For simple systems such as H$^+$ incident on a gas at energies where it is expected that double electron transfer is negligible, one expects only two possible charge states to be present after the beam has passed through the cell, namely H$^+$ and H. For this situation one can determine the cross sections for electron capture σ_{10} and loss σ_{01} from the equations

$$F_{1\infty} = \frac{\sigma_{01}}{\sigma_{01} + \sigma_{10}} \quad \text{and} \quad F_{0\infty} = \frac{\sigma_{10}}{\sigma_{10} + \sigma_{01}} \qquad (3.1)$$

This method has frequently been used to obtain information about charge transfer,[53] but its usefulness is generally limited to situations in which one can have no more than three charge components leaving the gas cell.

3.5.1.2. *Detecting Slow Ions.* In addition to the methods just described where the charge transfer events are observed by detecting the fast particles that have picked up an electron in the collision, one can also observe the relatively slow ions produced when the target atoms lose an electron. This method is best applied at rather low energies where the probability of ionization is small, since the ionization events also produce slow positive ions which makes the measurement of charge-transfer cross sections less straightforward. In this method, the beam passes through a gas cell where positive ions are produced with small kinetic energy and then collected by applying a small transverse electric field created by a pair of parallel plates, with the ions being collected on the negative plate. Care must be exercised that the measured current is not significantly influenced by secondary electron emission from the collecting plate caused by either photons, electrons, or the positive ions striking the plate. This method has been widely used with a variety of configurations for producing the collecting electric field.[54,55]

As a variation on this technique the slow ions can be extracted from the collision region and subsequently mass analyzed. The advantages of this method are most apparent when the target gas is composed of molecules so that a number of different species of positive ions might be produced.[56,57]

3.5.2. Differential Cross-Section Measurements

Experimental studies of the charge-transfer process in which the angular distribution of one of the collision products is measured generally provide more information about the nature of the process than does the measurement of total cross sections. Still more detail is obtained if some information can be gathered concerning the internal electronic-energy states that participate in the collision. As stated earlier, most of the incident ions that pick up an electron during the collision undergo only a small change in direction if the incident energy is high (e.g., >10 keV), making an experimental study of the angular distribution difficult. For this reason most of the differential cross-section measurements discussed in this section were performed at energies less than 10 keV (with a few exceptions where at the higher energies the small number of events that do result in angular deflections of several degrees were observed). One of the principal advantages to be gained from differential measurements is the fact that it may be possible to relate scattering angle to classical impact parameter and thus provide a firmer basis for comparing experiment with theory. The ability to relate angle to impact parameter depends on one's knowledge of the potential curves that participate in the collision and the positions at which transitions between these curves take place.

Let us now consider the charge-transfer process

$$A^+ + B \rightarrow A + B^+ \quad (3.2)$$

One possibility for obtaining differential information about this process is to observe the angular distribution of the neutral product A. The detection of these products is usually accomplished by the use of an electron multiplier,[58-62] but if A is in its ground electronic state, then it must have enough kinetic energy to produce a significant number of secondary electrons in the detector, thereby producing a lower limit on the energies for which such studies may be carried out. If A happens to be in a metastable state, then it is quite possible that sufficient secondary emission can be obtained even at low kinetic energies, but this of course limits the type of processes that can be studied to those events in which the electron is transferred to a metastable state. It should be emphasized that a critical factor in any study of differential charge transfer by observing the neutral products is a knowledge of the absolute efficiency of the detector for detecting neutral particles, whether ground state or not. Another possibility for detecting the results of charge transfer to metastable states is by quenching the atoms in these states by an electric field and observing the photons produced.[63] This has the added advantage of providing information about the states in which the charge transfer has left the neutral product.

In addition to the methods described in which the relatively fast neutral atom is observed, it is also possible to observe the angular distribution of the slow B^+ ions formed in process (3.2), since these charged particles are easily detected even though they have small kinetic energies.[64]

Let us now consider the charge-transfer process

$$A^{2+} + B \rightarrow A^+ + B^+ \qquad (3.3)$$

This type of collision has a large experimental advantage over those already discussed since both products are charged. Since A^+ usually has the larger kinetic energy, it is more common to detect it[65] rather than B^+. As in any experiment in which one is detecting charged products, when these products have a kinetic energy which is not sufficiently large to produce secondary electrons at the entrance of an electron multiplier, it is a simple matter to accelerate them with an appropriate potential difference before allowing them to strike the electron multiplier. The existence of the charged product A^+ provides the opportunity to use electrostatic analysis of its kinetic energy spectrum to provide information on the internal energy states to which the charge transfer has taken place.[66,67]

The measurements described by Siegel *et al.*[67] were performed with the system shown in Fig. 8. They were differential in angle and the kinetic energy of the products A^+ was measured with a cylindrical electrostatic analyzer. Thus, at different values of impact parameter, corresponding to different scattering angles, the relative population of the final intermediate states could to a large extent be determined by observing these energy spectra, since in many cases only one of the product ions was in an excited electronic state after the collision. A disadvantage to this method of observing the final internal states of the products is the limited energy resolution possible for electrostatic analyzers operated without deceleration of the ions before analysis. Values of $\delta E/E$ in the range 0.001–0.01 are commonly used, which corresponded in the experiments of Ref. 67 to peak widths of a few electron volts. If one achieves improved energy resolution by decelerating the ions before analyzing them, then the problem of losing ions in the region of deceleration creates difficulties in making measurements of absolute cross sections. One could achieve a greatly improved energy resolution by just looking at the optical spectra from transitions that follow the charge-transfer events, but at the expense of not being able to correlate observed transitions with a given impact parameter. This difficulty could in principle be obviated by observing coincidence between charged particles scattered through a known angle and the photons produced from the scattering event, but low-intensity experiments of this type are quite difficult.

For the system

$$He^{2+} + B \rightarrow He^+ + B^+$$

where B is a rare-gas atom, it is especially easy to assign internal states for all the participants in the collision since there is no question concerning the states of the incident and target particles, and one of the products, He^+, has a set of hydrogenic states with rather large and well-known separations.

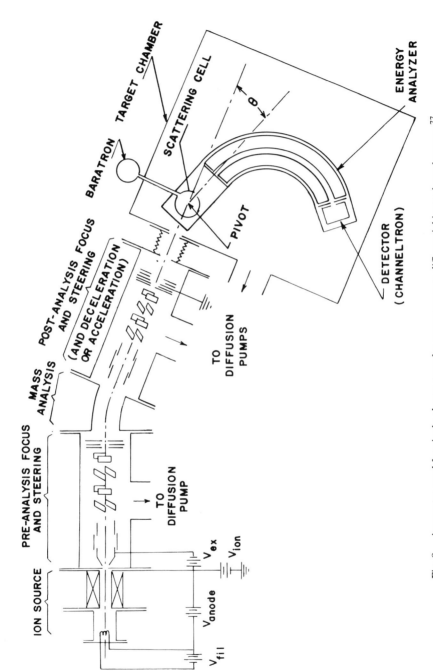

Fig. 8. Apparatus used for single-charge-transfer measurements differential in angle and energy.[77]

4. Experimental Data and Interpretation

4.1. Introduction

In this section we will present some characteristic experimental results and discuss them in light of the theoretical discussion given earlier. More extensive collections of experimental data and comparisons of that data with theoretical calculations are available elsewhere.[6,8,30,31,67] Since quantum mechanics can presumably be used to describe ion–atom collisions as accurately as one likes, the primary role of theory in this field is the interpretation of the experimental phenomena to gain an understanding of the physical processes at work. For this reason simple model descriptions which indicate the mechanism involved are still often preferred to detailed calculations since they can be used for predictive calculations on other systems. Here we hope to indicate how potential energy information, interaction mechanisms, and spectroscopic data may be obtained from cross-sectional data.

4.2. Integrated Cross Sections

Integrated cross sections for charge transfer with incident singly and doubly charged ions have well-known general characteristics. For strong coupling, the cross terms in Eq. (2.1) which depend on the difference in action generally go through many oscillations as a function of b (or τ) at intermediate and low velocities. These terms are often assumed to be self-canceling (the random phase approximation) out to some impact parameters b^* where the phase differences and/or the transition probabilities become very small. Using this, the integrated cross section is written in a form that lends itself to direct comparison with experiment:

$$\sigma(v) \approx \bar{\mathcal{P}}_{0,f} \cdot \pi b^{*2} \quad (4.1)$$
$$\scriptstyle 0 \to f$$

where $\bar{\mathcal{P}}_{0,f}$ is the transition probability in Eq. (2.5) averaged over the impact parameter from $b=0$ to $b=b^*$. In the two-state approximation, b^* is chosen to be the largest value of impact parameter for which the phase differences are equal to some small value (Firsov[68] used $2/\pi$ for this value). The velocity dependence of the cross section may be determined by either $\bar{\mathcal{P}}_{0,f}$ or b^*.

For symmetric resonant charge transfer, the phase difference between the g and u states is related to the exchange energy, which falls off exponentially at large R. Therefore, $b^* \approx \alpha^{-1}[\beta - \ln(v)]$, where β depends on the potential parameters, and α is the screening constant indicating the overlap of the atomic wave functions.[6,30] As $\bar{\mathcal{P}}_{0,f} \approx \frac{1}{2}$, this cross section increases monotonically with decreasing velocity at low and intermediate velocities (Fig. 9). The experimental cross section can therefore be used to obtain the parameters β and α. This procedure fails at high energies since the cosine of phase differences does not go through a number of oscillations, and, also, the excited states become important. At very low velocities b^* may be determined by the orbiting radius.[7]

For charge transfer due to a curve crossing, a reasonable approximation is obtained by setting $b^* \simeq R_x$. Then using the LZS transition probabilities of Eq. (2.9), one has

$$\bar{\mathscr{P}}_{0,f} \simeq \int_0^1 2P_x(1-P_x)\left(\frac{b}{R_x}\right) d\left(\frac{b}{R_x}\right) \tag{4.2}$$

which is velocity dependent. This expression has been applied to a large variety of charge exchange collisions involving curve crossings, particularly single charge transfer with incident doubly charged ions.[69] The general velocity dependence given by this prescription is characteristic of the observed velocity dependence for nonresonant charge-transfer collisions (Fig. 10). The cross sections usually go through a single broad maximum, falling off rapidly at low energies, and becoming orders of magnitude smaller than the maximum value of the cross section.

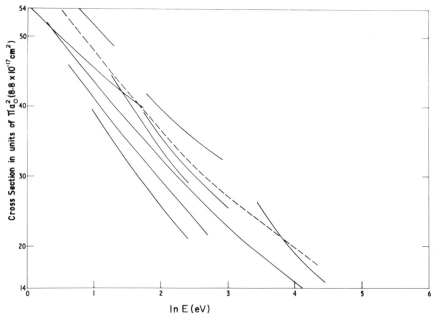

Fig. 9. Charge-transfer cross section (atomic units) as a function of incident-ion energy (E) for $Ar^+ + Ar$ collision showing dependence on $\ln(E)$ predicted by a simple model [Eq. (4.1)]. Solid lines are experiments; dashed line gives detailed impact parameter calculation.[27]

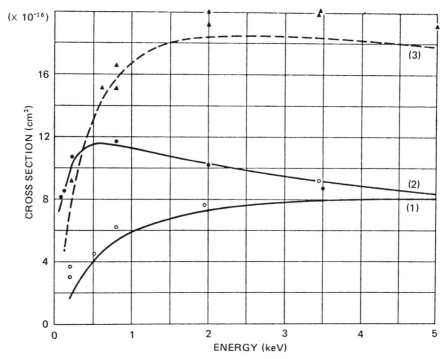

Fig. 10. Single-electron-capture cross section by doubly charged ions, from Ref. 8, p. 212. Results are typical of curve crossing mechanism. Curves are theoretical calculations based on LZS: (1) $Ar^{2+}(^3P) + Ne(^1S) \to Ar^+(^2P)$; (2) $N^{2+}(^2P) + He(^1S) \to N^+(^3P) + He^+(^2S)$; (3) $Ne^{2+}(^2P) + Ne(^1S) \to Ne^+(^1S) + Ne^+(^2P)$. Experimental results: $\bigcirc$, $Ar^{2+} + Ne$; $\bullet$, $N^{2+} + He$; $\triangle$, $N^{2+} + Ne$. These are favorable cases for the LZS approximation.[22]

The two-state, nonresonant charge-transfer reactions which do not involve a curve crossing, also go through a single broad maximum.[70–72] At low velocities $\bar{\mathscr{P}}_{0,f}$, obtained from averaging over the transition probability in Eq. (2.12), has the dominant velocity dependence, and therefore, as in the curve-cross case, one chooses $b^* \simeq R_x$, the region of strong coupling. However, at higher energies, $\bar{\mathscr{P}}_{0,f} \to \frac{1}{2}$ and b^* has a velocity dependence similar to the symmetric resonant case[71] (Fig. 11). For nonresonant charge transfer, two broad maxima have been found for some collisions for which the reaction may proceed by either crossing or noncrossing mechanisms.[31]

Using these prescriptions, potential energy information may be obtained directly from the measured cross sections.[67] The size of the maximum in the cross section is determined by the position of the transition region R_x, and the position of the maximum by the coupling strength. Obviously more information can be obtained with a more detailed parametrization of the problem. The position of the maximum $v_{\max}$, can be estimated

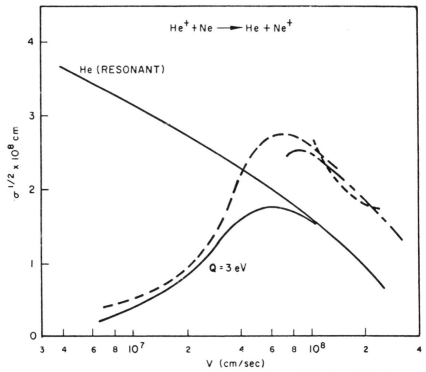

Fig. 11. Charge-transfer cross section plotted versus velocity for $He^+ + Ne$ collision. Results are typical of non-curve-crossing mechanism. Dashed lines are experiments; solid lines are model calculations based on Eq. (4.1). The energy defect Q is 3 eV. The line labeled resonant is the calculation for $He^+ + He$.[71]

from an argument based on the uncertainty principle,[30] yielding the equation

$$1 \simeq \frac{\Delta_{0,f} \cdot \delta R_x}{\hbar v_{max}} \qquad (4.3)$$

where $\Delta_{0,f}$ is the separation of the *adiabatic* potentials in the transition region and δR_x is the width of this region. For the noncrossing case where the distortion of the potentials is not important, one often estimates $\Delta_{0,f}$ as Q, the energy defect as in Eq. (2.13), and δR_x of the order of a few Bohr radii, obtaining the so-called adiabatic criterion of Massey.

For resonant or near-resonant charge transfer between unlike species, a common occurrence when the target is a molecule or when a number of close lying spin states are involved, e.g.,

$$H^+ + O(^3P_j) \rightarrow H(^2S_{1/2}) + O^+(^4S_{3/2})$$

similar concepts can be applied. However, the behavior of the long-range forces critically affects the position of the maximum as they split the degenerate or nearly degenerate initial and final states of the collision system.[73] In many cases, there is significant spin–orbit splitting in the final state multiplet. As the exchange interaction increases exponentially with decreasing R, it eventually dominates the spin–orbit interaction at some value of R. The concepts developed here are useful for determining the relative populations of products among the final state multiplet.[27,74] If b^* is less than that internuclear separation where the spin–orbit and exchange interactions are approximately equal, the fraction of products in a given atomic J state of the multiplet can be determined from a statistical

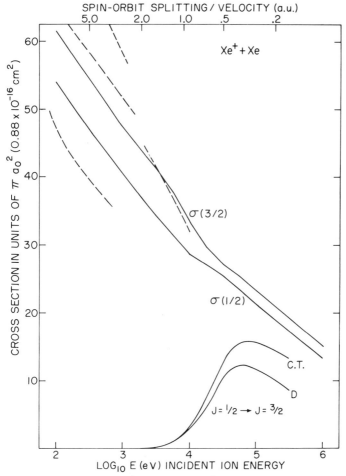

Fig. 12. Cross sections for $Xe^+ + Xe \rightarrow Xe + Xe^+$. Solid lines are theoretical curves: $\sigma(\frac{1}{2})$, $J=\frac{1}{2} \rightarrow J=\frac{1}{2}$; $\sigma(\frac{3}{2})$, $J=\frac{3}{2} \rightarrow J=\frac{3}{2}$; C.T., inelastic charge transfer; D, inelastic, no charge transfer. Dashed lines are experimental charge-transfer cross sections.[27]

weighting. If b^* is much greater than that internuclear separation, the different J states should be treated as distinct final states (Fig. 12).

Deviations from the general behavior described herein occur for a variety of reasons. For instance, if the difference in the potential functions and corresponding actions for two trajectories go through a maximum, the random phase approximation would break down in the vicinity of the maximum. Accounting for the constructive (or destructive) interference in this region will modify the cross section by superimposing an oscillatory structure on the general behavior already described. For instance, for the symmetric resonant two-state example discussed earlier, if a stationary point exists, Eq. (4.1) becomes[75]

$$\sigma_-(v) \approx \left\{ \frac{\pi}{2} b^{*2} - \left[\frac{\pi^{3/2} b}{\left| \frac{\partial^2}{\partial b^2} \frac{(A^{(g)} - A^{(u)})}{2 \cdot \hbar} \right|^{1/2}} \cos \left(\frac{A^{(g)} - A^{(u)}}{\hbar} + \frac{\pi}{4} \right) \right]_{b=b_0} \right\} \quad (4.4)$$

where b_0 is the position of the stationary point. A similar modification can be applied to the inelastic cases. The spacing and magnitude of the oscillatory structure[76] (Fig. 13) can be used to determine the behavior of the potential differences in the vicinity of the extremum.

4.3. Energy-Loss Spectra with Doubly Charged Ions

In the collision processes

$$A^{2+} + B \rightarrow A^+ + B^+$$

either product may be formed in an excited state. Since the kinetic energy of the scattered ions can be accurately determined, this provides a means for observing the states of the ions, as well as the mechanisms for charge transfer. Work with incident He^{2+} colliding on the rare gases and measuring the energy loss as a function of scattering angle has been carried out by Chen, Boring, and Siegel[77,67] and Cooks et al.[78] Because the states of the He^+ ion are well known a straightforward interpretation of the collision in terms of the excited states of the rare-gas ion can be made. The important parameter is the energy defect Q between the electronic binding energies of the particles in the incident and exit channels. Since the binding energies of the target atoms may differ considerably, each collision system must be interpreted separately according to its peculiar arrangement of ionic states.

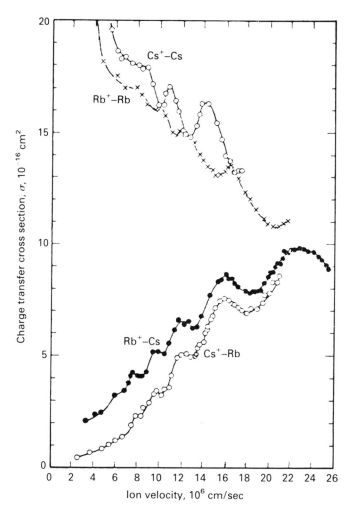

Fig. 13. Single-charge-transfer cross sections showing oscillatory structure superimposed on usual energy dependence for symmetric and nonsymmetric collisions.[76]

In Fig. 14, we give the energy-loss spectra for $He^{2+} + Ne$ as a function of laboratory angle for 200 eV laboratory energy from the data of Chen.[77] Because τ at small angles is roughly related to an impact parameter (Section 2.3), it can be seen in Fig. 14 that for distant collisions only one peak occurs, and this can be associated with Ne^+-excited states which are exothermic by 5–6 eV. There is another group of states with exothermicity of the order of 1.2–2.2 eV. Transitions to these states only occur at smaller impact parameters and the thresholds are easily observable. If the exothermic states

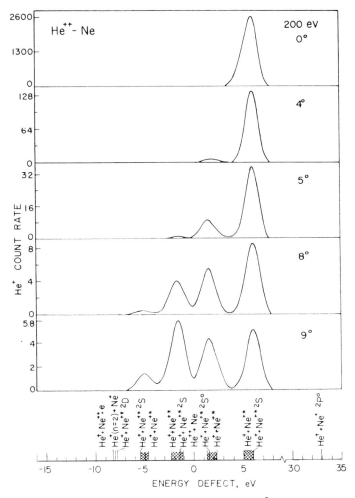

Fig. 14. Energy-change spectra for single-electron capture in $He^{2+}+Ne$. The energy scale is the inelastic energy loss, i.e., the measured energy change corrected for the elastic energy loss.[77]

discussed are treated as having a $1/R$ long-range repulsive force and the incident channel as having a polarization attractive force, then the crossings with the incident state can be deduced for both groups. For the state $He^+ + Ne^+[2s2p^6, {}^2S]$ with $Q = 5.94$ eV, this implies $R_x = 4.77a_0$, whereas for $He^+ + Ne^+[2p^42s, {}^2D]$ with $Q = 2.2$ eV, $R_x = 12.3a_0$. The transition probability at a crossing depends on the strength of the coupling [Eq. (2.9)], which falls off exponentially for the charge-transfer reaction. For this reason, transitions to states with higher exothermicity dominate at small scattering angle, and the channel with small exothermicity has an angular dependence similar to the *noncrossing* endothermic channels. Because of the finite angular resolution in the experiments, impact parameters equal to and greater than $R_x = 4.77$ are included in the 0° measurement, and the threshold for the $Q = 5.94$ eV channel is not seen. Further, no transitions to the exit channel with exothermicity of 32.8 eV are observed because the large energy defect at these collision energies results in rapid oscillations of the phase factors in the transition probabilities [Eq. (2.13)]. The quantity of importance is Q/v.

The incident channel $He^{2+} + Ne[2p^6, {}^1S]$ leads to a single molecular state of symmetry ${}^1\Sigma^+$. According to the coupling mechanisms discussed in Section 2, one would imagine that ${}^1\Sigma^+ \to {}^1\Sigma^+$ and ${}^1\Sigma^+ \to {}^1\Pi$ transitions would be allowed. Using the Wigner–Witmer coupling scheme[30] for the exit channels of exothermicity between 5 and 6 eV, $He^+[{}^2S] + Ne^+[2s(2p)^6, {}^2S]$ leads to ${}^{1,3}\Sigma^+$ molecular states and $He^+[{}^2S] + Ne^+[({}^3P)3s, {}^2P]$ leads to ${}^{1,3}\Sigma^-$ and ${}^{1,3}\Pi$. However, at small angles the peak has been identified[67,77] with the $Ne^+[2s(2p)^6, {}^2S]$ state. The fact that the other state is not observed is consistent with the fact that rotational coupling is weak for slow collisions at large internuclear separation.[6]

The spacing of the first and second peaks as a function of τ is shown in Fig. 15. It is clear that the position of the peak changes with τ or impact parameter. For small τ, i.e., large impact parameters, the second peak appears to be associated with the $Ne^+[({}^3P)3p, {}^2S^o]$ state, a transition conserving electronic orbital angular momentum. However, this leads to ${}^{1,3}\Sigma^-$ quasi-molecular states or transitions which are forbidden under the rules described earlier. This implies that spin–orbit coupling of the outer electron with the open shell of Ne^+ most likely plays a role, for there is a close-lying $Ne^+[({}^3P)3p, {}^2P^o]$ state which leads to a ${}^1\Sigma^+$ state, among others. For larger τ, i.e., smaller impact parameter, the peak appears to be a varying mixture of the two final state channels with exothermicity between 1.2 and 2.2 eV associated with ${}^1\Sigma^+$ molecular states. The oscillations for each energy appear to have some regularity. Neutral helium atoms associated with a double-charge transfer were not observed in these experiments. However, the molecular states associated with this reaction are felt to play an important role here. That is, the repulsive final states may be populated by crossings with the double-charge-transfer states. The results of Fig. 15 illustrate how measurements differential in energy and in scattering angle can identify reaction channels which could not be resolved by kinetic energy measurements alone.

The $He^{2+} + Ar$ system has very different energy characteristics. The entrance channel lies in the continuum of the Ar^+ ion for charge transfer to the He^+ ground state. As in the previous case, the possible endothermic channels at large R diverge in energy from the incident ${}^1\Sigma^+$ state. Therefore, for large impact parameters exothermic reactions are preferred. Three

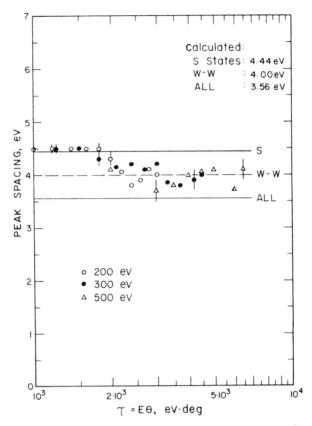

Fig. 15. Measured spacing between first and second peaks appearing in $He^{2+}+Ne$ energy-change spectra (Fig. 14) corresponding to two groups of single-charge-transfer states. S, average spacing between incoming and outgoing S states; W-W, average spacing between $^1\Sigma^+$ states; ALL, average spacing between all states in the two groups, each state weighted equally.

broad peaks, i.e., broader than the elastic peaks, are distinguishable[77] (Fig. 16). The largest is associated with the closely spaced Ar^+ single excited states. The next peak may be associated with doubly excited autoionizing states in the continuum of Ar^+. It is known that these doubly excited states may lead to well-defined quasi-molecular potentials.[23] Autoionizing states of doubly excited Ar^+ have been identified at 0.3 and 1.8 eV into the continuum.[79] The last peak is associated with single charge transfer into the continuum of states associated with $He^+(1s)+Ar^{2+}+e$ and those other bound states that lie in the continuum, e.g., $He^+(2s)+Ar^+$. This peak grows and broadens with increasing τ (decreasing impact parameter; see Fig. 17). Further, the maximum shifts toward the endothermic energy transfers as τ increases, as the transitions are taking place at smaller values of R where the biasing effect of the $1/R$ long-range potential has diminished. The peak can be roughly described by the weak coupling probability equation (2.7), modified to account for the density of states. Such a peak can be thought of as arising from the system relaxing into some statistical distribution of final states after the disturbing particle has passed and, in this case, removed an

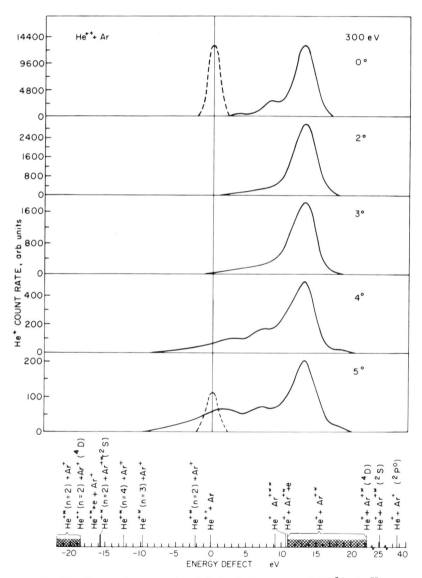

Fig. 16. Energy-change spectra of single-electron capture in $He^{2+} + Ar$.[77]

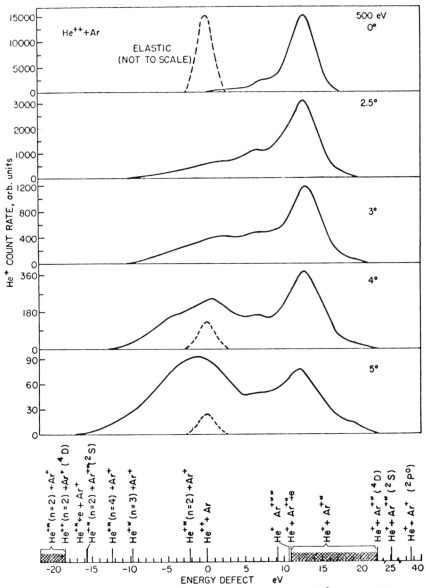

Fig. 17. Energy-change spectra of single-electron capture in $He^{2+}+Ar$, collisions at 500 eV.[77]

electron. For closer collisions the disturbance, and hence the spread in final states, is greater. In contrast, the maximum of the peak due to coupling at *large R* remains very nearly fixed in energy transfer over a broad range of τ. This peak at $Q \approx 12$ eV is broader than the elastic peak, indicating that a number of close-lying states are involved. For the $He^{2+}+Kr$ collision, Cooks et al.[78] find a sharp peak near the incident channel associated with $He^+(n = 2)+Kr^+$ which is nearly resonant, and a broad, weak peak associated with the excited states and continuum of Kr^+ with He^+ in the ground state.

4.4. Angular Differential Cross Sections

Angular differential cross-section data yield information not only on the relative likelihood of the possible transition but also on the interaction potentials, through the angular dependence. For purely elastic scattering on a monotonically decreasing potential with increasing R, straightforward mathematical schemes are available for inverting the data to obtain the potential.[6] For large τ values where there is a common repulsive potential for all processes such procedures have often been used to calculate an average, total differential cross section.[80] This is defined as the sum of all the elastic and inelastic cross sections, including charge transfer [see Eq. (2.5)]. For more complicated situations, the procedure is one of trial and error, based on an understanding of the physical phenomena. The theoretical procedure described in Section 2.2 has been employed with success by Smith and co-workers[16,19,81] to a number of collisions involving charge transfer. They find that even though the procedure has severe limitations near the angular threshold region, it gives an excellent starting point for interpreting the angular dependence and interference oscillations in inelastic and elastic cross sections. The quantities one wishes to find are the differences between potential energy curves, the position of transition regions, and the coupling strength.

The semiclassical relationship

$$\frac{\partial A^{(j)}}{\partial \theta} = L^{(j)} \qquad (4.5)$$

where $L^{(j)} = pb^{(j)}$, can be used to relate the oscillations in the differential cross sections, when they can be resolved, to the spacing between the potential curves for the different trajectories.

If the energy defect in the transition is small compared with the incident particle energy E, then using Eq. (4.5), the spacing between the maxima is related to the difference in impact parameters for two trajectories associated with the scattering into a given angle θ, by

$$\Delta b(\theta, E) \simeq 2\pi\hbar \left(\frac{E}{2\mu}\right)^{1/2} \frac{\partial n}{\partial \tau} \qquad (4.6)$$

where n is an indexing number for the maxima.[81] Therefore, a first step is to choose potentials which yield deflection functions producing the correct spacing. This does not uniquely determine the potential curves unless one

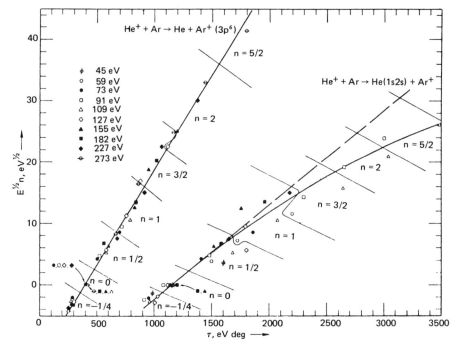

Fig. 18. $nE^{1/2}$ plotted versus τ for $He^+ + Ar \rightarrow He + Ar^+(3s, 3p^6)$ and $He^+ + Ar \rightarrow He(1s, 2s) + Ar^+$, at a variety of energies.[81]

knows the value of the potentials at some R from an accurate calculation or the trial potentials also yield good absolute values of the cross section. Since absolute differential cross-section measurements often have large errors associated with them, the latter check is sometimes difficult. For the unsymmetric charge transfer collision $He^+ + Ar \rightarrow He + Ar^+[3s3p^6, {}^2S]$, which takes place via curve crossing, Smith et al.[81] have plotted the position of the maxima in the charge transfer cross section as a function of τ (Fig. 18) for incident ion energies of 45–300 eV. It is seen that the slope of this plot is very nearly a constant, $\Delta b \simeq 0.51 a_0$, implying that the deflection functions are separated in b by a constant. This suggests that the potentials for the appropriate internuclear separations are nearly exponential functions of R with approximately the same screening constant but different multiplicative constants.

This procedure has been applied most successfully for high-energy, small-angle collisions, where the trajectories are simply approximated by straight lines and the phase differences have the form

$$\frac{1}{\hbar}|A^{(1)} - A^{(2)}| \simeq \frac{1}{\hbar v}\left|\int_{-\infty}^{\infty}[V_1(R) - V_2(R)]\,dZ\right| \quad (4.7)$$

where $R^2 \equiv b^2 + Z^2$ and τ, a function of b, is determined from some average interaction potential. In this limit, the oscillations have a simple v^{-1} dependence (Fig. 19). Everhart and collaborators[58,59,82] applied this concept to the symmetric resonant single-charge-transfer collisions $H^+ + H$ and $H^+ + He$ to obtain values for the exchange energy $V_g - V_u$ as a function of internuclear separation. At lower energies, the expression given in Eq. (2.3) should be used to account for the different trajectories on the gerade and ungerade states.[15] Here, if one of the states has a minimum, the oscillations may be superimposed on a rainbow structure.[15,83] At low energies, additional oscillations may occur if the nuclei are identical;[84] at both high and low energies, an effective shift in phase from this expression may occur due to coupling to close-lying excited states.[15] Equation (2.1) is the starting point for interpreting these phenomena.

Nonresonant capture cross sections at intermediate and high energies show a similar v^{-1} dependence in the charge-transfer probability. However, as can be seen for the $H^+ + He$ collision (Fig. 20), the oscillations are heavily

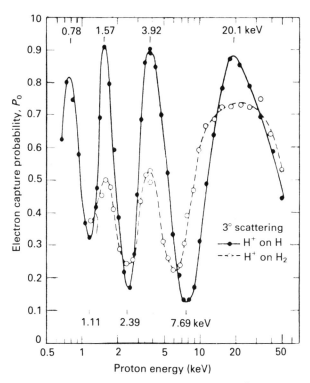

Fig. 19. Electron-capture probability versus ion energy for $H^+ + H$ collision showing the reciprocal velocity dependence of the oscillatory structure.[82]

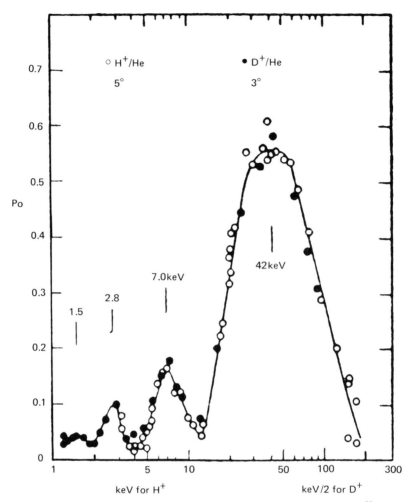

Fig. 20. Electron-capture probability versus ion energy for $H^+ + He$.[58]

damped, the damping being related to the collision time. This is an example of a non-curve-crossing, strong-coupling collision in the energy range shown. The damping has been shown to proceed roughly as $e^{-v_0/v}$, where v_0 is a constant, for both crossing and noncrossing cases.[10] This is easily deduced from Eqs. (2.9) and (2.13) for the transition probabilities. Here again the spacings between the oscillations are a measure of the separation of the two lowest $^1\Sigma^+$ molecule–ion interaction potentials of HeH^+.[13]

The position of a curve crossing can be experimentally estimated from either perturbations on the elastic scattering curves or the onset of the inelastic differential cross section. For the elastic scattering cross section for

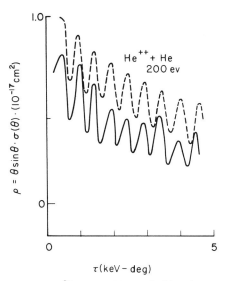

Fig. 21. Elastic scattering of He^{2+} + He at 200-eV incident ion energy from Ref. 27, solid curve. Calculation based on Eq. (2.3) using potentials chosen to fit oscillatory structure and approximate the absolute value of experiment,[86] dashed curve.

He^{2+} + He at 200 eV (Fig. 21), the competing double-charge-transfer reaction is evident from the oscillatory structure.[84] At 3400 and 4500-eV deg, two perturbations appear which can be associated with crossing of the incident channel $^1\Sigma_g^+$ state, with higher states of the same symmetry leading to

$$He^{2+} + He \rightarrow He^+(1s) + He^+(n=2)$$

(Fig. 4). This type of perturbation has been described by Smith and co-workers for a number of ion–atom collisions.[20,85] In Fig. 22, we represent hypothetical classical deflection functions corresponding to the two elastic

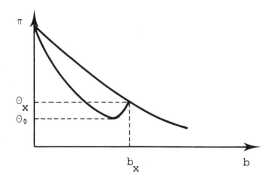

Fig. 22. Hypothetical classical deflection functions corresponding to the two *elastic* scattering trajectories involved in a curve crossing; b_x indicates impact parameter for which the distance of closest approach is R_x. Corresponding classical deflection functions exist for inelastic trajectories, Fig. 3. See Ref. 20.

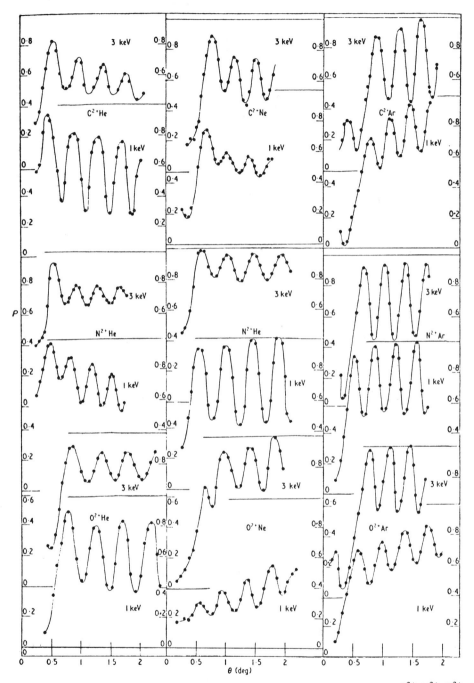

Fig. 23. Probabilities of single-electron capture as a function of scattering angle for C^{2+}, N^{2+}, O^{2+} on He, Ne, Ar at impact energies of 1 and 3 keV.[65]

scattering trajectories, where the crossing region is indicated by b_x. When the lower trajectory is smoothed over the region about b_x, it can be seen that it goes through two extrema which will result in the perturbations occurring between θ_0 and θ_x as indicated. Using potential functions in Eq. (2.3) which reproduce the oscillatory behavior of the elastic differential cross section, the crossing is estimated[86] to occur at $R_x \approx 0.97 a_0$. These perturbations will be followed, at larger τ, by a drop due to the inelastic process; however, the perturbations are often not resolved in an experiment.

In Fig. 23 are shown charge-transfer probabilities as measured by Hasted et al.[65] for a number of collisions of the type $A^{2+} + B \rightarrow A^+ + B^+$. The oscillations in angle are clearly observed since the data were deconvoluted to enhance the oscillatory structure. In some cases, the data indicate that more distant curve crossings contribute to charge transfer at small angles. On a τ plot, it is clear that the threshold regions seen here are very energy dependent. A calculation by Bates et al.[87] in the impact parameter approximation of the transition probability for $Be^{2+} + H \rightarrow Be^+ + H^+$ is illustrated in Fig. 24, and this energy dependence is apparent. Using the long-range interaction potential of the incident channel to describe the collision, Hasted et al.[65] have plotted transition probabilities versus impact parameter. In Fig. 25 we give the results for $C^{2+} + Ne$ collision where the dashed line indicates the crossing point. The values of b should not be taken too seriously, since in the threshold region a deflection function which is the average of the deflection functions for the incident and outgoing channel should be used. However, the general energy dependence exhibited in the sample calculation by Bates is evident. These figures also indicate one of the major flaws in the LZS transition probabilities. When the distance of closest approach equals R_x, the LZS transition probability is zero. Therefore, it should only be applied for impact parameters much smaller than R_x, as stated earlier. The problem is clearly that the transition region is finite in size. Recently the LZS transition region probabilities were employed in a partial wave calculation yielding improved agreement in this region, though not as good as the distorted wave approximation.[19] The energy dependence seen in Fig. 24 is much less than in Fig. 25, indicating that the crossing region may be better defined or narrower for the first case.

From the average value of the transition probability one can attempt to estimate the magnitude of the coupling matrix element based on the LZS approximation. It is found that coupling matrix elements determined from experiment may differ significantly from theoretically determined matrix elements for some single-charge-transfer collisions.[31] This has been attributed to coupling to excited states and the fact that the active electron does not start and finish in an s orbital,[22] upon which criteria the LZS approximation is derived. However, conceptually the LZS approximation continues to be useful.

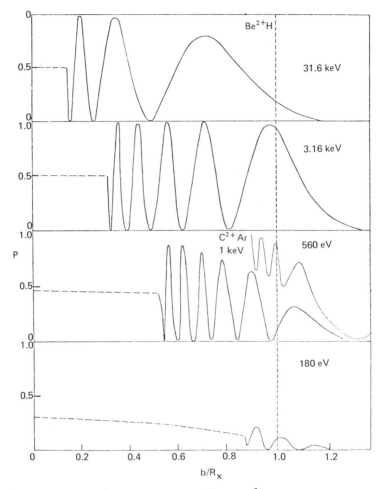

Fig. 24. Comparison of electron-capture probability in $Be^{2+} + H$, for a number of collision energies, as a function of b/R_x as calculated by Bates et al.[87] Plotted for reference are the experimental data for $C^{2+} + Ar$.[65]

4.5. Cross Sections Differential in Angle and Energy

A most valuable result is to obtain differential cross sections with final state analysis as discussed earlier. Cross sections differential in angle and energy obtained by analyzing the kinetic energy of the scattered particles provide such a result. Much of the preceding discussion applies to this case since in the analysis it was always presumed that the final state was known, and that only one final state was involved. In fact, calculations involving two states have been used for years for comparison with experiments in which the products were *not* state analyzed. In many of these cases the predominant final state of the products can be deduced, but this is not always reliable, and in other cases there is no single predominant final state.

The angular threshold for the charge-transfer differential cross sections, when it can be seen, yields information about the transition region and the coupling strength. At low energies for $He^{2+}+He$, the threshold in τ on the elastic channel, located as described earlier, may be shifted considerably

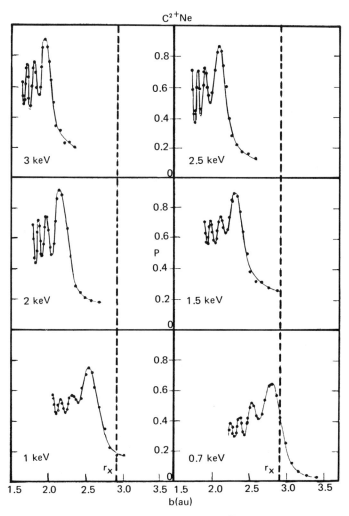

Fig. 25. Probabilities of single-charge transfer in $C^{2+}+Ne$ as a function of impact parameter b.[65]

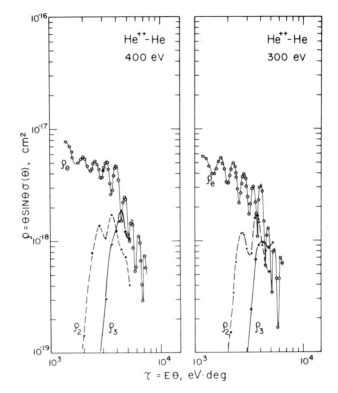

Fig. 26. Differential elastic scattering and single-electron capture cross sections for He^{2+} + He at 300 and 400 eV. Elastic scattering, ρ_e; single charge transfer to the $n = 2$ states, ρ_2; single-charge transfer to $n = 3$ and higher states, ρ_3.[86]

from that on the inelastic channel (Fig. 26) because the differences in the potentials of the final states are important. Since the outgoing potential for the reaction

$$He^{2+} + He \rightarrow He^+(1s) + He^+(n = 2)$$

is much less repulsive for R near the transition region (Fig. 4), the threshold in Fig. 26 seems to be consistent with the elastic perturbations discussed in the preceding section. Analysis of this cross section based on the LZS approximation yields a coupling strength of 0.17 a.u. and a value of 3.7 a.u. for the quantity $(\partial/\partial R)(V_f - V_0)|_{R=R_x}$. However, the two peaks in the cross section have been established to be associated with the separate crossing of the incoming channel $^1\Sigma_g^+$ state with the two $^1\Sigma_g^+$ states leading to the $He^+(1S) + He(n = 2)$ final states (Fig. 4). Using the half-height of the peaks and the elastic potentials discussed earlier, the second crossing occurs at about $R_x \approx 0.85 a_0$. The cross section ρ_3 in Fig. 26 is the sum of all unresolved higher single-charge-transfer processes. The onset of transitions to these higher states is about $R_x \gtrsim 0.8a_0$.

The single-charge-transfer spectra shown earlier (Fig. 14) lead to the differential cross sections for $He^{2+} + Ne \rightarrow He^+ + Ne^+$ shown in Fig. 27. Also shown is the elastic cross section and a theoretical average cross section obtained from a screened Coulomb potential with the screening constant chosen too be the same as used for $He^+ + Ne$ collisions. The impact parameters at the top of the figure are determined from this average potential and are useful only as a rough reference. The cross section represented by ρ_1 is that leading to the $He^+(1S) + Ne^+(2S, 2p^6)$, as discussed earlier, which is populated by transitions at a curve crossing. This threshold is not observed. However, thresholds for the higher-lying blocks of states are evident. Further, on ρ_1 and ρ_e, the elastic cross section, the interference pattern due to the first transition is evident prior to the onset of the higher transitions.

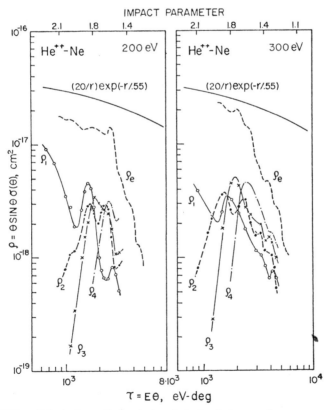

Fig. 27. Differential elastic scattering (ρ_e) and single-electron capture cross sections for $He^{2+} + Ne$ at 200 and 300 eV. Impact parameter is in atomic units.[77]

Using $R_x \approx 4.77 a_0$ as discussed earlier, and $(\partial/\partial R)(V_f - V_0)|_{R_x}$ determined from a Coulomb and polarization attractive potential, the coupling strength at R_x is estimated to be 0.14 eV. Even though the long-range force for the exit channels associated with ρ_2, ρ_3, and ρ_4 is stronger than that for elastic scattering, the threshold for these processes as observed on the elastic channel is displaced to larger scattering angle than on the inelastic cross sections. This indicates that the potential associated with the elastic channel is more deeply repulsive near the distance of closest approach, which is roughly the transition region for these energy transfers. It is quite clear also that the threshold for ρ_2 changes with energy, possibly implying a rotational coupling effect, while the half-heights of the ρ_3 and ρ_4 curves are roughly at 1450 and 2000 eV-deg for these energies.

It is clear from these figures that the energy loss cross sections for these single charge transfer processes are complex since only in very isolated circumstances can one reduce the problem to the simple two-state form. In all cases the analysis presented here is only useful in describing general features and as a first attempt to obtain values for potential parameters. Clearly if a very accurate understanding of the experimental results is needed, it can only be achieved through detailed calculations.

5. Summary

In the present chapter we have considered the principal experimental and theoretical methods for describing the charge-transfer reaction.

To summarize the analytical concepts—charge transfer, like other inelastic ion–atom collisions, can be described as transitions between diabatic molecular states, taking place at well-defined internuclear separations in most instances. The transitions can be grouped roughly into weak coupling, where the transition region is at the distance of closest approach, and strong coupling, where it is not. Simple methods exist for predictive calculations or for making preliminary analyses of experimental data. Two useful methods are the LZS and Rosen–Zener methods when the coupling is strong. These involve obtaining a transition probability based on some interaction potential. For low-velocity charge-transfer collisions, this potential generally has an exponential behavior related to the overlap of the electron clouds on the two centers. The transition probabilities are also determined by the differences in the behavior of the diabatic potential curves associated with the incoming and outgoing states of the colliding particles, as well as the energy defect Q. The principal selection rule is that transitions take place between diabatic states which involve no change in the component of electronic angular momentum along the internuclear axis. By changing the angle of observation of the scattered particles at fixed energy, or the incident-particle kinetic energy with fixed scattering angle, one changes the distance of closest approach and therefore scans the interaction potentials of the colliding particles. This procedure allows one to determine the transition regions for charge transfer. Since charge transfer results from a sharing of the electrons by the two centers, the region where the molecular states are formed, 3–5 a.u., is a region of considerable interest. Therefore, for single-electron transfer with incident doubly charged ions, the outgoing repulsive states, with the exothermicity appropriate to result in a crossing in this region, will have large transition probabilities when the distance of closest approach is comparable with or smaller than these values.

It has been pointed out in this chapter that in investigating charge transfer processes experimentally one must produce an ion beam, allow it to collide with some target atoms (usually in a gas cell), and detect the result of the events of interest. It is in this last category of detection that the experimenter's greatest ingenuity is required to achieve the maximum amount of information.

References

1. See, for example, A. Dalgarno, *Rev. Mod. Phys.* **39**, 858 (1967).
2. R. C. Amme and H. C. Hayden, *J. Chem. Phys.* **42**, 2011 (1965); J. Amdur and E. A. Mason, *ibid.* **25**, 632 (1956); J. W. Boring and R. R. Humphris, *AIAA J.* **8**, 1658 (1970).
3. R. C. Amme and N. G. Utterback, *Proceedings of the Third International Conference on the Physics of Electronic and Atomic Collisions*, North-Holland Publ., Amsterdam, 1963, p. 847.
4. M. L. Goldberger and K. M. Watson, *Collision Theory*, Wiley, New York, 1964.
5. B. H. Bransden, *Atomic Collision Theory*, Benjamin, New York, 1970.
6. N. F. Mott and H. S. W. Massey, *The Theory of Atomic Collisions*, 3rd ed., Oxford Univ. Press, London and New York, 1965; D. R. Bates (ed.), *Atomic and Molecular Processes*, Academic Press, New York, 1962; M. R. C. McDowell and J. P. Coleman, *Introduction to the Theory of Ion–Atom Collisions*, North-Holland Publ., Amsterdam, 1970.
7. E. W. McDaniel, V. Čermák, A. Dalgarno, E. E. Ferguson, and L. Friedman, *Ion–Molecule Reactions*, Wiley, New York, 1970.
8. R. A. Mapleton, *Theory of Charge Exchange*, Wiley, New York, 1972.
9. F. T. Smith, *Phys. Rev.* **179**, 111 (1969); T. F. O'Malley, *ibid.* **162**, 98 (1967).
10. W. Lichten, *Phys. Rev.* **131**, 229 (1963); **139**, 27 (1965); **164**, 131 (1967).
11. W. Kauzman, *Quantum Chemistry*, Academic Press, New York, 1960.
12. K. W. Ford and J. A. Wheeler, *Ann. Phys.* **7**, 259 (1959).
13. T. A. Green and R. E. Johnson, *Phys. Rev.* **152**, 9 (1966).
14. W. R. Thorson and S. A. Boorstein, in *Fourth International Conference on the Physics of Electron and Atomic Collisions, Abstracts*, Science Bookcrafters, Hastings-on-Hudson, New York, 1965, p. 218.
15. F. J. Smith, *Proc. Phys. Soc.* **84**, 889 (1964).
16. R. P. Marchi and F. T. Smith, *Phys. Rev.* **139**, A 1025 (1965).
17. F. T. Smith, R. P. Marchi, and K. G. Dedrick, *Phys. Rev.* **150**, 79 (1966).
18. J. B. Delos, W. R. Thorson, and S. K. Knudson, *Phys. Rev. A* **6**, 709 (1972); J. B. Delos and W. R. Thorson, *ibid.* **6**, 728 (1972).

19. R. E. Olson and F. T. Smith, *Phys. Rev.* **3**, 1607 (1971).
20. E. C. G. Stueckleberg, *Helv. Phys. Acta* **5**, 369 (1932); L. Landau, *Phys. Z. U.S.S.R.* **1**, 88 (1932); C. Zener, *Proc. R. Soc. (London) A* **137**, 696 (1932).
21. D. R. Bates, *Proc. Phys. Soc.* **73**, 227 (1959).
22. D. R. Bates, *Proc. R. Soc. (London) A* **257**, 22 (1960).
23. H. Feshback, *Ann. Phys.* **5**, 357 (1958); **19**, 287 (1962), for treating discrete states in a continuum.
24. R. E. Johnson, C. E. Carlston, and J. W. Boring, *Chem. Phys. Lett.* **16**, 119 (1972); see also V. Sidis, *J. Phys. B* **5**, 1517 (1972).
25. N. Rosen and C. Zener, *Phys. Rev.* **40**, 502 (1932); D. R. Bates, *Discuss. Faraday Soc.* **33**, 7 (1962).
26. R. Shakeshaft and J. Macek, *Phys. Rev. Lett.* **27**, 1487 (1971); *Phys. Rev. A* **6**, 1876 (1972); **7**, 1876 (1972).
27. R. E. Johnson, *J. Phys. B* **3**, 539 (1970); *J. Phys. Jap.* **32**, 1612 (1972).
28. J. I. Gersten, *Phys. Rev. Lett.* **31**, 73 (1973); W. E. Meyerhoff *et al.*, *ibid.* **30**, 1279 (1973).
29. I. K. Komarov and R. K. Yanev, *Sov. Phys. JETP* **24**, 1159 (1967).
30. E. E. McDaniel, *Collision Phenomena in Ionized Gases*, Wiley, New York, 1964; J. B. Hasted, *Physics in Atomic Collisions*, Butterworth, London, 1964.
31. J. B. Hasted, *Adv. At. Mol. Phys.* **4**, 237 (1968).
32. C. F. Barnett and H. B. Gilbody, *Methods Exp. Phys.* (L. Marton, ed.) **1**, 390 (1968).
33. P. H. Carr, *Vacuum* **14**, 37 (1964).
34. M. Rusch and O. Bunge, *Z. Tech. Phys.* **13**, 77 (1932).
35. W. Gaede, *Ann. Phys.* **46**, 357 (1915).
36. H. Ishii and K. I. Nakayama, *Trans. 8th Natl. Vacuum Symp., Washington, D.C. 1961*, Pergamon, Oxford, 1962, p. 519.
37. N. G. Utterback and T. G. Griffith, Jr., *Rev. Sci. Instr.* **37**, 866 (1966).
38. V. W. Hughes and H. L. Schults (eds.), *Methods Exp. Phys.* **4**, Part A (1967); B. Bederson and W. L. Fite (eds.), *ibid.* **7**, Part A (1968).
39. T. Jorgensen, Jr., C. E. Kuyatt, W. W. Lang, D. C. Lorents, and C. A. Sautter, *Phys. Rev. A* **140**, 1481 (1965).
40. P. M. Stier, C. F. Barnett, and G. E. Evans, *Phys. Rev.* **96**, 973 (1954).
41. E. S. Chambers, *Rev. Sci. Instr.* **35**, 95 (1964).
42. E. C. Crittenden and E. D. Spiel, *J. Appl. Phys.* **42**, 3182 (1971).
43. J. A. Hoyle, unpublished Ph.D. dissertation, University of Virginia (1973); J. A. Hoyle, R. R. Humphris, and J. W. Boring, *IEEE Trans. Magnetics* Mag.-II, **2** (March 1975).
44. A. I. Akishin, *Sov. Phys. Usp.* **66**, 113 (1958).
45. N. R. Daly, *Rev. Sci. Instr.* **34**, 1116 (1963).
46. G. W. McClure and D. L. Allensworth, *Rev. Sci. Instr.* **37**, 1511 (1966).
47. C. F. Barnett and J. A. Ray, Report ORNL-3836, 70, Oak Ridge National Lab., 1965.
48. C. F. Barnett and P. M. Stier, *Phys. Rev.* **109**, 385 (1958).
49. R. C. Amme and H. C. Hayden, *J. Chem. Phys.* **42**, 2011 (1965).
50. S. M. Trijillo, R. H. Neynaber, and E. W. Rothe, *Rev. Sci. Instr.* **37**, 1655 (1966).
51. Ya. M. Fogel, *Sov. Phys. Usp.* **3**, 390 (1960).
52. S. K. Allison, *Rev. Mod. Phys.* **30**, 1137 (1958).
53. S. K. Allison and M. Garcia-Munoz, Electron capture and loss at high energies, in *Atomic and Molecular Processes* (D. R. Bates, ed.), Academic Press, New York, 1962.
54. H. B. Gilbody, J. B. Hasted, J. V. Ireland, A. R. Lee, E. W. Thomas, and A. S. Whiteman, *Proc. R. Soc. A* **274**, 40 (1963).
55. R. Curran, T. M. Donahue, and W. H. Kasner, *Phys. Rev.* **114**, 490 (1959).
56. H. von Koch and E. Lindholm, *Ark. Fys.* **19**, 123 (1961).
57. D. V. Philipenko and Ya. M. Fogel, *Sov. Phys. JETP* **21**, 266 (1965).

58. F. P. Ziemba, G. J. Lockwood, G. H. Morgan, and E. Everhart, *Phys. Rev.* **118**, 1552 (1960).
59. H. F. Helbig and E. Everhart, *Phys. Rev.* **136**, A 674 (1964).
60. J. Bardon, M. Barat, and M. Abignoli, *J. Phys. B.* **1**, 1083 (1968).
61. M. Abignoli, M. Barat, J. Baudon, J. Fayeton, and J. C. Houver, *J. Phys. B* **5**, 1533 (1972).
62. S. W. Nagy, S. M. Fernandez, and E. Pollock, *Phys. Rev. A* **3**, 280 (1971).
63. D. H. Crandall and D. H. Jaecks, *Phys. Rev. A* **4**, 2271 (1971).
64. R. L. Champion and L. D. Doverspike, *J. Phys. B* **2**, 1353 (1969).
65. J. B. Hasted, S. M. Iqbal, and M. M. Yousaf, *J. Phys. B* **4**, 343 (1971).
66. K. E. Maher and J. J. Leventhal, *Phys. Rev. Lett.* **27**, 1253 (1972).
67. M. W. Siegel, Y. H. Chen, and J. W. Boring, *Phys. Rev. Lett.* **28**, 465 (1972); see also R. E. Olson, F. T. Smith, and E. Bauer, *Appl. Opt.* **10**, 1848 (1971), which contains extensive references to earlier work.
68. O. Firsov, *Zh. Eksp. Teor. Fiz.* **21**, 1001 (1951).
69. D. R. Bates and B. L. Moiseiwitsch, *Proc. Phys. Soc. A* **67**, 805 (1954); T. J. M. Boyd and B. L. Moiseiwitsch, *ibid.* **70**, 809 (1957).
70. D. R. Bates, *Discuss. Faraday Soc.* **33**, 7 (1962).
71. D. Rapp and W. E. Francis, *J. Chem. Phys.* **37**, 2631 (1962).
72. Yu N. Demkov, *Sov. Phys. JETP* **18**, 138 (1964).
73. R. E. Olson and F. T. Smith, *Phys. Rev. A* **8**, 1544 (1973).
74. E. E. Nikitin, *Opt. Spectrosc.* **19**, 91 (1965).
75. F. J. Smith, *Phys. Lett.* **20**, 271 (1966); R. E. Olson, *Phys. Rev. A* **2**, 121 (1970).
76. J. Perel, R. H. Vernon, and H. L. Daley, *Phys. Rev.* **138**, A937 (1965).
77. Y. H. Chen, Ph.D. Thesis, University of Virginia, 1972.
78. R. G. Cooks, J. H. Beynon, R. M. Caprioli, and G. R. Lester, *Metastable Ions*, Elsevier, Amsterdam, 1973; T. Ast, D. T. Terwilliger, J. H. Beynon, and R. G. Cooks, *J. Chem. Phys.* **62**, 3855 (1975).
79. J. W. McGowan and L. Kerwin, *Can. J. Phys.* **41**, 1535 (1963); A. S. Newton, A. F. Sciamanna, and R. Clampitt, *J. Chem. Phys.* **47**, 4843 (1967).
80. G. H. Lane and E. Everhart, *Phys. Rev.* **120**, 2064 (1960).
81. F. T. Smith, H. H. Fleischman, and R. A. Young, *Phys. Rev. A* **2**, 379 (1970).
82. F. P. Ziemba and E. Everhart, *Phys. Rev. Lett.* **2**, 299 (1959); G. J. Lockwood and E. Everhart, *Phys. Rev.* **125**, 567 (1962).
83. L. Newman, Thesis, University of Virginia, 1972.
84. S. K. Lam, L. D. Doverspike, and R. L. Champion, *Phys. Rev. A* **7**, 1595 (1973).
85. F. T. Smith, D. C. Lorents, W. Aberth, and R. P. Marchi, *Phys. Rev. Lett.* **15**, 742 (1965).
86. Y. H. Chen, R. E. Johnson, R. V. Humphris, H. W. Siegel, and J. W. Boring, *J. Phys. B* **8**, 1527 (1975).
87. D. R. Bates, H. C. Johnson, and I. Stewart, *Proc. Phys. Soc.* **84**, 517 (1964).

3

Inelastic Energy Loss: Newer Experimental Techniques and Molecular Orbital Theory

Q. C. Kessel, E. Pollack, and W. W. Smith

1. Introduction

The measurement of the inelastic energy losses which occur when heavy ions and atoms strike other atoms or molecules has provided information of unique importance to chemists and physicists. The inelastic energy loss in a collision is the portion of translational energy which is transferred to electronic, vibrational, and rotational excitation energy by the collision. After, or sometimes even during such a collision, the electron shells surrounding the collision participants may rearrange themselves so as to lose most of this excess energy. In atomic systems the ionization of electrons and the emission of photons are the primary mechanisms through which this

---

Q. C. Kessel • Department of Physics and The Institute of Materials Science, University of Connecticut. *E. Pollack and W. W. Smith* • Department of Physics, University of Connecticut, Storrs, Connecticut 06268.

energy is removed from the system. In collisions with molecular partners dissociation energies are important. In ordinary spectroscopy the energies of the ejected photons and electrons are measured, but such measurements are primarily concerned with *deexcitation* of the atom or ion. In contrast to this, inelastic energy-loss spectroscopy is primarily concerned with the *excitation* process. The inelastic energy loss Q represents the excitation energy of the collision system, and it follows that the sum of all deexcitation energies, together with any metastable excitation energy or dissociation energy resulting from that collision, must equal Q. For a very wide range of excitation processes, inelastic energy-loss spectroscopy has shown that *the excitations are molecular in nature* and have their origin in the transitory molecule formed during the collision. This chapter concentrates exclusively on collisions for which this is the case. Broadly speaking, these are the collisions in which the relative nuclear velocities are lower than the characteristic velocities of the electrons in question. When this is true, the Born–Oppenheimer approximation is valid and the nuclear and electronic motions may be considered separately.

The ideal experiment would measure the excitation energy of the collision as well as observe all of the participating deexcitation channels. Unfortunately such an all-inclusive experiment is not possible for complex collisions involving many electrons. Instead, one must settle for the separate measurement of the different collision parameters and by-products and attempt to develop the total picture through a correlation of these separate results. To do this, it is essential to understand the strengths and weaknesses of the various experimental techniques.

The importance of the Q value measurements is that they are inherently *differential* in nature. Inelastic energy losses are determined from a knowledge of the kinetic energies before and after collision. Usually it is also necessary to measure at least one of the scattering angles involved (see Section 3). Using this information, the distance of closest approach R_0 of the collision may be calculated and related to the minimum internuclear separation in the transitory molecule formed during the collision. Spectroscopic measurements, on the other hand, are generally *total* in nature. Although deexcitation energies may be measured very accurately through electron and photon spectroscopy, the observation of the total electron or photon emission yields limited information about the actual collision process itself.

Spectroscopic measurements provide numbers which are inherently averages over collisions having a wide range of impact parameters. It is often difficult to extract information about R_0, the resulting ionization states, or even specific excitation probabilities from total cross section measurements. A coincidence measurement can, in principle, resolve some of these difficulties: For example, the spectroscopic observation of a photon or electron in coincidence with the scattered incident particle can provide the desired impact parameter dependence of the excitation in question (Section 3.2).

Before going into details of the experimental work it is worthwhile to review the molecular orbital model and to demonstrate how it can be used to explain certain collision phenomena. This is done in Section 2. Section 3 describes the variety of experimental techniques available for the measurement of inelastic energy losses, together with a sampling of the corresponding data and an assessment of the relative strength and weakness of each technique. The techniques include energy analysis of the scattered fast particle, the method emphasized throughout this book, velocity analysis of fast particles by time-of-flight techniques, and coincidence measurements, which combine energy analysis with photon and electron emission data.

2. The Molecular Orbital Model

The history of the application of the molecular orbital model to atomic collisions is an interesting one. In 1927 Leonard Loeb addressed himself to the controversy surrounding the question of whether it was possible for positive ions to ionize atoms in binary collisions.[1] At the time he concluded that there must exist a mechanism by means of which the electrons surrounding the two nuclei could interact and thereby transform a portion of the collision's kinetic energy into electronic excitation energy. He also lamented that, "To date the new quantum mechanics has been unable to cope with the problem." Experiments resulting in ionization and thus showing the inelastic nature of such collisions were carried out during the next few years,[2,3] and almost immediately, Weizel and Beeck[4] explained the phenomenon as being due to electronic transitions between molecular orbitals of the transitory molecule formed during the collision. This explanation, which by now has been verified and refined, poses two questions: First, what are the most appropriate molecular wave functions for the description of a given collision, and second, what are the mechanisms which can cause transitions between energy levels of the quasi molecule formed during a collision? It is appropriate to discuss these questions separately. A related discussion on molecular orbital theory as applied to charge exchanging collisions is presented in Chapter 2, Section 2.

2.1. Molecular Orbitals for Collision Processes

For the collisions under consideration here, the Born–Oppenheimer approximation[5] is valid. This means that we are considering only those collisions for which the relative collision velocity of the nuclei is lower than the orbital velocities of the electrons in question. Under this assumption, the relative nuclear motion and the electronic motions may be considered separately. Furthermore, classical methods may be used for analyzing the trajectories of the relatively massive nuclei. This section will deal only with the electronic behavior, which must be treated quantally. The following section will deal with the coupling of the nuclear and electronic motions, which can give rise to inelastic scattering events.

In an adiabatic ion–atom or atom–atom collision, for which electronic motions are confined to particular potential surfaces, the collision partners may be viewed as the constituents of a diatomic molecule; and at any fixed value of the internuclear separation R, the allowed electronic states must be those of the molecule. However, during a collision, the value of R changes, as do the allowed electronic states. In fact, adiabaticity requires that the molecular states must change in a well-ordered manner as R proceeds from infinity (before the collision) to R_0 (the distance of closest approach, reached during the collision) and then back to infinity after the collision. For the case of slow-moving nuclei, the allowed electronic energies may be calculated, to a first approximation, by assuming the nuclei to be fixed and then repeating this calculation for closely spaced values of R. In this way, curves may be generated which describe the allowed electronic energies for all values of R. When the Coulomb repulsion between the nuclei is added to the electronic energy so calculated, one obtains the potential energy function for the nuclear motion in the Born–Oppenheimer approximation. Thus, a potential curve is calculated for each molecular electronic state: the set of so-called adiabatic molecular energy levels. These adiabatic molecular states are well known to molecular spectroscopists. Herzberg,[6] for example, describes the rules for relating the atomic states of the separated atoms to the molecular states of the molecule and to the atomic states of the united atom for R varying from infinity to zero. Correlation diagrams, which show these relationships, are very useful, and one is given in Fig. 1 for the one-electron hydrogen molecular ion (H_2^+).[7] This case was first treated in detail by Morse and Stueckelberg.[8] The molecular orbital notation used here is that introduced by Hund[8,9]; the lower-case Greek letters ($\sigma, \pi, \delta, \phi, \gamma, \ldots$) indicate Λ, the component of the electronic orbital angular momentum along the internuclear axis. The $1s, 2s, 2p, \ldots$, designations correspond to the united atom limit of the particular molecular orbital (MO). For symmetric molecules, the parity (reflection symmetry about the center of the molecule) of the MO is that of the united atom orbital, whereas the parity assignments are not unique in the separated atom limit.

The adiabatic correlation diagrams for molecules containing more than one electron may also be drawn, in which case the von Neumann–Wigner noncrossing rule[10] must be taken into account. This results in the "noncrossing," or avoidance of crossing, of adiabatic potential curves of the same symmetry. It turns out, however, that such multielectron adiabatic diagrams

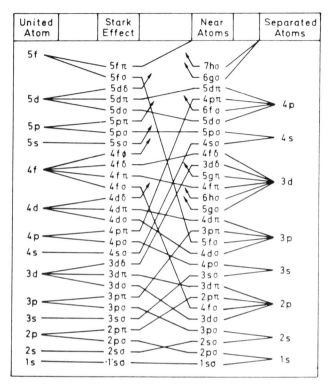

Fig. 1. Correlation diagram for the adiabatic molecular orbitals of the hydrogen molecular ion (H_2^+) tracing the electronic reorganization which accompanies the collisions of H^+ and H^-. The same correlation diagram can also be used to describe the diabatic one-electron molecular orbitals of any symmetrical molecule.[7] The correlation diagram is constructed by conserving Λ, the component of electronic angular momentum on the internuclear axis.

are usually inadequate for the explanation of collision phenomena which result in inelastic energy losses. In 1963 Lichten[11] noted that certain aspects of He^+–He collisions could not be explained using adiabatic energy-level diagrams. Instead, at certain energies, the collisions behaved as if the correlations between the united and separated atom limits were those given in Fig. 1; *i.e., the electrons were behaving as if they were single electrons moving in a two-center potential.* Lichten suggested using what are popularly termed the *diabatic* correlations and diabatic energy level diagrams for the explanation of collision processes. This approach simply neglects, to a first approximation, the electron–electron interactions, and writes the many-electron wave function for the collision system as an antisymmetrized product of one-electron wave functions. In this one-electron approximation, each electron behaves as if it were moving in an average potential due to the two nuclei and the other electrons. It is assumed that the appropriate MO correlations are identical to those of the H_2^+ molecular ion in Fig. 1. This concept has proved to be very useful in both symmetric and asymmetric collisions and Figs. 2 and 3 show one-electron correlation diagrams for two asymmetric cases.[12,13]

Computation of the correct orbitals is a much more difficult problem; however, these simple MOs are required at least as an initial step for any quantitative discussion. It should be strongly emphasized that only the many-electron molecular states, and not the single-electron MOs, correctly

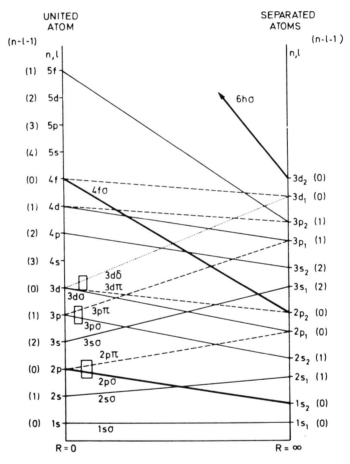

Fig. 2. MO correlation diagram constructed using the Barat–Lichten rule, which assumes that as R varies, the number of radial nodes $n-l-1$ remains constant for a given MO. This diagram is for the case of Z_1 slightly larger than Z_2. The solid lines correspond to σ states, the dashed lines to π states, and the dotted lines to δ states. The prominent MOs, $2p\sigma$, $4f\sigma$, and $6h\sigma$, are responsible for promotion of $1s$, $2p$, and $3d$ electrons, respectively. The rectangles show interactions for which rotational coupling between two MOs may be expected.[13]

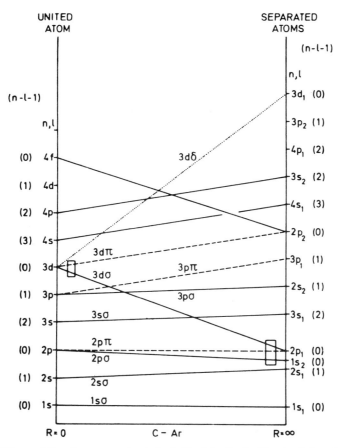

Fig. 3. MO correlation diagram constructed as in Fig. 2 for the Ar–C molecule for which there exists a near-resonance between the 1s level of carbon and the 2s and 2p levels of argon.[13]

describe the collision system. However, discussions in terms of single-electron MOs are useful in many cases. As two atoms approach each other in a collision, a particular electronic state of the molecule is formed and a transition to another molecular state is possible if the MOs describing these two states approach close enough to each other in energy so that coupling can occur. Knowledge of the exact behavior of the MOs and their dependence on R can be obtained only through computation. Morse and Stueckelberg[8] first accomplished this for the H_2^+ system (where the distinction between adiabatic and diabatic levels does not exist). For the molecular levels near $R = 0$, they perturbed He levels by allowing the nuclear charge to be split in two and separated by a finite distance. In a similar way they used perturbed hydrogenic levels near $R = \infty$. The energy levels between these extremes in R were extrapolated using the known correlations between states. With the advent of computers much more accurate determination of these levels has been possible.[14] The advances in computational technique have been such that there are now a large number of general methods and computer programs available. However, most of the calculations have been of the *adiabatic* state functions for multielectron systems,[14–21] and there is still no unique or generally accepted definition of states suitable for the inelastic collision problems for which diabatic orbitals are required.

The actual approaches used to calculate MOs are as varied as the methods and approximations used in calculating atomic energy levels, and often different basis sets are used for different ranges of R. Thus, Larkins[15] performed one of the first *ab initio* MO calculations for the heavy molecules, Ar–Ar and Ne–Ar, over the full range of R from zero to infinity, using atomic wave functions calculated from the Hartree–Fock equations. Mulliken[16] has done a self-consistent field computation for the N–N molecule using Slater-type wave functions. Given sufficient computer time, these methods can produce very accurate values for the MO energies. Briggs and Hayns[17] have used approximate methods to obtain Hartree–Fock MOs for the N–N molecule and have obtained good agreement with Mulliken's calculations without so great a use of computer time. Eichler and Wille[18] used a superposition of modified atomic Thomas–Fermi potentials to calculate the MOs of the Ne–Ne system. One advantage of this last method of calculation is that it may be extended to higher-Z molecules whereas a complete two-center Hartree–Fock calculation would require an inordinate amount of computer time. The results of the Eichler and Wille calculation are reproduced in Fig. 4 to serve as an example in discussing the differences between adiabatic MO energy levels and diabatic MO energy levels. Here the customary adiabatic MO designations are more useful because states of the same total symmetry never cross. Hence, $1\sigma_g$ corresponds to the lowest even symmetry (gerade) and $1\sigma_u$ to the lowest odd symmetry (ungerade) state. The $2\sigma_g$ is then the next lowest even symmetry state, and so forth.

The curves in Fig. 4 are adiabatic curves, calculated under the assumption that the internuclear separation R does not change with time. This is clearly a poor assumption to make when attempting to describe a situation so dynamic as a collision. As noted earlier, it was for this reason that Lichten[7,11] popularized what are termed diabatic molecular states based on one-electron MOs, which do not take into account the electron–electron interactions. The most apparent difference between adiabatic states and diabatic states is that the von Neumann–Wigner noncrossing rule is not applicable to the one-electron MOs or to antisymmetrized products of such states whose total energy is simply the sum of one-electron MO energies. The effect of the noncrossing rule, which forbids molecular state energy levels of the same parity from crossing in the adiabatic approximation, is

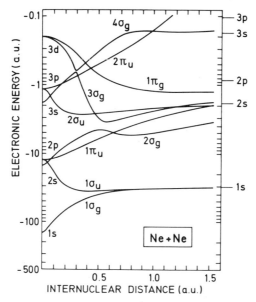

Fig. 4. Electronic energy correlation diagram calculated by Eichler and Wille[18] for the Ne–Ne system. The calculation has been restricted to $\sigma(\Lambda=0)$ and $\pi(\Lambda=1)$ states; the levels are labeled according to their "gerade" (g) or "ungerade" (u) character. At $R=0$, the electronic energies have been determined directly by solving the radial Schrödinger equation with the united-atom TF potential. a.u. = 0.529 Å.

apparent in Fig. 4. The $3\sigma_g$ MO and the $4\sigma_g$ MO both suffer sudden changes in their character when they approach each other. In the diabatic approximation these curves would pass smoothly through each other at this point. This example points up a conceptual advantage in employing diabatic curves when dealing with phenomena in which the internuclear distance changes rapidly with time. A rapidly moving atom will have a much larger probability of passing from the $3\sigma_g$ to the $4\sigma_g$ MO.

There is not, at present, an accepted unique formal definition for the diabatic states, as there is for the adiabatic states. As a practical matter, many interpretations of inelastic collision phenomena have used diabatic orbitals derived from adiabatic calculations. Landau and Zener might be considered the first to have used diabatic orbitals when they discussed the nonadiabatic crossing of energy levels in 1932.[22,23] In their articles, they recognize the limitation of the static adiabatic states and introduce the coupling between the nuclear and electronic motions as a perturbation in the adiabatic approximation. They considered only coupling due to the radial component R of the relative position vector and found the radial coupling to be most effective at high velocities. If two adiabatic potential curves experience a very close approach at some internuclear separation, the radial coupling can be sufficient to *ensure* a transition between states of like symmetry, thus showing the diabatic behavior of the collision. During a collision a given crossing will tend to behave either adiabatically or diabatically, depending on the nature of the crossing and the collision velocity. For example, it might be expected that the avoided $3\sigma_g$–$4\sigma_g$ crossing in Fig. 4 would behave diabatically even for relatively low collision velocities and that the $2p$ level of the separated atom would thus correlate with the $4f$ level of the united atom during a collision. On the other hand, the $2\sigma_g$ and $3\sigma_g$ levels also approach each other, but maintain a much greater separation, and therefore this "crossing" would tend to maintain its adiabatic character even for fairly high collision velocities. It is clear that to make quantitative judgments concerning a particular collision, the MOs must be known accurately for that collision combination. Predictions concerning specific excitations can be made from Fig. 4 that could only be guessed at on the basis of straight-line correlation diagrams such as those in Figs. 1–3.

Radial coupling, an example of which we have just discussed, is mainly due to residual electrostatic interactions between two electrons (i.e., the breakdown of the independent electron model). This kind of coupling is effective between orbitals having the same parity and same projection of the electronic angular momentum along the internuclear axis (e.g., two Σ states or two Π states). Another coupling mechanism is possible due to rotation of the internuclear axis: the molecular axis of quantization for the electronic states changes with time, leading to *rotational* coupling that mixes states with orbital angular momentum projection along the internuclear axis Λ differing by one unit of $\hbar$. Thus a Σ state can be coupled to a Π state by rotational coupling as discussed further in Section 2.2.

2.2. Transitions between Molecular Orbitals

During a collision an electron on one of the collision partners may be excited by its interaction with the approaching nucleus or by effects arising from changes in the entire electronic structure as the collision progresses. It was at first believed that the former process would prove to be the dominant one. The theory of this process was developed by Massey and Burhop[24] and its initial applications predicted that the maximum value of the total cross section for excitation would occur at energies well into the keV range (cf. also Chapter 1). The cross section was expected to rise to its peak value and then to decrease with further increase in energy. The energy E_m at which the maximum value of the cross section occurs is given conveniently by[25]

$$E_m = 36(\delta E)^2 m l^2 \qquad (2.1)$$

where m is the incident particle mass in atomic mass units (amu), l is the "range of the interaction" in units of the Bohr radius ($a_0 = 0.53 \times 10^{-10}$ m), and δE is the energy defect in electron volts of the transition in question (E_m is also in electron volts). In using Eq. (2.1) it was at times incorrectly assumed (in spite of Massey's warnings) that the energy defect after the collision, i.e., Q, was to be substituted for δE. As Q may be orders of magnitude greater than the appropriate δE, Eq. (2.1) then predicted that many higher-energy collisions would be elastic, when they were not. The criterion is still useful when properly applied as, for example, in Martin and Jaecks' investigation of the total cross section for charge transfer to H($2p$) in $H^+ + Ne$ collisions.[26] It is now known that inelastic scattering processes are also quite important even at energies well below those predicted by the Massey criterion (except in those special cases in which both partners have He-type closed electron shells). The low- and medium-energy scattering processes are best understood in terms of Landau–Zener-type level crossings. Until recently, most of the published theoretical work using the Landau–Zener approximation was concerned only with effects arising from the radial component of the relative motion, although it was recognized[27-29] that since the internuclear line rotates during a collision, rotational coupling effects may be quite important.

Following Russek[30] we now outline the basic theory of the rotational coupling process in the one-electron MO approximation. Assume a collision between two atoms having a vector internuclear separation $\mathbf{R}$, electron coordinates $\mathbf{r}$ as measured in the laboratory frame, and $\mathbf{r}'$, in a body-fixed frame attached to the center of mass and rotating with the internuclear axis. For the electronic Hamiltonian $\mathcal{H}$, the adiabatic states at a fixed internuclear separation $\mathbf{R}$ are defined by

$$\mathcal{H}\psi_i(\mathbf{r}', \mathbf{R}) = \varepsilon_i \psi_i(\mathbf{r}', \mathbf{R}) \qquad (2.2)$$

The coordinates $\mathbf{r}'$ are used because the component of the electronic angular momentum along the internuclear axis is quantized. Following the usual Landau–Zener[22,23] approximation we now assume that only two states are involved ($i = 1, 2$). In the two-state approximation ψ, the wave function during the collision, is written as

$$\psi = C_1(t)\psi_1(\mathbf{r}', \mathbf{R}) \exp\left[\left(\frac{-i}{\hbar}\right) \int^t \varepsilon_1 \, d\tau\right] + C_2(t)\psi_2(\mathbf{r}', \mathbf{R}) \exp\left[\left(\frac{-i}{\hbar}\right) \int^t \varepsilon_2 \, d\tau\right] \qquad (2.3)$$

with the collision trajectory described by $R(t)$. Since $\psi_1, \psi_2, \varepsilon_1$, and ε_2 are known from Eq. (2.2), only $C_1(t)$ and $C_2(t)$ are needed to determine Ψ. These coefficients are found by solving the time-dependent Schrödinger equation

$$\left(\mathcal{H} - i\hbar \frac{\partial}{\partial t}\right)\Psi = 0 \qquad (2.4)$$

Within the context of the adiabatic two-state model, however, the coefficients are determined by the weaker conditions

$$\left\langle \psi_1 \left| \mathcal{H} - i\hbar \frac{\partial}{\partial t} \right| \psi \right\rangle = 0$$
$$\left\langle \psi_2 \left| \mathcal{H} - i\hbar \frac{\partial}{\partial t} \right| \psi \right\rangle = 0 \tag{2.5}$$

where the integrations are carried out only over the electronic coordinates. Using Eqs. (2.2) and (2.5) it may be shown that

$$[(\dot{C}_1 + C_1 \langle \psi_1 | \dot{\psi}_1 \rangle)] \exp\left[\left(\frac{-i}{\hbar}\right) \int^t \varepsilon_1 \, d\tau \right] + C_2 \langle \psi_1 | \dot{\psi}_2 \rangle \exp\left[\left(\frac{-i}{\hbar}\right) \int^t \varepsilon_2 \, d\tau \right] = 0$$
$$[\dot{C}_2 + C_2 \langle \psi_2 | \dot{\psi}_2 \rangle] \exp\left[\left(\frac{-i}{\hbar}\right) \int^t \varepsilon_2 \, d\tau \right] + C_1 \langle \psi_2 | \dot{\psi}_1 \rangle \exp\left[\left(\frac{-i}{\hbar}\right) \int^t \varepsilon_2 \, d\tau \right] = 0 \tag{2.6}$$

Since the electronic coordinates are defined with respect to the rotating internuclear line

$$\dot{\psi}_i = \dot{R} \frac{\partial \psi_i}{\partial R} + \frac{i}{\hbar} \dot{\theta} L_\perp \psi_i \tag{2.7}$$

where $\dot{R}$ is the relative radial velocity, $\dot{\theta}$ is the angular velocity of the internuclear axis, and $L_\perp$ is the component of the electronic angular momentum operator, in the body-fixed frame, which is perpendicular to the plane of the collision trajectory. Since $\partial \psi_i / \partial R$ has the same symmetry as ψ_i, only states of identical symmetry are coupled by the first term in Eq. (2.7). This term gives rise to the usual Landau–Zener transition with allowed couplings between the many-electron molecular states, such as Σ to Σ and Π to Π, having the same component of the total electronic angular momentum along the internuclear axis. Since

$$L_\perp = \frac{L_+ - L_-}{2i} \tag{2.8}$$

where L_+ and L_- are raising and lowering operators for the angular momentum component along the internuclear line, the term containing $L_\perp$ in Eq. (2.7), which is responsible for rotational excitation, allows transitions between states such as Π to Σ, Π to Δ, etc. It should be noted that for symmetric collisions in which the electronic wave function remains unchanged (gerade, g) or changes sign (ungerade, u) when reflected about the center, each term in Eq. (2.7) allows transitions only between states g to g or u to u. It is well known that transitions from g to u states and from u to g states are forbidden for both the radial and rotational coupling in this type of problem.

Since the states are normalized,

$$\langle\psi_i|\dot\psi_i\rangle = \frac{1}{2}\frac{\partial}{\partial t}\langle\psi_i|\psi_i\rangle = 0 \tag{2.9}$$

and in the case of rotational coupling where only the second term of Eq. (2.7) contributes, Eqs. (2.6) reduce to

$$\dot C_1 + i\dot\theta\hbar^{-1}C_2\langle\psi_1|L_\perp|\psi_2\rangle \exp\left[-i\hbar^{-1}\int^t(\varepsilon_2-\varepsilon_1)\,d\tau\right] = 0$$

$$\dot C_2 + i\dot\theta\hbar^{-1}C_1\langle\psi_2|L_\perp|\psi_1\rangle \exp\left[-i\hbar^{-1}\int^t(\varepsilon_1-\varepsilon_2)\,d\tau\right] = 0 \tag{2.10}$$

It is reasonable to assume that near a crossing, $R = R_x$, of the electronic energy curves the energy difference varies linearly with internuclear separation. Therefore,

$$\frac{\varepsilon_2-\varepsilon_1}{\hbar} = B(R-R_x) = \alpha t \tag{2.11}$$

with

$$\alpha = \frac{dR}{dt}\frac{1}{\hbar}\frac{d}{dR}(\varepsilon_2-\varepsilon_1)|_{R=R_x} = \frac{v_R}{\hbar}\frac{d}{dR}(\varepsilon_2-\varepsilon_1)|_{R=R_x} \tag{2.12}$$

In addition, the matrix element

$$V = \langle\psi_1|L_\perp|\psi_2\rangle \tag{2.13}$$

remains essentially constant over the range of internuclear separation where there is a strong interaction between the levels. Equations (2.10) may now be rewritten as

$$\dot C_1 + i\omega V\hbar^{-1}C_2 \exp\left[-i\hbar^{-1}\int^t(\varepsilon_2-\varepsilon_1)\,d\tau\right] = 0$$

$$\dot C_2 + i\omega V^*\hbar^{-1}C_1 \exp\left[+i\hbar^{-1}\int^t(\varepsilon_2-\varepsilon_1)\,d\tau\right] = 0 \tag{2.14}$$

where ω is the angular velocity at $R = R_x$.

For a straight-line trajectory with impact parameter b and initial relative velocity v

$$\omega = \frac{vb}{R_x^2} \quad \text{and} \quad v_R = \frac{v(R_x^2-b^2)^{1/2}}{R_x} \tag{2.15}$$

If the system is in state ψ_2 prior to collision, then $|C_2(-\infty)|=1$ and $|C_1(-\infty)|=0$. The probability P of making a transition to ψ_1 as a result of the collision is given by $P=|C_1(+\infty)|^2$. This can be evaluated (at least numerically) from Eqs. (2.14) and represents the solution of the problem. Equations (2.14) have essentially the same mathematical form as those solved by Zener (here ωV replaces the constant coupling matrix element in Zener's work) and his results may be transferred intact. However, it should be noted that the two interacting adiabatic levels do actually cross in the rotational coupling case but do not cross for a radial coupling. As pointed out earlier, Landau and Zener introduced what were later termed "diabatic" states, which do cross, and they solved the problem for these states. In the Landau–Zener solution, C_1 and C_2 are expressed in terms of Weber functions of complex argument, which are not tabulated but whose asymptotic behavior is known. The solutions valid for a single pass through the crossing, and for the initial conditions $|C_1(-\infty)|=0$ and $|C_2(-\infty)|=1$, are

$$|C_1(+\infty)|^2 = 1 - e^{-2\pi\gamma} \tag{2.16}$$

$$|C_2(+\infty)|^2 = e^{-2\pi\gamma} \tag{2.17}$$

where

$$\gamma = \frac{\omega^2|V|^2}{\hbar(d/dt)(\varepsilon_2-\varepsilon_1)|_{t=0}} = \frac{\omega^2|V|^2}{\hbar v_R(d/dR)(\varepsilon_2-\varepsilon_1)|_{R=R_x}} \tag{2.18}$$

Using Eqs. (2.15)

$$\gamma = \frac{v|V|^2}{\hbar(d/dR)(\varepsilon_2-\varepsilon_1)|_{R=R_x}} \cdot \frac{b^2}{R_x^3(R_x^2-b^2)^{1/2}} \tag{2.19}$$

For the conventional Landau–Zener approximation

$$\gamma_{LZ} = \frac{\beta^2}{\frac{v}{\hbar}\left(1-\frac{b^2}{R_x^2}\right)^{1/2} \frac{d}{dR}(\varepsilon_2-\varepsilon_1)|_{R=R_x}} \tag{2.20}$$

where β is the constant coupling matrix element. Of particular importance is the fact that γ for a rotational excitation varies linearly with v whereas it varies inversely with v for the radial case. The probability of making a transition in a single pass through a crossing is given by

$$p = 1 - e^{-2\pi\gamma} \tag{2.21}$$

where γ is calculated from Eq. (2.19) or (2.20) in the respective cases. It should be emphasized that p is not observable in a collision since the crossing will occur twice (if it is reached at all). The Landau–Zener formulation takes account of this by assuming that the experimentally observable probability P for making a transition as a result of the collision is given by

$$P = 2p(1-p) \tag{2.22}$$

Bates has shown that there is an interval ΔR on either side of the crossing radius given by

$$\Delta R = \left(\frac{\pi \hbar v_R s}{B}\right)^{1/2} \tag{2.23}$$

in which the probabilities of the two states achieve the major portion of their changes from initial to final values [B has been defined in Eq. (2.11) and s is a number of the order of but greater than unity]. Russek has performed a direct numerical integration of the coupled equations for radial excitation, demonstrated the validity of Eq. (2.23), and concluded that if the turning point R_0 of the nuclear trajectory lies within an interval ΔR of R_x, then the simple model represented by Eq. (2.22) breaks down. If $R_x - R_0 \leq (\pi \hbar v_R s/B)^{1/2}$, Eq. (2.22) is invalid, but if $R_x - R_0 \gg (\pi \hbar v_R s/B)^{1/2}$, Eq. (2.22) is valid if phase interference oscillations are averaged over. The breakdown of Eq. (2.22) was shown to be particularly bad for a rotational excitation since the numerical results show that considerable excitation can occur for impact parameters greater than R_x. These large impact parameter contributions are heavily weighted in the total cross section and the Landau–Zener approach is inadequate except for cases in which $R_x \gg \Delta R$.

Russek solved Eq. (2.14) numerically, taking $R_x = 1.0$ a.u. and an angular momentum coupling $V = 0.6$ a.u. He showed that contrary to what would be expected from the idea that the two passages through R_x are independent, the main change in probability occurs when the nuclei are in the region between the two points when $R = R_x$. There was no indication of a sharp transition at each crossing. The transition takes place over the entire region between the two crossings. Figures 5 and 6 show the rotational excitation probability as a function of impact parameter for two velocities

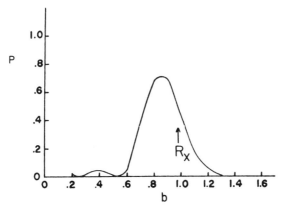

Fig. 5. Rotational excitation probability P versus impact parameter b. The curve shows P versus b for $v/B = 0.1$ and angular momentum coupling $V = 0.6$, in atomic units.[30]

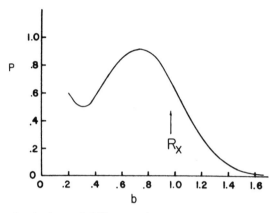

Fig. 6. Rotational excitation probability versus impact parameter. The curve shows P versus b for $v/B = 0.3$ and $V = 0.6$ in atomic units.[30]

$v/B = 0.1$ and $v/B = 0.3$. P is seen to be quite large at $b = R_x$ and does not drop off to negligible values until b is substantially larger than R_x. Smaller values of v/B result in a P which is sharply peaked at a b value just smaller than R_x whereas the large v/B results yield a broad-peaked P having its maximum value at smaller b values but extending further out beyond $b = R_x$. The total cross sections are obtained by multiplying the curves of Figs. 5 and 6 by $2\pi b$ and integrating over b. It is now clear that impact parameters greater than $b = R_x$ contribute significantly to the total cross section for rotational excitation as a function of incident velocity, assuming straight-line trajectories, for several values of V. Since the maximum value of P in Eq. (2.22) is 0.5 and the geometric cross section, for getting inside R_x, is πR_x^2, the total cross section of the two-independent-crossings picture is

$$\sigma \leq \tfrac{1}{2}\pi R_x^2 \tag{2.24}$$

A comparison of Eq. (2.24) and Fig. 7 shows the shortcomings of the simple two-independent-crossings model.

The two-state theory just discussed requires the crossing or pseudo-crossing of the electron potential curves for excitation. Transitions can also occur between two close-lying curves that do not cross.[31,32] Here again both rotational and radial couplings between the states are possible. These mechanisms are particularly useful for understanding charge exchange between initial and final states that lie close together at large internuclear separation such as in $He^+ + Ne \rightarrow He + Ne^+$ and in nonsymmetric alkali ion–alkali atom-charge exchange.[31,33]

The importance of the molecular states (rather than the one-electron MOs) in inelastic collisions was pointed out by Sidis and is discussed in two recent papers by Brenot and co-workers.[34] The first of these papers summarizes the known inelastic scattering mechanisms and the characteristics associated with the symmetric rare-gas collision systems. It is pointed out that a series of quasi-diabatic state crossings (resulting in inelastic scattering) is generated when a vacancy is present in an inner MO. In some cases (rare-gas–ion–rare-gas–atom systems) this vacancy is initially present and causes inelastic scattering without MO crossings. In the rare-gas-atom–atom case, no vacancies in inner MOs are initially present and an MO crossing is required before inelastic effects are seen. Of particular significance is that within this model a large number of molecular state crossings may be generated by the production of an inner MO vacancy.

3. Experimental Techniques and Results

Currently several techniques are available for studying inelastic scattering, including electron and photon spectroscopy, energy-loss measurements on scattered and recoil (target) species, and delayed coincidence measurements on the scattered and recoil particles. Each of these methods has characteristic advantages and disadvantages. In electron and photon spectroscopy experiments there is generally some uncertainty about the states originally excited in the collision since cascade effects (see Chapter 1, Section 3) may contribute to the measured signal. Furthermore, electron and photon spectroscopy cannot provide information about R_0, the distance of closest approach in the collision. This is a particular disadvantage since

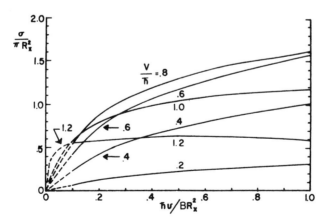

Fig. 7. Collision cross section for rotational excitation. The reduced total cross section $\sigma_{\rm rot}/\pi R_x^2$ for rotational excitation is plotted as functions of reduced velocity $\hbar v/BR_x^2$ for collisions with several values of angular momentum coupling V, given in atomic units.[30]

both ionization and excitation processes are expected to be strong functions of R_0. Direct measurements on the scattered ions or atoms, however, are not complicated by cascade effects and do provide information on R_0. On the other hand, electron and photon spectroscopy techniques are superior in energy resolution. In addition to broadening effects resulting from finite angular resolution and from the motion of the target atoms, the unavoidable energy spread in the incident beam limits the ultimate resolution in scattered ion (or atom) energy-loss measurements. This spread does not affect the ejected-electron or photon-flux measurements so greatly since the cross sections are generally slowly varying functions of the incident beam energy. It should also be noted that the resolution of electron and ion energy analyzers is usually a fixed percentage of the electron or ion energy, thereby allowing higher resolution in measurements on electrons since these energies are smaller than scattered ion energies. Deceleration of the ion beam prior to analysis can somewhat offset this disadvantage. Finally, measurements by electron or photon spectroscopy are limited to those collisions which result in the emission of an electron or photon. With the addition of time-of-flight techniques for analyzing the energy of neutral scattered atoms, scattered ion or atom spectroscopy can, in principle, be used to study any inelastic process.

The inelastic energy loss is derived from a knowledge of the kinetic energies of the collision participants before and after the collision. Figure 8 is a schematic diagram defining the important collision parameters. Here ion or atom A with charge i strikes a target atom B and is scattered through the angle θ while B leaves the collision region at the angle ϕ. The masses M,

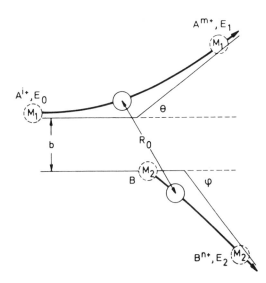

Fig. 8. Schematic of the collision $A^{i+} + B \rightarrow A^{m+} + B^{n+} + (m + n - 1)e$ in laboratory coordinates. The ion energies E, the scattering angles θ and ϕ, and the impact parameter b are indicated. The solid circles show the positions of the ions at the distance of closest approach R_0.

energies E, and final charge states m and n are indicated. The distance b is the impact parameter of the collision and R_0 represents the distance of closest approach for the two particles.

The inelastic energy loss Q is then given by

$$Q = E_0 - E_1 - E_2 \tag{3.1}$$

This inelastic energy represents the total energy lost from the translational motion of the nuclei and must equal the sum of the energies of the emitted photons and electrons plus the ionization energies of the electrons removed and any energy metastable species might carry away from the collision. To determine Q experimentally, any three of the five parameters E_0, E_1, E_2, θ, and ϕ must be measured independently; then through the conservation of energy and momentum, Q may be calculated. This divides the experiments into three categories.[13] First, the scattered-particle experiment in which E_0, E_1, and θ are measured. For this case

$$Q = 2\gamma(E_0 E_1)^{1/2} \cos\theta + (1-\gamma)E_0 - (1+\gamma)E_1 \tag{3.2}$$

where $\gamma = M_1/M_2$ is the mass ratio of the incident particle to the target particle. Second is the recoil-particle experiment in which E_0, E_2, and ϕ are measured. Here,

$$Q = \left(\frac{E_0 E_2}{\gamma}\right)^{1/2} \cos\phi - \left(1+\frac{1}{\gamma}\right)E_2 \tag{3.3}$$

Finally, there is the delayed coincidence experiment wherein the quantities E_0, θ, and β ($\beta \equiv \theta + \phi$) are measured and for which

$$Q = E_0\left(1 - \frac{\sin^2(\beta-\theta)}{\sin^2\beta} - \frac{\gamma \sin^2\theta}{\sin^2\beta}\right) \tag{3.4}$$

Of these three methods, the scattered-particle technique provides the best resolution. This is because the thermal motion of the target particles can cause considerable broadening of the spectra obtained by the latter two techniques.

The thermal velocities, superimposed upon the velocity given to the target atom by the collision itself, contribute in a major way to the observed line widths. With all three of these methods for measuring Q, the observed line widths δQ_{obs} have three important contributions: the natural line width δQ_{nat}, the instrumental line width δQ_{inst}, and the thermal width δQ_T. As these contributions are independent of one another

$$(\delta Q_{obs})^2 = (\delta Q_{nat})^2 + (\delta Q_{inst})^2 + (\delta Q_T)^2 \tag{3.5}$$

Fastrup and co-workers[35] have derived expressions for the thermal width

δQ_T (at $1/e$ height) for each of these types of experiments. Their results, with δQ_T in electron volts, are as follows:

Scattered particle: $$\delta Q_T = 2(kTE_2)^{1/2} = 10(E_2)^{1/2} \text{ eV} \qquad (3.6)$$

Recoil particle: $$\delta Q_T = 2\left(\frac{kTE_1}{\gamma}\right)^{1/2} = 10\left(\frac{E_1}{\gamma}\right)^{1/2} \text{ eV} \qquad (3.7)$$

Coincidence: $$\delta Q_T = 2\left(\frac{kTE_0}{\gamma}\right)^{1/2} = 10\left(\frac{E_0}{\gamma}\right)^{1/2} \text{ eV} \qquad (3.8)$$

In these equations, k is the Boltzmann constant, T is the absolute temperature, and the energies E_0, E_1, and E_2 are expressed in keV. In most experiments the angle θ is less than 20° and thus E_2 is much smaller than either E_0 or E_1. Because of this, the thermal contribution to the measured widths is the smallest in the scattered-particle experiments. A typical example is that of 3.0-keV aluminum ions being scattered through 4.5° by argon atoms. In this case $E_0 = 3.00$ keV, $E_1 = 2.99$ keV, and $E_2 = 0.01$ keV. For these values these equations for the thermal width yield values of 1.1, 21, and 21 eV, respectively. Under most conditions, instrumental effects limit the resolution in scattered-particle experiments whereas thermal broadening is the limiting factor in recoil-particle and coincidence experiments. Afrosimov and co-workers[47] have performed experiments in which they measured not only E_0, θ, and ϕ (a coincidence measurement) but E_1 and E_2 as well. By doing this they combined the advantages of coincidence measurements with the superior resolution of the scattered-particle method.

3.1. Energy-Loss Measurements

3.1.1. A Typical Scattered-Particle Apparatus

Much has been learned about the dynamics of collisions from measurements, including coincidence measurements, on the scattered particles. Figure 9 shows an apparatus developed for this purpose by Pollack and co-workers.[36-39] Atomic particles, generated in the source, make collisions in the scattering cell and some of the scattered particles are detected by an electron multiplier in the detector chamber.

The production of both keV-energy ions and neutral atoms requires the initial ionization of the parent atoms. There is a vast literature on sources and in addition many commercial units are available. For investigations of inelastic energy losses by measurements on scattered ions and atoms it is essential to use a source of ions that introduces the least possible energy spread in the beam since this spread will degrade the ultimate resolution of

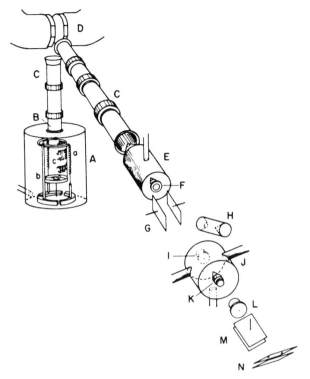

Fig. 9. Schematic diagram of the apparatus of Pollack and co-workers for studying inelastic collisions.[39] A, Ion source; B, extractor; C, lenses; D, magnet; E, charge-exchange cell; F, I, K, L, collimating holes; G, M, ion deflector plates; H, Faraday-cup monitor; J, scattering chamber; N, electrostatic energy analyzer.

the experiment. In experiments involving measurements on ejected electrons or photons the energy distribution in the incident beam is not as critical. It is possible to obtain a combined source and analyzer resolution of better than 0.5 eV/1000 eV, full width at half-maximum (FWHM). The source must also produce predominantly ground state ions. In spite of the fact that radiofrequency and glow discharge sources yield the largest intensities of ions, the inherent energy spread or excited-state population in the ions from such sources makes them unacceptable for energy-loss measurements at low energies. The sources of choice are the electron-impact ion sources except for the alkali ions for which a surface ionization source is superior.[40]

The source shown in Fig. 9 is of the electron-impact type. It is operable over a wide range of gas pressures and filament power to yield ion beams with desirable characteristics. One such characteristic is beam intensity. Total cross section measurements require low beam intensities to avoid saturation of the pulse counting detection system, whereas differential angular measurements require high-intensity beams. The beam intensities obtained from such a source increase very rapidly with increasing filament-to-grid (anode) voltage. This voltage determines the number as well as the maximum energy of the ionizing electrons. Unfortunately the beam intensity is small when the filament-to-grid voltage is low enough to ensure against the production of excited-state ions. Data can be taken at filament-to-grid voltages above the threshold for excited state ion production, but it should be demonstrated that the experimental results are not affected. As an example, Nagy and co-workers[36] used 150-eV electrons in their ion source for a study of charge exchange in He^+–He collisions, but demonstrated that essentially the same cross sections are obtained with 50-eV electrons. Since 65-eV electrons are required to excite the metastable ion state of He^+, it may be concluded that either these ions were not produced by the source in significant numbers or that the presence of the metastable state does not influence the measurement of the cross section. Independence of the experimental results on the ion-source gas pressure should also be established.

The source in Fig. 9, used by Nagy and co-workers[37] for He^+ beams and Savola and co-workers[38] for He-scattering studies, is typical of electron-impact ion sources. It consists of a stainless-steel water-cooled jacket and a removable flange on which are mounted a cylindrical grid structure (c) and a molybdenum support structure (b) holding two (one a spare) filaments (a) in place. The grid is 2 in. long and $\frac{5}{8}$ in. in diameter and is formed from medium-fine tungsten screen mesh. A spoked molybdenum disk attached to the base of the grid provides support for it. The filaments are made of 0.015-in. tungsten wire and are wound in helices 0.125 in. in diameter and 1.25 in. long. The electrode structure is mounted on pins passing through a ceramic disk positioned on the removable flange. Typical ion-source operating pressures are in the range of 20 mtorr; after mass analysis and the collimation described below, about 10^{-10} A of He^+ enters the target gas cell.

Ions are extracted from the source by an extractor (B) which, in conjunction with six additional cylindrical lenses (C) and a permanent magnet (D), form the ion optical and mass analysis system. The lenses range in length from 1.0 to 1.5 in. and have a 0.5-in. diameter bore. Shielding from stray fields is accomplished by having the lenses telescope as shown in the figure. The ion optical system design was established with the aid of characteristic curves found in the book by Spangenberg.[41]

For neutral beam production the ions pass through a charge-exchange cell (E) which is supplied with a suitable gas. Savola and co-workers[38] used a cell 0.8 in. in diameter and 1.2 in. long. Both ends were capped with inserts containing small holes at their centers. The entrance hole was 0.063 in. in diameter and the exit insert (F) had a 0.0044-in.-diameter hole at its center.

With charge-exchange gas in the cell, a beam containing both ions and neutrals emerges from F. A neutral beam is obtained by placing a voltage across the pair of parallel plates (G) and deflecting away any residual ions. Using a charge-exchange gas pressure of about 30 μm, ~30% of a low-keV He$^+$ beam is neutralized by symmetric charge-exchange in this cell.

The possible presence of long-lived metastable neutral atoms in the beam is of concern. In charge exchange between He$^+$ and He, metastable state production is kinematically possible in forward scattering. Even a small metastable population in the neutral beam could have serious consequences since the scattering cross sections for metastable He atoms may be large compared to those for ground-state atoms. It is essential to demonstrate that the neutral beam is primarily in its ground state (this may be done directly by a time-of-flight energy measurement on the neutral beam). In the case of He, the metastable state is excited by a curve-crossing mechanism.[42] This mode of excitation requires a relatively hard collision (in the case of He) and if small enough beam collimation holes are employed both before and after charge exchange, the metastable population in the beam will be negligible. In the studies of Savola and co-workers[38] the angular width of the incident He beam was determined by the hole in F and a second collimating hole at the entrance of the scattering chamber. These holes were sufficiently small to block any metastable component generated by a narrow He$^+$ beam incident on the charge exchange cell. The incident He$^+$ beam was indeed shown to be narrow by placing a high-angular-resolution collimating system, consisting of two holes of diameters 0.0108 and 0.0088 in., separated by a distance of 1.17 in. between the last lens and the charge exchange cell. The intensity and angular profile of the resulting beam were found to be unchanged, which showed that the beam originally incident on the charge-exchange cell was paraxial. Furthermore, to within the accuracy of the measurements, no differences in the total cross-section values were observed between those obtained with the ordinary and the additionally collimated beams. In addition the consistency between the cross-section values measured by Savola,[38] Amdur,[43] and Belyaev[44] and their co-workers, where the He beams are generated under quite different conditions, seems to indicate that metastables do not have a large effect (at least in the total cross sections that were measured).

The actual scattering measurements that may be made with such apparatus fall into three categories: total cross sections measured in the forward direction, total differential cross sections measured over a range of scattering angles, and doubly differential cross sections. The inelastic energy-loss measurements to be discussed here fall into this last category. The term "doubly differential" refers to the measurements being differential in energy as well as angle. To determine Q from Eq. (2.26) it is clear that both E_1 and θ must be measured for the scattered particle. For completeness, though, we will begin with a brief mention of total cross sections measured in the forward direction. Total cross sections are most easily observed by measuring the attentuation of the incident beam after passing through a gas target. If the beam's initial intensity is I_0 and it traverses a gas cell of length l, containing n particles per unit volume, then the intensity of the emergent beam is given by

$$I = I_0 e^{-n\sigma l} \tag{3.9}$$

where σ is the total cross section for scattering a beam atom beyond the acceptance angle of the detector. It is assumed in the derivation of this equation that if a particle is scattered out of the beam it does not scatter back into the beam as a result of subsequent collisions. Since the value of the cross section σ measured this way may be a function of detector acceptance angle, care must be taken in interpreting the data. Figure 10 shows the effect of the detector's angular resolution on the total cross section as a function of energy for the scattering of He by He. The curves labeled F give the results of Jordan and Amdur[43] (effective angular resolution 0.57°), A of Savola[38]

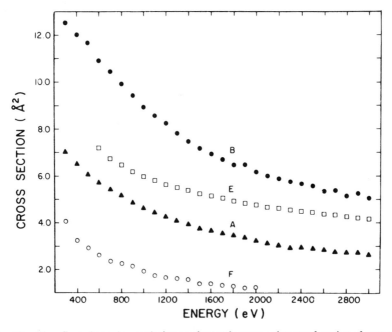

Fig. 10. The effect of angular resolution on the total cross section as a function of energy for the scattering of He by He. Starting with the lowest curve the angular resolutions are 0.056°,[38] 0.11°,[44] 0.26°,[38] and 0.57°.[43]

(angular resolution 0.260°), E of Belyaev[44] (effective angular resolution 0.11°), and B of Savola[38] (angular resolution 0.056°). A "total cross section" determined in this type of measurement contains contributions from the elastic and all the inelastic channels that scatter particles beyond the detector acceptance angle. Generally it should not be assumed that the detected signal measures only the attenuated incident beam. Inelastic scattering into the forward direction is possible even for those channels excited in a crossing of the molecular potential curves if at least one of the curves is attractive. If inelastic effects are small and the detector's angular resolution is not high enough to require a quantum mechanical treatment of the scattering, then a classical analysis can provide information on the interaction potential. In the case of He–He scattering, the minimum angle to which a classical analysis is applicable is estimated to be 0.5° at an energy of 1 keV.[45] The classical analysis leads to a formalism for extracting the short-range interaction potential, as a function of internuclear separation, from the energy dependence of the total cross section. The necessary techniques were pioneered by Amdur and his co-workers and successfully applied to numerous systems (see Ref. 43 for a discussion of experiments with neutral beams and for additional references to the literature). Several of the important problems relating to the interpretation of experimental data (including quantum scattering corrections and inelastic scattering effects) are discussed in the review article by Mason and Vanderslice.[45]

The total *differential* cross section (for a particular process) $\sigma(\theta) \equiv d\sigma/d\Omega$, where $d\Omega$ is the element of solid angle, is the effective area presented by a target particle in scattering a beam particle into a unit solid angle at the required scattering angle. In the single collision approximation it is found from

$$I = I_0 n \sigma(\theta) l \, \Delta\Omega \tag{3.10}$$

where $\Delta\Omega$ is the solid angle of acceptance of the detector. This equation will give the cross section if the scattered signal increases linearly with n (indicating a predominance of single collisions) at the angle in question, if l is well defined, and if the cross section does not vary appreciably over the detector's acceptance angle. It is also understood that the efficiency of the detector is the same for the incident and for all components of the scattered beam. Since the detector must have a finite angular resolution, the scattered signal is determined by an integrated (over the detector acceptance angle) value of $\sigma l \, \Delta\Omega$ at the "scattering angle." An approximation for $\sigma l \, \Delta\Omega$ has been given by Jordan and Brode,[46] which is useful at the larger scattering angles where $\sigma(\theta)$ may be taken outside the integral. The most accurate cross-section values are obtained if circular, rather than rectangular, apertures are used for the required collimation since they maintain the same angular resolution in both the horizontal and vertical directions. Rectangular slits are often used at large scattering angles where there is difficulty in obtaining sufficient particle intensities. Since differential cross sections do not vary as rapidly with angle at the large angles this does not necessarily degrade the angular resolution to a great extent. Slits have been successfully employed by Everhart and co-workers who used combinations of slits and holes,[48] by Lorents and Aberth,[49] and recently by Crandall and co-workers.[50] Problems associated with scattering geometry are discussed in these references as well as in numerous other works. Of particular interest is a paper by Filippenko[51] which analyzes the systematic errors in calculations of differential scattering cross sections that are due to the finite resolving power of an apparatus.

A total differential cross-section experiment is performed by measuring the intensity of the beam as a function of scattering angle. Particle-counting techniques are necessary for these measurements since very low intensities are generally encountered in differential-scattering work. The usual measurement involves a determination of the number of particles detected in a given time interval. Since the count rates are small at the larger angles, the measurements take longer and may be severely affected by variations in the incident beam intensity or fluctuations in the scattering gas pressure. These errors may be compensated for by using a reference signal which is proportional to the number of collisions occurring. One possible reference is the electron or ion signal reaching a monitor which "sees" a fixed interaction region in the scattering cell. A photomultiplier positioned near the scattering volume has provided a suitable reference signal for some ion–molecule studies.[52] The measurements were made by simply determining the number of particles that reached the detector (at each angle) for a predetermined number of photons coming from the scattering volume. This number depends on the product of beam intensity and scattering gas density and is therefore proportional to the number of collisions. At small scattering angles Bray[52] used an $x-y$ recorder to measure the scattered ion signal as a function of scattering gas pressure. At each angle the differential cross section is then obtainable from the slope of the curve. An additional advantage of this method is that for single collisions, the resulting curve is a straight line and any departure from single-collision conditions is immediately obvious. Bray's curves for He^+-H_2 are displayed in Figs. 11 and 12. The scattered-particle intensity versus gas-pressure curves, at several scattering angles but at a fixed incident beam energy, are shown in Fig. 11. The total differential cross sections for various energies are shown in Fig. 12.

In order to determine the *doubly differential* cross sections, required for determination of the inelastic energy loss, the energy spectrum of the scattered particle must also be measured. If the scattered particle has a charge, then it is a fairly simple matter to pass it through an electrostatic

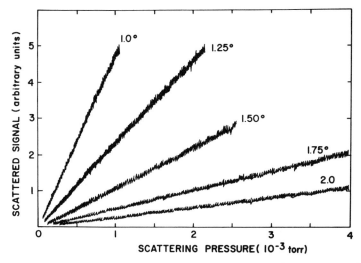

Fig. 11. Scattered intensity versus target-gas pressure curves for $He^+ + H_2$ collisions.[52] The differential cross sections are obtained from the slopes of the lines.

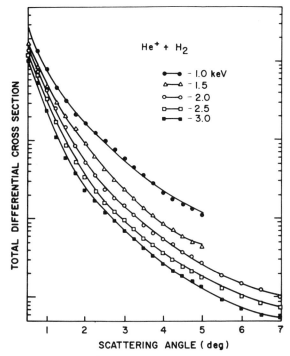

Fig. 12. Total differential cross section plotted versus scattering angle for $He^+ + H_2$ collisions derived from intensity versus pressure curves such as those in Fig. 11.[52]

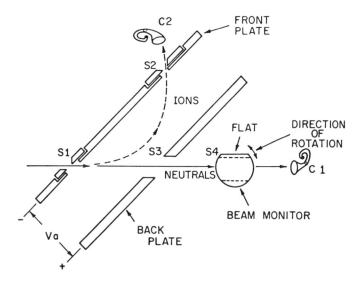

Fig. 13. Schematic diagram of the parallel plate electrostatic analyzer.[39]

analyzer and to measure its energy this way. Because of their simplicity and focusing properties, cylindrical analyzers and parallel-plate analyzers are most frequently used, and there exists extensive literature on these analyzers.[53-57] An analyzer of the parallel plate type, used by Fernandez, Eriksen, and co-workers,[39,58] is outlined in Fig. 13. With no voltage applied to the plates, all particles are intercepted either by the beam monitor or particle detector C1. With voltage applied to the plates only neutral particles will be detected at C1. By adjusting the voltage, the charged particles may be deflected through slit S2, and detected with C2. The detectors shown are electron multipliers of the solid-state variety, of which the Bendix Channeltron is an example. The beam monitor shown is a Faraday cup which may be rotated into position and used for measurements at small scattering angles (where the beam is too intense for the Channeltron) and also to plot the incident beam profile. The analyzer is operated with its front plate grounded and with a variable voltage on the rear plate (it is also possible to float the entire system and thereby increase the resolution by retarding the beam). The energy profile of the scattered beam is determined by measuring the beam intensity as a function of the voltage across the plates. Examples of typical spectra obtained this way are shown in Fig. 14. Through the use of Eq. (3.2) the abscissa has been converted from scattered ion energy E_1 to Q, the inelastic energy loss. It is seen that for the Ar^+–Ar collisions a fairly large number of events correspond to elastic scattering; i.e., $Q = 0$ for one of the peaks displayed. The other peaks at 14 and 27 eV result from collisions for which excitation of electrons has taken place.[39] These specific excitations have also been discussed in detail by Barat and co-workers.[34,59]

The energy-loss spectra were determined by counting the number of ions arriving at C2, with a given voltage across the plates, for a preset number of neutral atoms detected by C1. Thus the neutral beam serves as a reference signal. After detection of the required preset number of neutrals, the ion count was stored in a multichannel analyzer, the voltage incremented by a small amount, and the cycle repeated. Measurements utilizing the detection of a preset number of neutral atoms (to indicate that a fixed number of collisions have occurred) minimize the fluctuations in ion signal normally associated with incident beam fluctuations and scattering gas pressure variations. The data acquisition system employed in these measurements is described by Bray et al.[60] In some experiments it is essential to know the detection efficiency for the scattered particles. This is particularly important when a neutral and an ion signal must be compared as, for example, in determining charge-exchange probabilities. At energies above a few keV the detection efficiencies for ions and neutrals are usually equal. This fact has been used and demonstrated by several workers.[61-64] As the beam energies decrease the relative detection efficiency of neutrals to ions

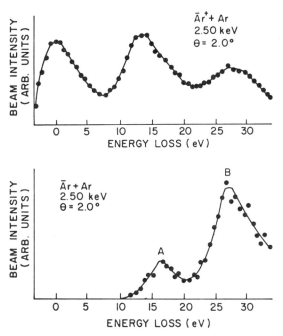

Fig. 14. Typical inelastic energy-loss spectra obtained by Eriksen and co-workers[39] for Ar^+-Ar and Ar-Ar collisions, product Ar^+ ions being analyzed in a parallel-plate energy analyzer.

decreases so that significant errors may result if this is not taken into account. A considerable amount of work has been done by Hagstrum on the number of electrons ejected by various projectiles incident on surfaces.[65] These results, although useful, must be applied with care in any particular case since the efficiencies depend on the quality of the surface.

A novel technique for determining the relative efficiency of neutral to ion detection is described by Nagy et al.[37] It is known that in resonant charge exchange such as occurs in $He^+ + He$ collisions, interference between Σ_g and Σ_u states causes the intensity of the scattered He beam to exhibit a marked oscillatory behavior when plotted as a function of angle. A similar structure, differing in phase, is obtained for the He^+ signal. At small angles where inelastic effects are negligible, the combined ion and neutral beam intensities should decrease monotonically with increasing scattering angle. An oscillatory structure will result in the combined signal when the detection efficiencies of the neutral atom and ion are different. At those angles where the neutral signal predominates, a low-relative-detection efficiency for neutral atoms would result in a scattered signal which falls below the expected smooth curve. The technique involves a high-angular-resolution measurement of the combined neutral and ion beam intensities as a function of angle. The envelopes of the maxima (where the ions predominate) and minima (where the neutral atoms predominate) are compared and yield the relative detection efficiencies. Using this technique Nagy et al.[37] found that at 1.0 keV the detection efficiency of He is about 85% that of He^+. At 3.0 keV the efficiencies are essentially equal. Another interesting technique for determining detection efficiencies is described by Jaecks et al.[63,64] for the case of protons incident on various gases. This method makes use of $2s$ hydrogen atoms produced in these collisions. By electrostatic quenching of these states and then measuring the number of photons found in coincidence with the fast particles a rather accurate determination of the detector efficiency was made.

A scattering event may result in the excitation of the projectile, the target, or of both collision partners. For a given process the excitation energy is well defined. As an example, the Ar^*3p^44s4p state, which is excited in Ar + Ar collisions,[39] lies approximately 29 eV above the ground state. In decaying to $Ar^+3p^5 + e^-$ the electron acquires a kinetic energy of $29 - 16 = 13$ eV relative to a frame of reference fixed on the Ar. In a collision experiment the Ar projectile is moving in the laboratory frame and the ejected electrons have laboratory kinetic energies which depend on the velocity of the Ar as well as on the angle of ejection. This kinematic energy shift of the ejected electrons was first reported by Rudd et al.[66] and has more recently been discussed by several other investigators.[67,68] It should be realized that related shifts (but much smaller in energy) of the resulting ion must occur, and indeed, the effect is analogous to that associated with the kinetic energy release occurring on fragmentation of polyatomic ions which is discussed extensively in Chapters 6 and 7. Since the direction of electron ejection is generally not known in a noncoincidence type of experiment it is impossible to correct the measured inelastic energy losses to determine the exact excitation energy. For spherically symmetric ejection the profiles are simply broadened about the proper energy loss. In the nonspherical case a measured energy shift is possible. This would be most severe in the case of light projectiles.

The second spectrum in Fig. 14 demonstrates a deficiency of electrostatic analyzers—they are unable to measure the energy of scattered neutral particles. For the Ar + Ar collisions, the elastically scattered projectiles are neutral and cannot be energy-analyzed electrostatically. Thus, no peak corresponding to the elastically scattered neutrals appears in the second spectrum of Fig. 14. Since not all of the scattered neutral particles are elastically scattered, it is clear that an apparatus which cannot measure the energy loss of collisions having neutral atoms among their final products is inappropriate for many investigations. This deficiency can be met, either by using the coincidence method of Eq. (3.4)[69,70] or time-of-flight measurements on the scattered particle. Since the latter technique is not so affected by the thermal broadening of Eqs. (3.6)–(3.8), it has been the most widely used. The following section is limited to a discussion of time-of-flight measurements.

3.1.2. Time-of-Flight Measurements

The feasibility of making high-resolution energy-loss measurements on neutral scattered atoms having keV energies was recently demonstrated by Barat and co-workers.[34,71,72] Time-of-flight (TOF) techniques employed in their measurements were found to yield resolutions comparable to those obtained with conventional electrostatic-energy analyzers. In principle the technique is quite simple and depends on the fact that the time t required for an atom of mass m and energy E to traverse a distance L is given by

$$t = \frac{L}{v} = \frac{(m/2)^{1/2}L}{E^{1/2}} \qquad (3.11)$$

Therefore an atom undergoing an energy loss ΔE before traversing the same distance is delayed, in reaching the detector, by an amount of time Δt, where

$$\Delta t = \frac{m^{1/2}L\,\Delta E}{(2E)^{3/2}} \qquad (3.12)$$

TOF measurements can be made by electrically pulsing the incident ion beam as it moves between the plates of a parallel-plate capacitor and measuring the time interval between the detection of a particle and the

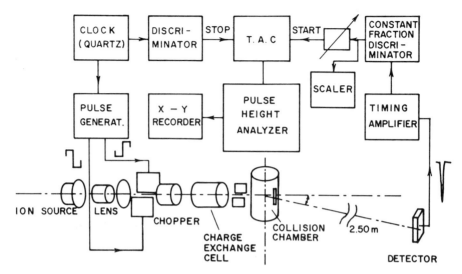

Fig. 15. Schematic description of the time-of-flight apparatus of Barat and co-workers[34] for kinetic energy analysis of fast neutrals.

arrival of a suitably delayed signal generated by the chopping pulse. Figure 15 outlines the experimental arrangement employed by Barat and co-workers[34] for TOF measurements. An ion beam is extracted from a discharge type of ion source at an energy of 250–300 eV (this low extraction voltage yields well-focused beams having a small energy spread). Following this, the beam is accelerated, focused by a three-element cylindrical lens, passed through some shim fields, and collimated by a slit. The beam then passes between electrostatic plates where it is chopped by a time-varying voltage. The transmitted fraction continues through a second shim field region followed by a slit, enters a charge-exchange cell where it is partially neutralized, and further collimated by a third slit. Electrodes behind the last slit deflect the residual ions, leaving only neutral atoms in the incident beam. This beam enters a scattering cell, undergoes collisions, and traverses a 2.5-m flight path to an electron multiplier. A "beam cleaner" placed in the flight path separates the ions and neutrals prior to detection. Because of the large scattering cell-to-detector distance required for high-resolution TOF measurements it was found convenient to change the observed scattering angle by moving the source rather than the detector.

The chopper is in the form of a simple parallel-plate capacitor having a plate length of 1.0 cm. On these plates are applied square wave pulses of amplitude ±10 V (rise time typically 3 nsec) in a frequency range of from 0.1 to 1.0 MHz (stabilized by a 20-MHz quartz oscillator). The system is designed so that the detected particle initiates a start pulse at the time-to-amplitude converter (TAC) and the arrival of the delayed chopper voltage pulse then provides the stop pulse and determines the pulse height analyzer channel into which the event is placed. The energy distribution in the scattered beam is then determined with the help of Eq. (3.12). Although it is possible to use the chopper voltage pulse as the start pulse with the detected particle providing the stop pulse, this method is less desirable because not all chopper pulses are followed by a detector pulse. Care must be taken to minimize the system's dead time to be certain that the faster particles are not counted preferentially. It is apparent that the scattering cell-to-detector distance is quite important in determining the energy resolution. It should also be pointed out that the distance between the scattering region and the detecting surface must be well defined in order to obtain an optimum energy resolution (care must be exercised to ensure that the particle-detection surface is flat). Figure 16 shows the measured energy resolution ΔE as a function of beam energy for He^+ beams from 200 to 3000 eV for values of L equal to 1.25 and 2.50 m. Experimental results obtained with this apparatus are discussed later, but first it is appropriate to review some earlier work on scattered *ions* in a more general way.

The $He^+ + Ne$ collision has been the subject of extensive experimental and theoretical work. It is still of current interest in many laboratories, and provides a good example of the types of information which scattering studies can yield. In addition to elastic scattering it is possible to have charge-exchange processes leading to final ground-state or to excited-state configurations. Direct scattering with only projectile or target excitation (or ionization) can occur as well as collisions in which both partners are excited. In the energy range of from several hundred to several thousand electronvolts, the collision is best understood in terms of the quasi-molecular model (cf. Section 2).

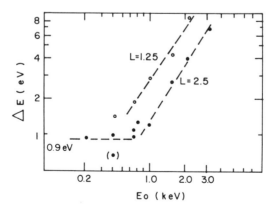

Fig. 16. Measured values of the energy resolution ΔE, plotted versus incident beam energy for lengths L of 1.25 and 2.5 m for the TOF system shown in Fig. 15.

In an early paper Aberth and Lorents[73] demonstrated the importance of inelastic scattering in $He^+ + Ne$ collisions. The inelastically scattered fraction of the beam was found to increase strongly with increasing scattering angle and two inelastic peaks were reported at energies close to the first excited levels of Ne and He^+. The collision was studied in greater detail in later papers published by the S.R.I. group.[74-76] Particular attention was given to the excitation of the $Ne(2p^5 3s)$ state since the 16.8-eV energy-loss peak could be uniquely resolved. Figure 17 shows the reduced cross section as a function of reduced scattering angle for exciting this state at an initial He^+ center-of-mass energy of 71 eV. This excitation is seen to be absent in forward scattering, and the reduced cross section has pronounced Stueckelberg interference oscillations caused by the fact that there are two semiclassical trajectories of different impact parameter that scatter the ions to a common angle (cf. Section 3.2.4). Analysis of the data suggested that the upper state was attractive, and in addition a curve crossing was identified as the cause of an observed perturbation in the elastic scattering cross section. The behavior of the reduced cross section at the first Stueckelberg peak gave a transition probability peaking at an energy of about 25 eV, which corresponds to a velocity of approximately 2.6×10^6 cm/sec at the crossing. Potentials were then inferred from the elastic and inelastic $He^+ + Ne$ scattering results and provided an early example of the information obtainable from a detailed analysis of the differential scattering data. Figure 18 shows a more recent example of the types of potential curves that account for the results of He^+–Ne scattering.[34]

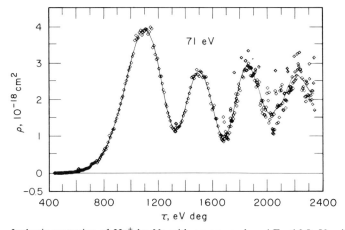

Fig. 17. Inelastic scattering of He^+ by Ne with an energy loss $\Delta E = 16.8$ eV, with initial center-of-mass energy 71 eV. Reduced cross section ρ plotted against reduced scattering angle τ (beam energy × scattering angle).[75]

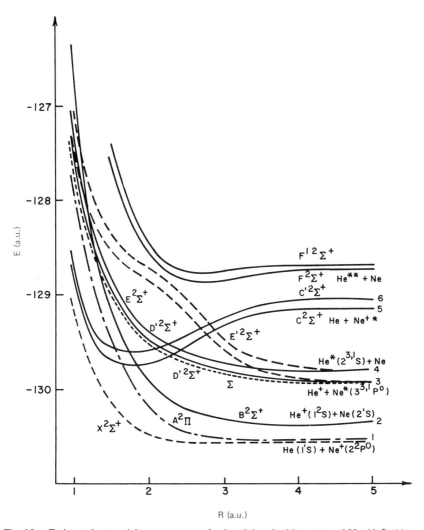

Fig. 18. Estimated potential energy curves for low-lying doublet states of He–Ne$^+$. Above about 30 eV other states are known to cross $B\ ^2\Sigma^+$.[34]

Elastic and inelastic scattering in He$^+$ + Ne collisions was also studied and analyzed by Baudon et al.[77] The measurements were made directly on the scattered He$^+$, and Fig. 19 shows the relative elastic and inelastic reduced cross sections as a function of reduced scattering angle for laboratory energies of 700 and 1580 eV. The curve labeled ρ_e represents the elastic scattering, ρ_1 is an average value of the cross section taken over all the excited levels of Ne, and ρ_2 represents the excitation of autoionizing states in the target Ne. The elastic scattering was found to have a similar behavior for

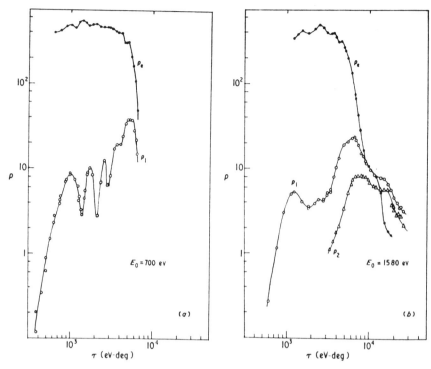

Fig. 19. He$^+$–Ne collision. Elastic (ρ_e) and inelastic (ρ_1 and ρ_2) reduced cross sections in arbitrary units, as function of the reduced scattering angle $\tau = E_0\theta$ for two impact energies.[77]

all the energies in the range investigated (0.5–3.0 keV). Of particular interest was the abrupt falloff in ρ_e at $\tau \approx 400$ eV-deg, which agrees well with the results of Aberth and Lorents.[73] This falloff, which is characteristic of a strong absorption (attributed to the rapid opening of inelastic channels), suggests the possible use of a complex optical type of potential to explain the scattering results. Below energies of 1.5 keV and for $\tau \leq 3500$ eV-deg, ρ_1 has a similar behavior for all energies. It is negligible for $\tau < 350$ eV-deg and increases rapidly to a first maximum at $\tau_0 = 1.1$ keV-deg. For $\tau > \tau_0$, ρ_1 is found to oscillate and attains its maximum value for $\tau \approx 6000$ eV-deg at all energies. For $\tau > 3500$ eV-deg the oscillations become damped, and at energies above 2 keV they disappear. This is a result of the increasing number of channels which contribute to the scattered signal. Work currently in progress on He$^+$ + Ne, by Barat et al.,[78] is of sufficiently high energy resolution to allow a more detailed analysis of the interactions that occur during the collision. Figure 20 shows an energy analysis of a 1.5-keV He$^+$ beam scattered at an angle of 4.4° and indicates the significant amount of information that is obtainable from such measurements.[78]

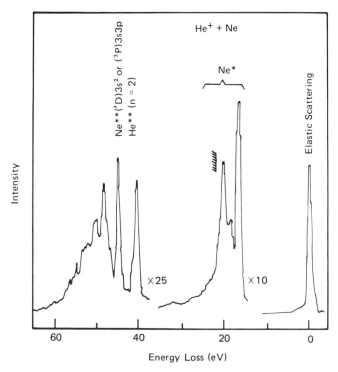

Fig. 20. High-resolution inelastic energy-loss spectrum for He^+–Ne 1.5 keV, $\theta = 4.4°$ collisions.

All the work on $He^+ + Ne$ discussed so far was done by making energy-loss measurements on the scattered He^+, and the results may generally be interpreted in terms of curve crossings in the quasi molecule. There is, of course, much work to be done in constructing valid potential diagrams for systems of interest. Charge-exchange processes (cf. Chapter 2) are also important in the low-keV energy range and contribute significantly to the inelastic scattering in the system. However, these processes are not directly represented in data taken with the scattered He^+ ion. Direct measurements on the scattered neutrals were made and Fig. 21 shows typical 2-keV collision data. The curve labeled N is a plot of the relative count rate of the detected He as a function of scattering angle, and the curves labeled I and P represent the relative count rate of the He^+ and the charge-exchange probability (the ratio of the scattered neutral to scattered total signals), respectively. The marked oscillation of the scattered neutral signal was found at all the energies investigated. Figure 22 shows the probability of charge exchange as a function of scattering angle in the energy range of 1000–3000 eV. A most interesting feature in the plots is that the

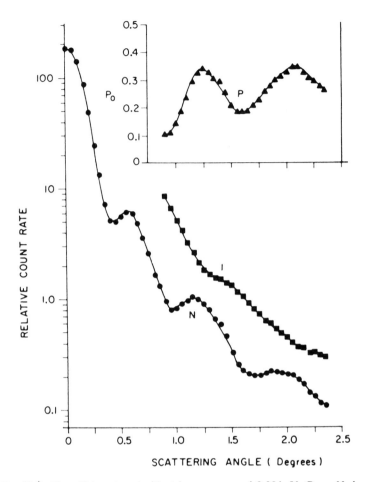

Fig. 21. He$^+$ + Ne collision at an incident beam energy of 2.00 keV. Curve N shows the neutral He signal as a function of scattering angle and curve I shows the He$^+$ signal. The insert shows curve P which represents the charge-exchange probability P_0 obtained by dividing N by I. Results for curves I and P are not shown for angles below 0.90°, since the incident beam may influence the data in this angular region. The data shown are normalized with respect to the neutral-atom data point at 3.00 keV and 1.00°.[37]

locations of the maxima and minima in the probability curves are generally independent of energy. Similar measurements at higher energies gave P_0 values which depended only on energy and not on scattering angle.[79] This latter behavior is typical of resonant charge exchange (as, e.g., in He$^+$ + He) and has been successfully explained by a two-state approximation. However, the results of Fig. 22 cannot be interpreted in terms of a simple curve crossing and resulting Stueckelberg oscillations; but they are due to transitions between the two close-lying states of the same symmetry (in this case between $B\ ^2\Sigma$ and $X\ ^2\Sigma$) that are coupled by an exponential matrix element.[78,80,81] The cross sections for these processes are characterized by a constant threshold value for τ at low to intermediate energies; at high energies oscillation features resembling those found in resonant charge exchange are observed.[81] Although the target and/or the projectile may be excited in a charge-exchange collision, the $He^+ + Ne \rightarrow He + Ne^+ + 3.021$-eV channel is dominant at energies ≥ 1 keV as has been shown in recent TOF energy-loss measurements on scattered He atoms by Barat and co-workers.[82] The TOF technique for determining the energy loss suffered

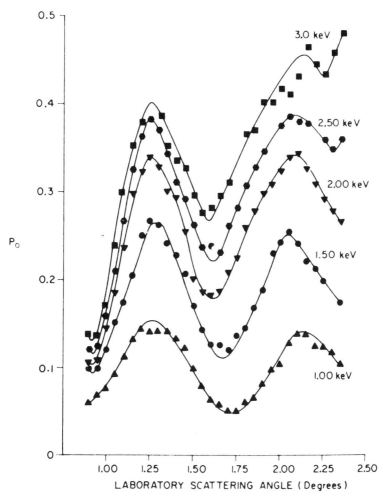

Fig. 22. Plots of the charge-exchange probability P_0 versus laboratory scattering angle for the $He^+ + Ne$ collision.[37]

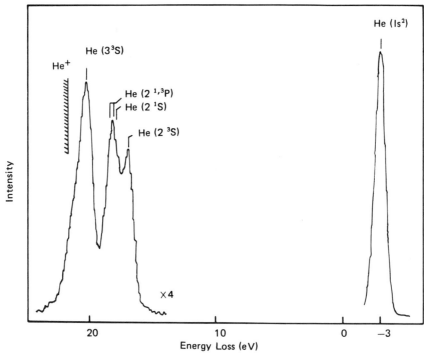

Fig. 23. Time-of-flight energy spectrum of He atoms formed in 750 eV, 2.1°, He$^+$–Ne collisions. The peak at +3 eV is due to the exothermic reaction giving ground-state products (He + Ne$^+$).

by scattered neutrals provides data which allow a more thorough analysis of the collision process than would be possible from ion data alone. The currently attainable energy resolution is sufficient to resolve relatively close-lying states. This is demonstrated in Fig. 23 which shows a TOF spectrum of He when a 0.75-keV He$^+$ beam is scattered at 2.7° by Ne.[78]

Differential scattering experiments with neutral incident beams are currently in progress at several laboratories. In this work energy analysis is performed on both the scattered neutrals and the fast ions created by the collision. Studies on the Ar + Ar system are typical of this type of work. Eriksen and co-workers[39,83] determined P_1 (the probability of ionization defined as the ratio of the scattered ion to scattered total signal) as a function of energy and scattering angle for this collision. Figure 24 is a plot of P_1 as a function of $\tau = E\theta$ at small values of τ, and Fig. 25 is a plot for larger τ values. Little ionization is seen to occur at small τ. Of particular significance is the threshold behavior of P_1 at $\tau \approx 3$ keV-deg. This behavior (common onsets at different energies) is characteristic of an (initial) interaction that occurs at a

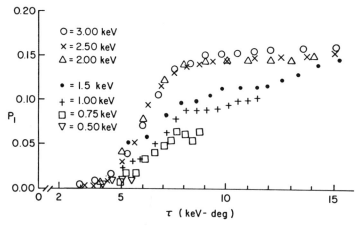

Fig. 24. Probability of ionization (P_I) as a function of reduced scattering angle for Ar–Ar collisions.[39]

fixed internuclear separation. Additional data taken at energies of 1.5, 1.0, 0.75, and 0.50 keV yield ionization-probability curves that maintain the general shape of the higher-energy results but show a decrease in P_I as the energy is reduced. This velocity dependence merits further study. Energy-loss measurements on the scattered Ar$^+$ over a large range of angles show several peaks whose relative intensities vary rapidly with angle. Figure 26 displays the evolution of the inelastic energy-loss spectrum at an incident Ar energy of 3.0 keV and scattering angles of 4.0, 6.0, 8.0, and 10°.[39] The peaks at 18 and 29 eV were studied in detail with higher-energy resolution, from threshold out to about 12 keV-deg, and it was found that in this limited

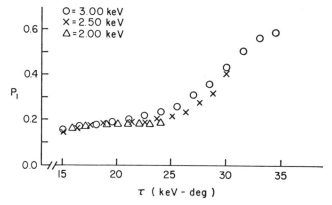

Fig. 25. Probability of ionization as a function of reduced scattering angle for Ar–Ar collisions.[39]

range the ratio of intensities of the 18–29-eV peaks was 0.25 and remained fairly constant. The 18-eV peak is attributed to autoionization in the quasi molecule. These processes occur when the collision excites electrons to orbitals that at infinite separation lead to states such as $Ar(3p^54s)+Ar(3p^54s)$, which lie in the continuum, and have sufficiently short lifetimes to ionize before the "collision" has ended. The resulting energy loss is simply the difference in energy (at the point of electron ejection) between the initial excited-state curve and the one leading to $Ar^+(3p^5)+Ar$ at infinite separation. If the lifetime for autoionization is long enough to allow the system to decay at an internuclear separation where the curves are essentially parallel, the energy-loss spectrum will be sharply peaked. If the decays occur at separations where the energy difference between the curves varies rapidly, then the peak will be broadened. An energy-loss spectrum peaked at the ionization potential suggests that the decay occurred near the region where the excited-state curve entered the continuum. Sharp peaking here indicates a well-defined region of decay. The 29-eV peak is due to the excitation of an atomic autoionizing state such as $Ar^*(3s^23p^44s4p)$ which decays after the collision is over. Although autoionizing states of the type $(3s3p^6ns)$, $(3s3p^6np)$, etc., also would lead to similar energy losses, their contributions are not thought to be important here since larger τ values are required to reach the internuclear separations where the $4p\sigma$ molecular orbital curve crosses those leading to final states having excited $3s$ electrons.[39] Higher-energy-loss peaks are attributed to the excitation of autoionizing states with simultaneous excitation or ionization of the target.

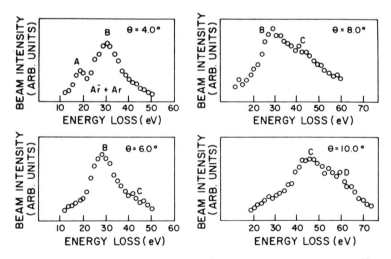

Fig. 26. Inelastic energy loss spectra of Ar^+ from 3.0-keV $Ar+Ar$ collisions.[39]

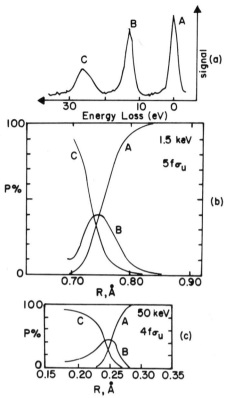

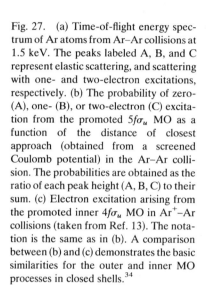

Fig. 27. (a) Time-of-flight energy spectrum of Ar atoms from Ar–Ar collisions at 1.5 keV. The peaks labeled A, B, and C represent elastic scattering, and scattering with one- and two-electron excitations, respectively. (b) The probability of zero- (A), one- (B), or two-electron (C) excitation from the promoted $5f\sigma_u$ MO as a function of the distance of closest approach (obtained from a screened Coulomb potential) in the Ar–Ar collision. The probabilities are obtained as the ratio of each peak height (A, B, C) to their sum. (c) Electron excitation arising from the promoted inner $4f\sigma_u$ MO in Ar^+–Ar collisions (taken from Ref. 13). The notation is the same as in (b). A comparison between (b) and (c) demonstrates the basic similarities for the outer and inner MO processes in closed shells.[34]

Recently[34] TOF techniques have been employed to provide energy analysis of neutral atoms scattered from Ar + Ar collisions. Figure 27 shows a spectrum taken at an incident-beam energy of 1.5 keV and scattering angle of 3.2°. The peak labeled A represents elastic scattering and peaks B and C correspond to the excitation of one and two electrons, respectively. Figure 28 shows the reduced cross sections for each of the peaks as a function of τ at a beam energy of 2.0 keV. Inelastic scattering is seen to be the dominant process for $\tau > 5$ keV-deg.[34]

3.1.3. The Use of Molecular Targets

For the examples discussed in the preceding sections only atomic targets were considered and all contributions to the inelastic energy loss Q had to be related to electronic excitation. When a molecular target is used, there exists the additional possibility of vibrational and rotational excitation, and a careful measurement of Q may allow the measurement of the vibrorotational energy given to the target molecule during the collision. An

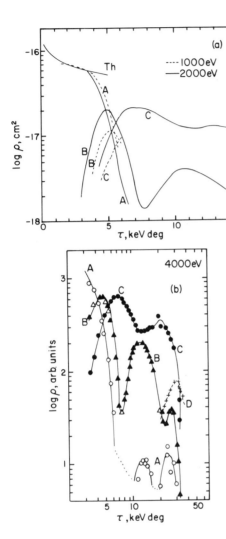

Fig. 28. Reduced differential cross sections for Ar–Ar collisions as a function of $\tau = E_0\theta$, the reduced scattering angle. (a) Results are normalized to theory at small angle. (b) Unnormalized reduced cross sections at a 4.0-keV laboratory energy over a large range of τ values. The elastic peak A is seen to exhibit a "diffraction" pattern in the region of strong absorption. Peaks B and C are seen to oscillate out of phase. The minimum separation (at $\tau = 12$ keV-deg) between B and C is found to decrease rapidly with increasing energy. At large τ values (30 keV-deg) process D is seen to dominate.[34]

obvious question to be asked is how does one apply Eq. (3.2) if the target is a molecule? In Eq. (3.2) $\gamma = M_1/M_2$, the ratio of the projectile mass to the target mass. Is M_2 to be taken as the mass of the molecule or the mass of a target atom within the molecule? Clearly, for very gentle collisions where dissociation does not occur, it is correct to use the total molecular mass. However, for more violent collisions where the binding energy of the molecule is small compared to Q the collision is more atomic in nature and it is correct to use only the mass of the individual atom which was struck by the projectile. Such high-energy collisions provide an interesting extension of the present work and are discussed later in this section. It may also be noted

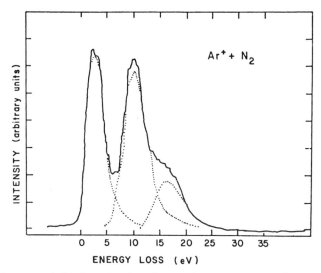

Fig. 29. Scattered particle intensity plotted versus $\Delta E = E_0 - E_1$ for $Ar^+ - N_2$ collisions. The peak on the left corresponds to those events for which there is no electronic excitation.[84]

that for very small scattering angles the distinction between these types of collisions is unimportant (cf. Fig. 30).

Figure 29 shows the energy spectrum[84,85] of 2.0-keV Ar^+ ions scattered through 1.25° by N_2 molecules. The relative intensity is plotted versus the measured energy loss of the fast ion, $\Delta E = E_0 - E_1$. The three peaks represent different degrees of electronic excitation of the N_2 molecule. The "elastic peak" nearest zero energy loss corresponds to events for which there is no electronic excitation. Fernandez and co-workers have obtained similar spectra for $Ar^+ - N_2$ collisions for energies of 1.0–3.0 keV and scattering angles of 0–4.0°. If these energy-loss data are plotted versus $E_0\theta^2$, the answers to several questions become obvious.

The reduced variable $E_0\theta^2$ is valuable for small-angle collisions because for small θ, Eq. (3.2) reduces to

$$Q \cong E_0 - E_1 - \gamma E_0 \theta^2 \qquad (3.13)$$

or

$$\Delta E \cong \left(\frac{m_1}{m_2}\right) E_0 \theta^2 + Q \qquad (3.14)$$

From Eq. (3.14) it is clear that for approximately elastic collisions ($Q \cong 0$), a plot of ΔE versus $E_0\theta^2$ will be a straight line from whose slope we can determine the appropriate mass to be used in Eq. (3.2) for the exact calculation of Q. Such a plot, from the work of Fernandez and co-workers,[84,85] is shown in Fig. 30. Line A shows the energy losses ΔE for the

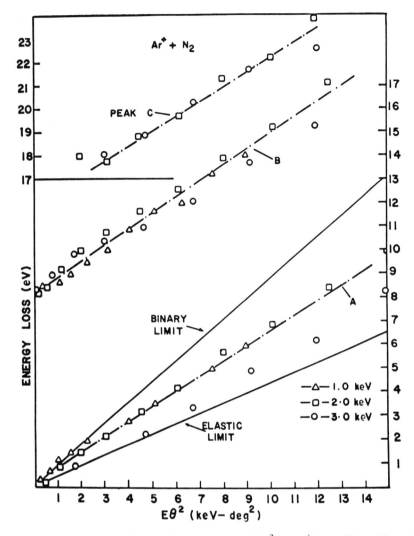

Fig. 30. The energy loss $\Delta E = E_0 - E_1$ plotted versus $E_0 \theta^2$ for Ar^+–N_2 collisions. Curve A represents the elastic data points from a large number of curves similar to those in Fig. 29. Curves B and C represent the corresponding inelastic peaks.[84]

"elastic peaks" of spectra similar to those in Fig. 29 plotted versus $E_0 \theta^2$. The solid "binary-limit" curve represents the minimum value ΔE can have (i.e., when $Q = 0$) if the mass m_2 in Eq. (3.14) is taken to be the mass of an N atom rather than the total mass of the N_2 molecule. The "elastic-limit" curve is the corresponding minimum-value curve obtained when m_2 is taken to be the mass of the molecule as a whole. Since the data fall below the binary-limit curve, it is clear that the collision is one of the projectile with the whole molecule and not with just one of the atoms within the molecule. Furthermore, the difference between the data points and the elastic-limit curve represents the Q value for each data point, i.e., the vibrorotational energy transferred to the N_2 molecule by the collision. It is observed that for the gentle, small-angle collisions, the data points fall near the elastic limit, whereas for the larger angles investigated this energy is 2 or 3 eV above the elastic-limit curve.

It is because the lowest excited electronic state of the N_2 molecule is 6.2 eV above the ground state that Q for peak A may be attributed solely to vibrorotational excitation. It is not so simple to explain the electronically excited inelastic peaks of Fig. 29. The collected data points for these peaks are displayed along curves B and C in Fig. 30. These curves show exactly the same slope as does curve A. Although it is possible that this slope is due to the excitation of higher-and-higher-lying electronic states as $E_0\theta^2$ is increased, Fernandez and co-workers argue that the lines being parallel indicate that the vibrorotational contribution to Q is the same for each of these three electronic excitation states. If this is correct, it would mean that there is negligible coupling between the mechanisms causing electronic excitation and the vibrorotational excitations. In a recent second-order Born approximation treatment of these collisions, Russek[86] does find the coupling between vibrational and electronic excitation to be small.

Figure 31 shows a similar ΔE versus $E_0\theta^2$ plot, this time for He^+-H_2 collisions obtained by Bray and co-workers.[52,87] Again, the upper and lower solid curves represent the expected values for elastic scattering either from an H atom ($\Delta E_{m=1}$) or from an H_2 molecule ($\Delta E_{m=2}$). Although Bray also observed electronic excitation the data points in Fig. 31 are only those corresponding to events for which electronic excitation did not take place. These data are of note because they clearly show where the onset of vibrorotational excitation occurs. Below values of $E_0\theta^2 \cong 6$ keV-deg^2, $Q = 0$ and the collision is elastic. For larger values of $E_0\theta^2$ the data depart from the elastic limit, and the difference between the data points and the $\Delta E_{m=2}$ curve represents vibrorotational excitation.

3.1.4. Inner-Shell Excitations

Investigations of more violent collisions, those in which the inner electron shells of heavy ions are made to interpenetrate, are also of great interest from the molecular point of view. Although it may be short-lived, a diatomic molecule is formed during an ion–atom collision. As long as the nuclear motion is slow compared with the pertinent electron velocities, the Born–Oppenheimer approximation is still valid and molecular states may be considered. Even though such a molecule may exist for less than 10^{-15} sec,

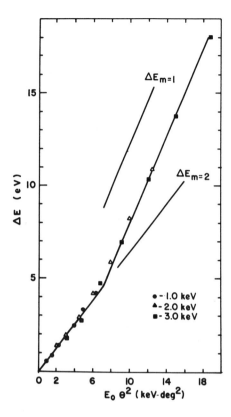

Fig. 31. The energy loss $\Delta E = E_0 - E_1$ plotted versus $E_0\theta^2$ for H^+-H_2 collisions which do not involve electronic excitation. $\Delta E_{m=1}$ corresponds to the binary limit and $\Delta E_{m=2}$ to the elastic limit.[52,87]

the radial and rotational coupling of the translational energy with the electronic energies (Section 2.2) can provide for excitation of electrons in the inner electron shells. When the resulting inner-shell vacancies are filled, usually following the collisions, characteristic x-rays and Auger electrons are emitted by the ions. Inner-shell excitations in single, heavy ion–atom collisions, were first observed by Morgan and Everhart,[88] then investigated more carefully by Afrosimov and co-workers[89] and Kessel and co-workers,[69,90] and finally explained in terms of MO promotions by Fano and Lichten.[7] Reviews of this early work[70] and later investigations of the related x-ray and Auger electron emission are available.[13,92,93] The following discussion is limited to K-shell excitations, first in ion–atom collisions and then ion–molecule collisions.

The straight-line correlation diagrams of Figs. 1 and 2 are suitable for a qualitative discussion of K-shell excitation in 50–500-keV collisions of combinations such as Ne^+-Ne, N^+-Ne, N^+-N_2, and the like. As an example of this, consider the application of Fig. 1 to Ne–Ne collisions. Prior to collision, the separated $1s$, $2s$, and $2p$ atomic levels of both atoms are completely filled. As the atoms approach each other the corresponding MOs are also completely filled. For a 1s, or K, electron to be excited or promoted during the collision, there must be a coupling between its MO and another one containing a hole into which the K electron might go. For Ne–Ne collisions there is no such hole and so K-shell promotion of this type is forbidden. On the other hand, for Ne^+–Ne collisions there is an initial $2p$ vacancy which will become a hole in one of the corresponding MOs. Lichten[7] first pointed out this relationship between the incoming ionization state and the probability for K-shell promotion and predicted that K-shell ionization should be twice as probable for Ne^{2+}–Ne collisions as the Ne^+–Ne collision. Inelastic energy-loss experiments allow these probabilities to be measured directly. From energy-loss spectra similar to those in Fig. 29, determination of the relative number of events contributing to each peak gives the probability for the specific excitation as a function of E_0 and θ. Figure 32 shows such K-shell ionization probabilities P_{II}, as a function of both $E_0\theta$ and R_0.[13,35] It is seen that P_{II} for the 300-keV Ne^{2+}–Ne collision is twice as large as that for the 300-keV Ne^+–Ne collision as predicted by Lichten.

Briggs and Macek have written a series of papers dealing with the production of K vacancies in symmetric ion–atom collisions.[94] They assume that the transition is due to a rotational coupling between the $2p\sigma$ and the $2p\pi$ MOs and they proceed in two steps: First they calculate the probability of a $2p$ vacancy in the separated atom limit becoming a vacancy in the $2p\pi$

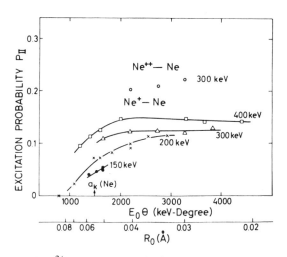

Fig. 32. Ne^+–Ne and Ne^{2+}–Ne: Probability (P_{II}) for producing one K-shell vacancy as a function of the reduced scattering angle $E_0\theta$. The K-shell radius for a neon atom $a_K(Ne)$ is indicated. The abscissa is also given in terms of R_0, the minimum distance of approach of the nuclei.[35]

MO. Next they perform a two-state perturbed stationary-state calculation of the rotational coupling and then calculate the resulting probability for the coupling to promote a $2p\sigma$ electron into the $2p\pi$ hole. Their results agree well with the Ne^+–Ne data of Fig. 32, but their N^+–N calculations do not agree so well with the measured probabilities for N^+–N_2 and N^+–NH_3 collisions.

For asymmetric collisions it is appropriate to use the correlation diagram of Fig. 2. The subscripts 1 and 2 in the separated-atom limit correspond to the higher-Z and lower-Z nucleus, respectively. Thus, the $1s_1$ level is more tightly bound than the $1s_2$ level. As a result of this, we see that such collisions are unlikely to promote any electrons from the K shell of the heavier atom. These $1s_1$ electrons become $1s\sigma$ electrons and this level would have difficulty interacting with any other MOs. The $1s_2$ electrons, on the other hand, enter the $2p\sigma$ MO and are eligible for promotion into the $2p\pi$ MO, providing other conditions are met. As in the symmetric case, these conditions are twofold. The rotational coupling must be sufficient and there must be a vacancy present in the $2p\pi$ MO. From this latter condition an important observation can be made: Since the $2p\pi$ MO correlates with the $2p$ level of the separated atom, K-shell ionizations, which are only likely for the lower-Z atoms, will not occur unless there is an initial $2p$ vacancy in the higher-Z atom. To a first approximation, experiments confirm these predictions. Figure 33 shows the probability for producing K vacancies in nitrogen in N^+–Ne, Ne^+–N_2, and Ne^+–NH_3 collisions.[13] Although there does exist a small probability for K excitation in N^+–Ne collisions, it is much smaller than for the Ne^+–N_2 and Ne^+–NH_3 collisions for which there is an initial $2p$ vacancy in the higher-Z collision partner. Auger electron and x-ray emission experiments have also confirmed these properties of asymmetric collisions for a wide range of ion–atom combinations. Fastrup and co-workers have also found a similar behavior for the case of L-shell excitations in Z_1–Ar collisions.[35]

In calculating the inelastic energy losses corresponding to the molecular target data in Fig. 33, it was apparent that the correct mass to use in Eq. (3.2) was the mass of the N atom and not the mass of the whole molecule. This was clear because the Q value has to be approximately equal to the K-binding energy of nitrogen (400 eV) and only by using the mass of a single N atom were values in the correct range obtained. However, this is not to say that the molecular structure of the target has no influence at all. Fastrup and Crone[35] have studied the influence of the molecular nature of the target on the probability of producing K vacancies in ion–molecule collisions, and some of their results are shown in Fig. 34. As in the atomic target case, the K electron is promoted from the lower-Z atom, but it is observed that when the higher-Z collision partner is part of a molecular target the promotion probability is affected. For the case of C^+ ions incident upon N_2 and NH_3

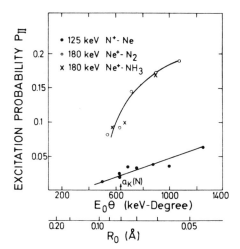

Fig. 33. N^+–Ne, Ne^+–N_2, and Ne^+–NH_3: Probabilities (P_{II}) for producing one K-shell vacancy in nitrogen at the same relative collision velocity $v = 1.3 \times 10^8$ cm/sec as a function of the reduced scattering angle $E_0\theta$ and R_0. The K-shell radius for a nitrogen atom $a_K(N)$ is indicated.[35]

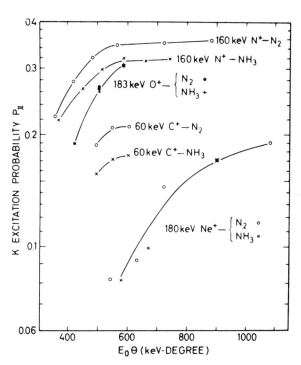

Fig. 34. The probability of producing a K vacancy in various ion–molecule collisions plotted as a function of the reduced scattering angle $E_0\theta$. Assuming an exponential screened coulomb potential, $E_0\theta = 600$ keV-deg corresponds to $R_0 = 0.058$ Å for C^+–N_2, NH_3; $R_0 = 0.063$ Å for N^+–N_2, NH_3; $R_0 = 0.071$ Å for O^+–N_2, NH_3; and $R_0 = 0.085$ Å for Ne^+–N_2, NH_3.[35]

molecular targets the effect was as large as 30%. The reason for this is unclear. However, since the promotion probability was shown to depend strongly on the number of vacancies in the $2p\pi$ MO for ion–atom collisions, it may be speculated that the molecular structure of the target somehow influences the number of $2p\pi$ vacancies available.

3.2. Ion–Photon and Ion–Electron Coincidence Measurements

3.2.1. Introduction

The past few years have seen refinements in methods of measuring inelastic energy losses that approach the "ideal experiment" referred to in Section 1. Measurements of the energy lost by an incident ion colliding with an atomic target in a differential scattering experiment give a great deal of information about collision dynamics and the excited states produced in inelastic collisions. However, these measurements are limited in energy resolution by the inherent energy spread of the beam and sometimes by considerations such as the experimental geometry, thermal motion of the target atoms, and broadening due to recoil of projectile or target following autoionization (Section 3.1). In the case of closely spaced or degenerate energy levels, ion–photon and ion–electron coincidence techniques have been developed which are capable of distinguishing between the states that are excited, achieving much higher energy resolution in practice than has been possible with conventional energy-loss spectroscopy.

With this coincidence technique, individual photons (or electrons) emitted as a result of excitation of a specific state of a target atom by the incident ion beam are detected in delayed coincidence with the corresponding scattered ions. Thus differential scattering information is preserved, with low-resolution optical (or electron) spectroscopy of the photons (or electrons) emitted after the collision serving as a substitute for higher-resolution energy analysis of the scattered ions. The basic idea is outlined in Fig. 35.

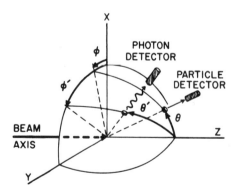

Fig. 35. Schematic diagram showing the variables in a scattered-particle–photon coincidence experiment. Primed and unprimed coordinates indicate the photon and particle detector positions.[101]

The coincidence techniques can differentiate between the two energy-degenerate inelastic processes of direct excitation and charge exchange in a symmetric collision such as He$^+$ on He, and can also be used to give information on angular distributions and polarization in differential scattering.[95]

In an inelastic scattering process, structure is sometimes observed in the total optical-excitation cross section. For He$^+$ on He collisions, the cross sections reach their maximum values only a few tens of electron volts above the inelastic threshold, in apparent violation of the Massey adiabatic criterion.[96] The *total* cross sections for each excited state reveal a rich oscillatory structure that is different for each L–S multiplet: Examples are shown in Fig. 36. For some time it was not clear why the oscillations were present in

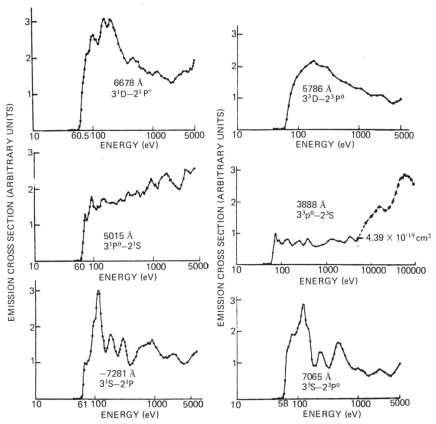

Fig. 36. Excitation functions for all the visible HeI lines originating from the $n = 3$ levels where He is bombarded by He$^+$ ions.[96] The cross section for the $3\,^3P$ state is normalized at 5 keV to the absolute data of de Heer, Wolterbeek-Muller, and Geballe [*Physica* **31**, 1745 (1965)]. Their value of the cross section rises to $\approx 7.7 \times 10^{-19}$ cm^2 at 90 keV. No cascade corrections have been applied.

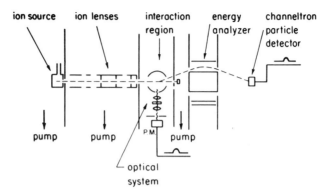

Fig. 37. Schematic diagram of the ion–photon coincidence apparatus used by Smick and Smith,[97] for studying $He^+ + He$ collisions.

the total cross section. Further, it was not possible to tell from the total excitation functions whether the structure in the cross sections came from the direct or charge-exchange collisions in this symmetric system. These observations provided the motivation for some of the first experiments designed to measure differential inelastic cross sections of optically excited states by the ion–photon coincidence method.[97]

3.2.2. Experimental Principles

The apparatus used by Smick and Smith at the University of Connecticut is illustrated schematically in Fig. 37. It is designed to have high efficiency for the collection of both the scattered ions and the emitted photons. A simple electron-bombardment ion source is used to form a He^+ beam,[98] which is focused and collimated by a differentially pumped system of electrostatic lenses. The collision region is filled with He gas at a pressure of at most 10^{-3} torr, to ensure single-collision events. Ions differentially scattered into a cone at a fixed angle of $11 \pm 0.5°$, as determined by a pair of annular slits coaxial with the ion beam, enter a coaxial cylindrical-mirror electrostatic energy analyzer. This low-resolution analyzer serves the dual purpose of refocusing the inelastically scattered He^+ ions back onto the beam axis for detection by an electron multiplier and also blocking the unwanted background of elastically scattered ions. The design of the cylindrical-mirror analyzer is similar to those discussed previously in the literature,[99], except for the fact that the entrance and exit slits for the ions are on the front and back flanges of the analyzer, rather than on the cylindrical surfaces as in Zashkvara's design.* The primary ion beam is

---

*Electric field end effects cause no problem in a low-resolution device. The analyzer need not even be designed for second-order angular focusing, though this would improve the resolution for any given acceptance angle.

collected on axis, after passing through the scattering chamber, by a Faraday cup with the inner cylinder of the ion-energy analyzer. Photons emitted in the collisions are counted by means of an approximately $f/0.7$ optical system that incorporates an infrared blocking filter, a narrow-band interference filter to isolate the helium spectral line of interest, and a cooled EMI 9558 photomultiplier. The optical system focuses the "collision volume" in the scattering chamber onto a mask in front of the photocathode of the photomultiplier. Two spherical mirrors are used, one containing a hole in the center to let the light through, so that it may be focused into a parallel beam by an auxiliary 16-mm movie-projection lens, before passing through the interference filter. A second projection lens is then used to focus the parallel beam onto the photomultiplier. It is important that the scattering volumes "seen" by the scattered-ion analyzer and by the photomultiplier coincide as far as possible, to assure a good signal-to-background ratio.

The electronic system for observing ion–photon coincidences is built entirely from commercially available "NIM" modules and is arranged as shown in Fig. 38. The use of a time-to-amplitude converter (TAC) makes it possible to use a multichannel pulse height analyzer to record the delayed coincidence spectrum: A sample spectrum is shown in Fig. 39. The multichannel analyzer can, if desired, be replaced by two less expensive single-channel analyzers (SCAs), one to record "true" coincidences by responding with an output pulse when photon and ion pulses are received with the correct delay time corresponding to the difference between the photon and

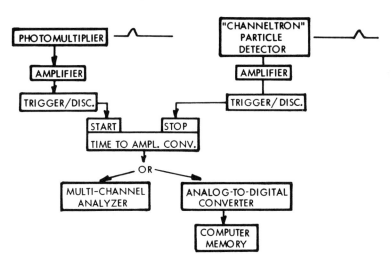

Fig. 38. Block diagram of electronic circuitry for detecting delayed coincidences between emitted photons and scattered ions.

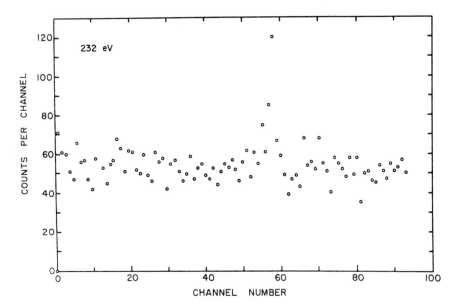

Fig. 39. Ion–photon coincidences plotted as a function of delay time for 232 eV $He^+ + He$ collisions. The photons are from the decay of the $3\,^3D$ state of the target at a laboratory scattering angle of 11.2°. The coincidence peak in channel 58 corresponds to a time delay of approximately 2.5 μsec. The data required several hours of accumulation time at a single beam energy.

ion flight times, and another SCA to record "chance" coincidences between ion and photon pulses that are not correlated. For true ion–photon coincidence events, the measured delay-time intervals will all fall in a small range about the expected delay time (typically a few microseconds). The delay-time spectrum for uncorrelated events will be essentially flat. The observed time-delay spectrum shown in Fig. 39 consists of the true events in the peak, superimposed on a background which must be subtracted from the coincidence data. The range of delay times (10–50 nsec, compared with an inherent electronic time resolution of a few nanoseconds) is determined by the spread in TOFs due to the finite size of the coincidence region, the spread in beam energy, and the mean decay time (negligible in this case) of the collisionally excited atoms. Chance coincidences (the background) are caused by the detection of ions subsequent to the detection of a photon but not from the same scattering event.

The true coincidence rate R_C [Eq. (3.15)] is proportional to the product of the differential inelastic cross section $d\sigma_j/d\Omega_i$ for exciting a particular optical state j, the ion flux I, the target gas particle density n, the effective length of the coincidence region along the beam L, the solid angle for detecting scattered ions $\Delta\Omega_i$, the photon detection efficiency $(\Delta\Omega_\gamma/4\pi)\eta_\gamma$, and the ion detection efficiency η_i. (Note that $\eta_i \simeq 1$ for a Channeltron electron multiplier if the ions are postaccelerated into the detector so that they arrive with energy > 1 keV.)

$$R_C = \frac{d\sigma_j}{d\Omega_i} InL \, \Delta\Omega_i \left(\frac{\Delta\Omega_\gamma}{4\pi}\right) \eta_\gamma \eta_i \qquad (3.15)$$

The random coincidence rate R_R [Eq. (3.16)] is proportional to the coincidence resolving time τ multiplied by the product of the *uncorrelated* ion and photon counting rates (R_i and R_γ, respectively):

$$R_R = R_i R_\gamma \tau \qquad (3.16)$$

Since each "singles" rate is proportional to the product of ion flux and gas density, random coincidences will be proportional to the *square* of this product, whereas true coincidences depend linearly on the nI product. In the limit of very large ion flux or gas density, chance coincidences will predominate over real ones. Thus, for a given geometry and collection efficiency, there will be an optimum nI product such that the coincidence rate is as large as possible without being swamped by the background. (It is desirable to be able to verify visually on the multichannel display that the apparatus is producing coincidences during a run.) Equation (3.16) indicates that, making use of $\sqrt{N}$ Poisson counting statistics, the signal-to-*noise* ratio (not the same as the *contrast*, or signal-to-*background* ratio) increases as the square root of the counting time t, in the absence of apparatus drifts:

$$\frac{S}{N} = \frac{R_C t}{(R_R t + R_C t)^{1/2}} \propto \sqrt{t} \qquad (3.17)$$

3.2.3. Charge-Exchange Measurements Using Coincidence Techniques

The use of scattered-ion–emitted-photon coincidence methods makes possible the clear distinction between direct and charge-exchange excitation in degenerate and near-degenerate cases where the channels cannot be resolved easily by energy-loss measurements alone. In the He^+ on He excitation already discussed, there is always a degeneracy between the direct and charge-exchange channels because of the charge-symmetry. For scattering angles much less than 45°, direct and exchange scattering can be approximately separated by whether the scattered (faster) particle is an ion or a neutral. The coincidence technique then makes it possible to "tag" an emitted photon with scattering of an ion ("direct" scattering) or a neutral ("charge-exchange" scattering). At low beam velocities, the coincidence technique may be the only practical way to distinguish these two processes. At higher beam velocities, they may be distinguished by looking at the Doppler shift of the emitted line radiation.[100]

The first successful application of the scattered-particle–photon coincidence technique we have been describing was not to direct collisional excitation but rather to investigate the differential cross section for charge-exchange excitation of the $2p$ state of hydrogen in $H^+ + He$ collisions. This work, by Jaecks et al.,[101] uses coincidences to distinguish between the charge-exchange excitation process yielding $H(2p) + He^+$ and the direct excitation process yielding $H^+ + He(2s, sp)$. These two channels involve nearly the same inelastic energy loss (within 1 eV) and so are very difficult to distinguish without coincidence, for incident protons in the multi-kiloelectron volt energy range. This experiment differs from those previously described in that neutral scattered particles are detected instead of ions.*

Lyman-α photon-counting in the experiment of Jaecks et al.[101] was accomplished by using a "solar-blind" photomultiplier (insensitive to the large light flux in the visible region from the collision chamber) equipped with a lithium fluoride ultraviolet transmitting window. The photon detector was placed close to the scattering center for high efficiency. Even though the particle detector looked at small-angle scattering (2–3°), the geometrical and photon-counting efficiency factors are so low that, with target-gas pressures in the millitorr range and 10^{-8}-A beam currents, typical coincidence counting rates are 20 and 200 hr^{-1}, for real and chance (random) coincidences, respectively.

These long counting times represent the major disadvantage of the ion–photon coincidence method: Usually the emitted photons have a broad (nearly isotropic) angular distribution which, coupled with the low quantum efficiency of most photomultipliers in the visible, results in very long counting times to attain reasonable statistics. Since most apparatus will not work unattended for very long periods without drift, these coincidence methods are worth using only for critical experiments, particularly where other methods such as total cross-section measurements, conventional spectroscopy, and conventional scattered-particle energy-loss measurements are inadequate. Usually it is wise to apply these simpler methods first before beginning a detailed coincidence study on any given collision system.

Figure 40, from Jaecks et al.,[101] shows both the differential cross section for charge-transfer excitation to the $2p$ state in the $H^+ + He$ case and the

---

*Counting of the neutral particles is accomplished with a 13-stage open electron multiplier with Cu–Be dynodes. Jaecks et al. measured the efficiency of this neutral detector as a part of the experiment. It was found to be $\sim 15\%$ for 15-keV ground-state hydrogen atoms, by measuring coincidences between the neutral detector and Lyman α photons from either $H(2p)$ or electric-field-quenched $H(2s)$ produced by collisions. This detection efficiency is considerably less than that quoted in the literature for Channeltron detectors, probably because of the sensitivity of the secondary-electron coefficient of open Cu–Be multipliers to vacuum system contaminants and to past history.

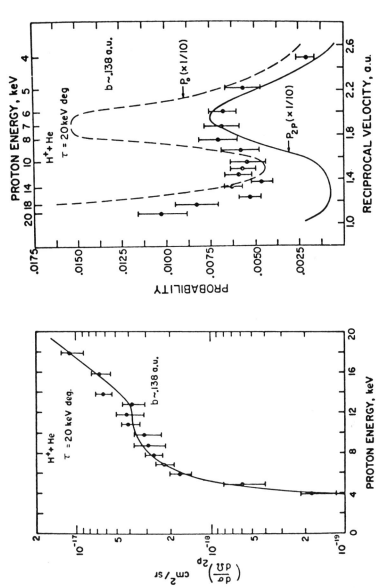

Fig. 40. Differential cross section ($d\sigma/d\Omega$) and probabilities for forming H($2p$) by charge transfer in H$^+$ + He collisions.[101] $\tau = E_0\theta$.

experimental probability of charge-transfer excitation to the $2p$ state, with the impact parameter b held constant. For comparison, the theoretical calculations of Sin Fai Lam[102] at the same impact parameter are shown in the curve marked P_{2p}. Although the phase and shape of the calculated curve agree with the experimental points fairly well, the calculation overestimates the observed probability by a factor of nearly 10. This is probably because these impact-parameter method calculations[102] did not include the nearly degenerate $H^+ + He(2\,^1S, 2\,^1P)$ channels, which should be strongly coupled to the observed charge transfer channel.

Jaecks' group at the University of Nebraska have applied their technique to other measurements, e.g., $H^+ + Ne \rightarrow H(1s, 2s, 2p) + Ne^+$.[103,104] These coincidence data shed light on whether the charge-transfer $2p$ excitation of the projectile occurs predominantly through a "direct" coupling between initial and final states, or through an intermediate channel involving charge transfer to the ground state of the hydrogen projectile, followed by excitation to the $2p$ state, in a two-step process. These mechanisms are distinct in the context of a second-order time-dependent perturbation theory calculation. When the two-step process is dominant, it can be shown that the probability for $H(2p)$ formation in the collision is the product $(\frac{1}{4})P_{0\rightarrow 1}P_{1\rightarrow 2}$, where the states 0, 1, and 2 are $(H^+-Ne$, elastic), $(H(1s) + Ne^+)$ and $(H(2p) + Ne^+)$, respectively. In the $H^+ + He$ case it was found that the two-step process dominated whereas in the $H^+ + Ne$ case the direct and the two-step processes have approximately equal importance.

The coincidence experiments involving scattered particles and emitted photons are designed to probe the impact-parameter dependence of excitation and charge-transfer processes. As we have indicated, these experiments give information similar to the kinds obtained from the inelastic energy-loss measurements discussed in Section 3.1, usually with higher resolution offset by longer counting times. The experiments so far described by no means exhaust the collision parameters accessible to a "complete" differential measurement. Optical polarization measurements in coincidence with scattered particles can given information on the partial cross sections for populating the various magnetic sublevels (M_L or M_J) of an excited state, with the impact-parameter dependence specified as well. A calculation of the final state wave function would specify the relative *phases* between coherently excited sublevel amplitudes—information also accessible through polarization-sensitive coincidence measurements.[95]

3.2.4. Stueckelberg Oscillations

Coincidence data, normalized to the number of emitted photons, are presented for the excitation of the $3\,^3D$ state of helium in He^+ on He collisions for a fixed angle of $11.5 \pm 0.5°$ as a function of $1/v^*$ (inverse beam velocity in the excited state) in Fig. 41. The total excitation function for the same state (total emission cross section) is shown for comparison in the inset, as a function of beam energy. The data show qualitatively the expected drop-off of the differential cross section at high energy and, in addition, striking oscillations of nearly constant frequency in $1/v^*$. The oscillation frequency is thus proportional to a characteristic "time of collision" over a considerable range. The high contrast (peak-to-valley ratio) of these oscillations is typical of an essentially two-state quantum-mechanical interference effect known as Stueckelberg oscillations (cf. p. 180).

Stueckelberg oscillations can occur in inelastic scattering whenever excitation occurs predominantly over a small range of internuclear distances via coupling between the ground state of a colliding ion–atom system and an electronically excited state of the quasi-molecular system, i.e., at a crossing of two "diabatic" molecular potential curves, already discussed in Section 2. The oscillations arise from the interference between two different scattering "trajectories" for two different impact parameters in the collision, both of which lead to the same final state of the system (the same excited state and scattering angle).[105] In inelastic trajectory II (Fig. 42), a transition from molecular potential curve V_1 to excited-state curve V_2 occurs at crossing distance R_x on the way in as the nuclei approach each other; the system then remains in state V_2 as the crossing point is traversed a second time while the nuclei separate from each other. The interfering inelastic trajectory I is one in which the transition from V_1 to the same final curve V_2 occurs on the way out instead of on the way in. The semiclassical phase ϕ (or classical action integral) differs for these two trajectories:

$$\phi_{I,II} = (2\mu)^{1/2}\hbar^{-1} \int_{R_t(V_{1,2})}^{R_x} \left[E - V_{1,2}(R) - \frac{Eb^2}{R^2} \right]^{1/2} dR \qquad (3.18)$$

where μ is the reduced mass and E is the center-of-mass energy. The impact parameter b is related to the angular momentum l by $b = \hbar(l+\frac{1}{2})$ and R_t is the classical turning point on curve V_1 or V_2, where the integrand goes to zero. The phase difference that shows up in the oscillations of the differential inelastic cross section to a particular excited state, such as the $3\,^3D$ of Fig. 41, is proportional to the difference between the classical action integrals [Eq. (3.18)] for the two alternative trajectories I and II. The action integrals can be calculated from the two potential curves if they are known, or can be used to find the parameters of one of the curves (say V_2) by fitting the data, when the other curve (V_1) is known.[107] Energy-loss data showing similar Stueckelberg oscillations in $He^+ + Ne$ inelastic scattering are shown in Fig. 17.

The methods of Olson and Smith[107] can be applied to the analysis of Stueckelberg oscillation patterns, whether from coincidence or energy-loss

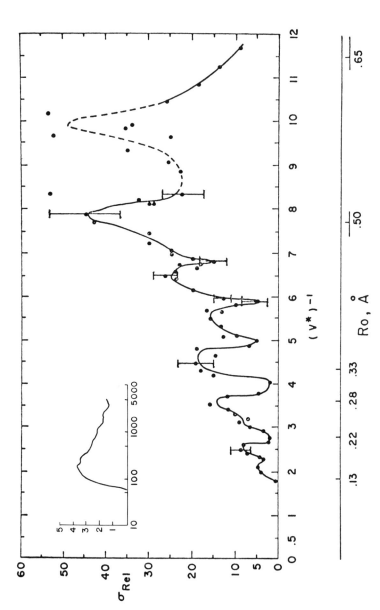

Fig. 41. Relative differential inelastic cross section at 11.2° (laboratory angle) for excitation of the $3\,^3D$ state of helium in $He^+ + He$ collisions. The abscissa $(v^*)^{-1}$ is a scaled reciprocal velocity parameter proportional to a characteristic collision time: $v^* = (E_0 - 80)^{1/2}/79$, where E_0 is the beam energy in the laboratory in electron volts, and R_0 is the approximate distance of closest approach for the collision, calculated from a screened coulomb potential. Data have not been corrected for gradual changes in ion analyzer transmission with energy. This curve represents approximately 1400 h of accumulation time (data of Smick and Smith[97]). The inset shows the energy dependence of the $3\,^3D$ total emission cross section.

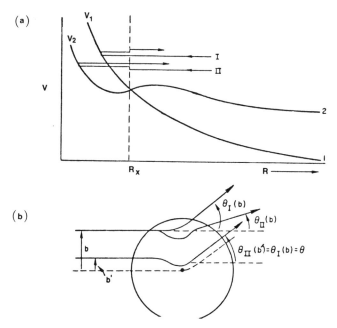

Fig. 42. Origin of Stueckelberg oscillations in the differential inelastic cross section.[105] (a) Two quasi-molecular potential curves V_1 and V_2 with a curve crossing at internuclear distance R_x. The two most probable inelastic paths are designated I and II. (b) The scattering trajectories I and II correspond to two different parameters b and b'. Interference occurs when trajectories I and II lead to the same scattering angle.

measurements. Ideally, one would like to start with a complete experimental contour map of maxima and minima of the differential cross section in the E_0–θ (energy-scattering angle) plane. Several such contours are plotted from inelastic energy-loss data on the excitation of the $2\,^3S$ state of He in $He^+ + He$ collisions in Fig. 43. Shown for comparison are maxima and minima of the $3\,^3D$ differential cross section from the much more time-consuming coincidence measurements of Fig. 41.

From these data information can be obtained about the shape of the molecular potential curves for two unbound excited states of the He_2^+ quasi-molecule, the states which go over into $He^+ + He(2\,^3S)$ and $He^+ + He(3\,^3D)$, respectively, at infinite internuclear separation. Figure 44 shows the corresponding diabatic potential curves for the elastic and two inelastic $He^+ + He$ channels. Because the available coincidence data for $3\,^3D$ are limited to one fixed laboratory scattering angle, the potential curve V_3 for this state is not determined uniquely but is a fit to a similar function to that

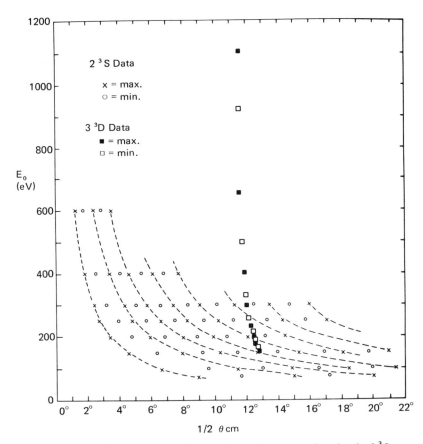

Fig. 43. Maxima and minima in the differential inelastic cross sections for the $2\,^3S$ state of helium and the $3\,^3D$ state in $He^+ + He$ collisions, plotted in the E_0–θ planes. [$2\,^3S$ data from D. C. Lorents, W. Aberth, and V. Hesterman, *Phys. Rev. Lett.* **17**, 849 (1966); $3\,^3D$ data from Fig. 41.]

for the potential curves V_2 and V_4, which serve as lower and upper bounds to the curve V_3. The analysis which leads to these potentials will break down at both low energies (below the inelastic threshold—adiabatic Born–Oppenheimer potentials will then be valid) and at high energies and scattering angles. The breakdown at high energies is due to breakdown of both semiclassical scattering theory and of the simple two-channel analysis using well-defined diabatic states. Nevertheless, there is a wide range of E_0 and θ over which these excited-state potential curves are useful in describing the scattering data. For example, the curve V_3 for the $3\,^3D$ state is useful from a few tens of volts above the inelastic threshold of 46 eV to $E_0\theta \simeq 4$ keV-deg (with E_0 in the kiloelectronvolt range). For more violent collisions than about 4 keV-deg, the potential V_3 no longer suffices to describe the excitation of the $3\,^3D$ state, probably because of the onset of new excitation channels through rotational coupling.

3.2.5. Rosenthal Oscillations and Long-Range Couplings

The explanation for the interference effects in the *total* excitation cross sections of Fig. 36 involves more than the two-state interference between

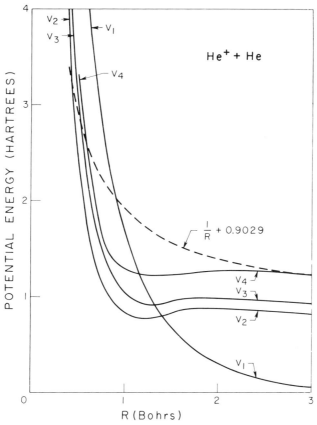

Fig. 44. Approximate diabatic potential curves for three unbound scattering states of the $He^+ + He$ system and for $He^+ + He^+$. Curves V_1, V_2, and V_3 are optimized functions of the form $V(R) = (4/R + A)\exp(-BR) + C$. V_1 is Olson's estimated lowest-energy curve of $^2\Sigma_g$ symmetry.[107] V_2 is the next highest diabatic $^2\Sigma_g$ curve, based on fits to energy-loss data for the excitation process $He^+ + He \rightarrow He^+ + He^*(2\,^3S)$.[107] V_3 is a good fit to the ion–photon coincidence data of Fig. 41 for direct excitation of $He^*(3\,^3D)$ from 0.9 to ~5 keV-deg.[97] The Rydberg series limit, i.e., the ground-state (adiabatic) potential curve for He_2^{2+}, is also shown [from Kolos and Roothaan, *Rev. Mod. Phys.* **32**, 219 (1960)] as V_4, together with a coulomb potential (dashed) leading to $He^+ + He^+$. Parameters: $V_1(A = 2.4, B = 1.33, C = 0)$; V_2 $[A = -3.2284, B = 2.03, C = 0.7284 + 0.18 \exp(-\alpha(R - 1.7)2)]$; V_3 $[A = -3.0, B = 1.63, C = 0.8484 + 0.18 \exp(-\alpha(R - 1.7)2)]$; $\alpha = 6.5$ if $R < 1.7$ and $\alpha = 0.325$ if $R \geq 1.7$.

collision trajectories found in the Stueckelberg oscillations. It has been shown by Rosenthal[109] that the oscillations in the total cross section in He$^+$ excitation of He are due to a long-range coupling between at least two *excited* states of the He$_2^+$ molecular system at internuclear separations of 20–50 Bohr radii, through the Stark effect and associated exchange interactions. The incoming ion interacts strongly with the highly polarizable target atom once it is in an excited state. These long-range interactions produce collisional coupling among at least three molecular potential curves and give rise to the possibility of several interfering inelastic trajectories and semiclassical phases rather than just two. Some of these will depend weakly on impact parameter and will produce oscillations in the total as well as the differential cross section.[110]

An elegant analysis of the long-range coupling between the 4 ^1S and 4 ^3S states of helium, based on ion–photon coincidence data for He$^+$ + He scattering, has been given by Rahmat and co-workers.[111] Figure 45, adapted from this work, shows a *double* oscillation with scattering angle of the

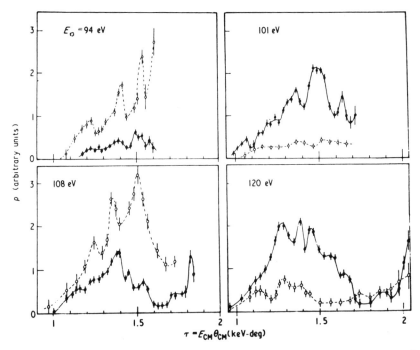

Fig. 45. Differential cross section (ρ) plotted against reduced scattering angle (τ) for 4 ^1S (dashed curve) and 4 ^3S excitation (solid curve) in He$^+$ + He collisions. The impact energies E_0 (lab) are such that the 4 ^1S total cross section is either a maximum (94, 108 eV) or a minimum (101, 120 eV). Arbitrary units are the same for all curves.

Inelastic Energy Loss 213

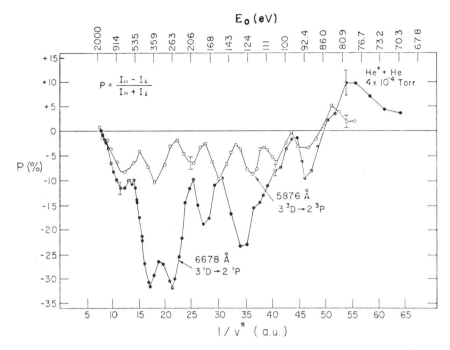

Fig. 46. Linear polarization P (with respect to ion beam direction) of light emitted at 90° from $3\,^1D$ and $3\,^3D$ states of He in $He^+ + He$ collisions as a function of $1/v^*$ and of beam energy E_0 in electron volts. Here v^* is the velocity (in atomic units) of the collision partners at large separation after excitation.[113]

differential cross section for both the $4\,^3S$ and $4\,^1S$ excitation: a rapid oscillation modulated with a slower oscillation.

From the fact that the frequency of the slower oscillation (in angle) agrees closely with the frequency of the Stueckelberg oscillations of the $n = 2$ and $n = 3$ states, Baudon and co-workers have concluded that the rapid oscillations have another origin. They are attributed to the effect of nuclear symmetry[112]: the indistinguishability of projectile and target for $^4He^+ + ^4He$ scattering, and would presumably disappear in the ion–photon coincidence data if $^3He^+$ were used on a ^4He target.

Long-range Rosenthal coupling between the $4\,^3S$ and $4\,^1S$ states is detected through the antiphase oscillations in the energy dependence of the *total* cross sections for the two states, but this effect is difficult to detect in the differential cross sections because of the complexity of the other interference effects present.[111] In some cases, the effects of long-range couplings between Rydberg-type states of the quasi molecule may not even be readily apparent in the total cross section but may be brought out by looking at the *polarization* of the emitted light as a function of collision energy. Strong oscillations in the polarization not easily seen in the total cross section have recently been observed for the $3\,^3D$ state of helium excited by He^+ (Fig. 46).[113]

3.2.6. Polarization and Angular Distribution Measurements in Coincidence

Some of the most recent ion–photon coincidence experiments have included linear polarization or angular distribution measurements on photons emitted from collisions. Figure 47, taken from a paper of Vassilev and co-workers,[114] shows how the rate for observing photons (from the transition $3\,^3P \to 2\,^3S$ in $He^+ + He$ collisions in coincidence with ions scattered at a fixed angle of 13.5°) varies with polarizer angle α. The ion beam direction is taken as $\alpha = 0$. Figure 48 shows the apparatus. The presence of strong polarization indicates that the excited atomic state is produced in the collision with some "alignment" or "orientation" with respect to the beam

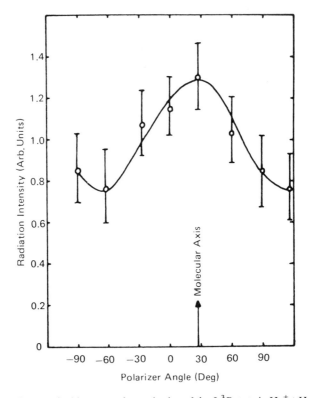

Fig. 47. Ion–photon coincidence rate for excitation of the $3\,^3P$ state in $He^+ + He$ collisions as a function of polarizer angle. Laboratory energy $E_0 = 150$ eV, lab scattering angle 13.5°.[114]

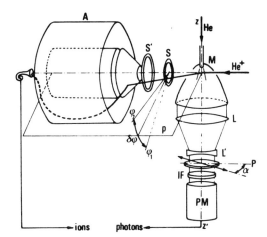

Fig. 48. Ion–photon coincidence apparatus of Vassilev et al.[114] with provision for detection of the linear polarization of the coincidence light. M, parabolic mirror; p, collision plane; LL', quartz lenses; IF, interference filter; P, polarizer. The ion scattering direction is defined by slits SS'. A is a cylindrical mirror energy analyzer.

direction. This means that the various degenerate magnetic substates labeled by the quantum number m_J are unequally populated[95] (where the axis of quantization is the final internuclear axis of the quasi-molecule). This unequal population, in turn, reflects preferential population of the various electronic angular momentum states of the diatomic quasi-molecule: Σ-, Π-, and Δ-symmetry states for example. Thus, polarization measurements, with or without coincidence, are of great help in deducing the angular-momentum quantum numbers of the specific alternative molecular states that may be involved in the inelastic scattering. This kind of detailed angular information on the excited states produced is not available in a conventional energy-loss measurement that looks only at the scattered and/or recoiling heavy particles (cf. the discussion on coincidence techniques in Chapter 6).

Extensive and very time-consuming polarization measurements in coincidence have been made by Jaecks et al.[115] for 3 keV He$^+$ + He collisions leading to He($3\,^3P$) excitation in a charge-transfer process, where the fast neutral He atom emerging at a small, known angle to the incident beam is detected and counted in coincidence with 3889-Å photons from the $3\,^3P$ excitation. These results show that at certain scattering angles, the excitation of the $3\,^3P$ state proceeds via a coherent superposition of Σ and Π symmetry states of the quasi molecule, with an approximately constant phase difference of 90° between the two amplitudes, independent of impact parameter. They interpret this interesting result as being due to a localized radial coupling between the $1s\sigma_g(2p\sigma_u)^2\,^2\Sigma_g$ and the $(1s\sigma_g)^2\,4d\sigma_g\,^2\Sigma_g$ molecular-orbital configurations of the He$_2^+$ molecule in the collision. On the plausible assumption that the $4d\sigma_g$ MO maintains the *space-fixed* orientation it had at the time of its impulsive excitation at the curve crossing, a coherent superposition of $4d\sigma$ and $4d\pi$ MOs with a 90° phase difference is automatically produced in the *body-fixed* molecular frame due to rotation of the internuclear axis, as the collision partners separate. This model for rotational coupling is analogous to that used by Briggs and Macek[94] to describe the coupling between $2p\sigma$–$2p\pi$ MOs, leading to K-shell vacancy production (and K x-rays) in $Ne^+ + Ne$ collisions (see Section 3.1.4 on inner-shell vacancies).

When an ion–photon coincidence experiment involves a noble-gas atom or ion as one of the collision partners, the emitted photon is often in the extreme ultraviolet range of the spectrum. In this case there is no simple way to measure the polarization of the emitted photons, but, in analogy with the practice of nuclear physics, one can get the same kind of information as one would derive from a polarization measurement by measuring the *angular distribution* of the emitted light relative to the ion beam axis. Vassilev *et al.* have done this for the excitation of the $2\,^1P$ state of helium in $He^+ + He$ collisions which results in emission of the 584 Å vacuum ultraviolet helium resonance line, by using the apparatus illustrated in Fig. 49.[114] A Channeltron electron multiplier serves as an excellent UV photon counter if there is no possibility of ions entering the detector and giving spurious counts. No wavelength selection was needed in this experiment because the ion-energy analyzer had sufficient resolution to separate the energy-loss peaks corresponding to $n = 2$ from those with $n \geq 3$. Thus any UV photons emitted in coincidence with scattered particles in the $n = 2$ energy-loss peak must necessarily correspond to $2\,^1P$ excitation.

As we have seen in Section 3.1.4, excitation of *inner-shell* vacancies in ion–atom collisions at low-keV energies and up appears to be well described as a molecular phenomenon. Many qualitative insights as to the most important inelastic processes can be gleaned by using molecular–orbital correlation diagrams and the electron-promotion model.[12] The sharp onset of inelastic energy losses at certain values of $\tau = E_0\theta$ makes it desirable to

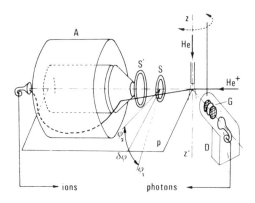

Fig. 49. Experimental arrangement for measuring the angular correlation of photons emitted from the $2\,^1P$ state of He (584 Å resonance line) with the scattered ion direction in coincidence.[114] The optics of Fig. 48 are replaced by an electron multiplier (UV detector) with grids G to prevent the detection of charged particles.

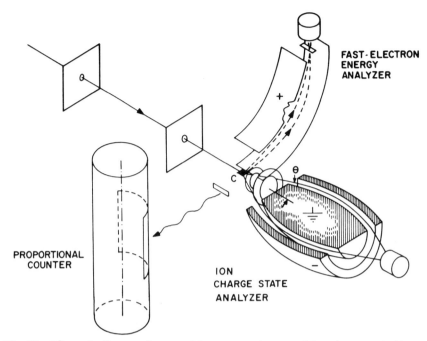

Fig. 50. Schematic diagram of scattered-ion–x-ray and scattered-ion–electron coincidence apparatus. Collisions occur at C.[116,117]

supplement energy-loss measurements with the use of scattered-ion–photon coincidence measurements so that the specific excitations can be identified spectroscopically as a function of impact parameter. In such cases, the photons are in the x-ray region of the spectrum[116] and are emitted in competition with Auger (autoionization) electrons.[117] Except for the details of the detectors used, ion–photon and ion–electron coincidence experiments used to study the details of inner-shell excitation processes are, in principle, similar to the type of coincidence experiments we have just described for studying the impact-parameter dependence of outer-shell excitations. Figure 50 shows an apparatus used to study x-ray or Auger electron emission in coincidence with scattered ions that have produced an inner-shell vacancy in a collision. Typical data showing the dependence of coincidence x-ray yield on scattered-ion final-charge state are illustrated in Fig. 51. A more detailed review of this field can be found in the recent article on ion-induced x-rays by Smith and Kessel.[93]

3.2.7. Ion–Molecule Coincidence Measurements

Although coincidence techniques have been used for the most part to study ion–atom collisions, there has recently been some interest in using

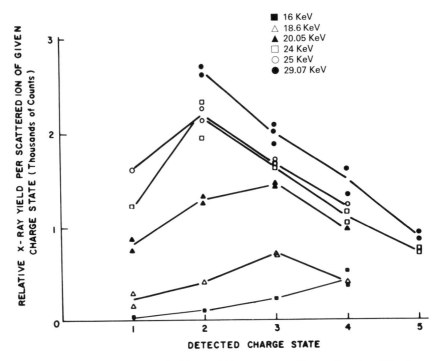

Fig. 51. Relative Ar LM x-ray coincidence yields per scattered ion as a function of the charge state of the ion, at $\theta_{lab} = 20 \pm 2°$ for Ar^+–Ar collisions. Statistical errors are less than the size of the points (integration time 1—20 hr/point). From Thoe and Smith.[116]

these methods for investigation of inelastic ion–diatomic-molecule collisions. This section concludes with a brief discussion of this largely unexplored research frontier.

Preliminary investigations have been made of the dissociative excitation of the simplest diatomic molecule H_2^+ in collisions with the rare gases in the low kiloelectron volt energy range[118] (cf. Chapter 6, Section 4). The simplest "direct" dissociative excitation process proceeds via the reaction

$$H_2^+ + A \rightarrow H_2^{+*} + A \rightarrow H(2p) + H^+ + A \quad (3.19)$$

which occurs in competition with dissociative excitation via charge transfer:

$$H_2^+ + A \rightarrow H_2^* + A^+ \rightarrow H(2p) + H + A^+ \quad (3.20)$$

The observation of coincidences between the hydrogen Lyman α photons (from decay of the $2p$ state) and scattered protons *or* neutral H atoms serves to distinguish the direct and charge-transfer reactions. Furthermore, the angular distribution of the molecular fragments from the collision can be measured in the coincidence experiment. This depends on the orientation of the H_2^+ internuclear axis. In the experiment described, the velocity of the H_2^+ ions was sufficient to freeze the rotational and vibrational motion during the collision. The conclusion from this work is that in H_2^+ + He collisions at 10 keV there is very little charge transfer, whereas in H_2^+ + Ar the charge-transfer process is approximately as likely to occur as the direct dissociation process.

Another type of coincidence experiment, involving total inelastic cross sections rather than differential scattering, is the photon–photon coincidence experiment of Young et al.[119]:

$$H^+ + N_2(X\ ^1\Sigma_g^+, \nu = 0) \rightarrow H(3p, 3d) + N_2^+(B\ ^2\Sigma_u^+, \nu = 0) \quad (3.21)$$

The coincidence rate between the 6563-Å Balmer α photon from excitation of the H(3p) and H(3d) states and the 3914-Å photon from excitation of the N_2^+ B state, is proportional to the total cross section for simultaneous charge transfer to the excited hydrogen with N_2^+ excitation. The energy dependence of this cross section was measured and showed a pronounced peak at a proton energy of $\sim$7.5 keV.

Another type of experiment, again involving the $H^+ + N_2$ collision, measures the 3914-Å photons from the decay of the B state of N_2^+ in coincidence with either forward-scattered protons or H atoms.[120] If the angular acceptance is a few degrees for the forward-scattered particles, almost all the scattering events making up the total cross section can be included. Thus, one has the capability of comparing the total cross sections for two processes: (a) charge transfer with excitation of N_2^+, and (b) ionization with excitation of N_2^+. This experiment would not work well much below a few hundred electron volts because of the difficulty in detecting the neutral particle by secondary electron emission.

4. Summary and Conclusion

Energy-loss measurements in ion–atom and ion–molecule collisions have become a widely used technique in recent years for investigating inelastic heavy-particle collision processes and reactions. These studies have been made over a wide range of energies from the inelastic thresholds of a few electron volts (for electronic excitation of valence electrons), up into the MeV range (where inner-shell excitation and x-ray production are of interest). Measurements in which the angle of scattering or the collisional impact parameter is specified are time-consuming but allow a much more critical comparison with theoretical collision models than do total cross-section measurements, which represent an integral over all angles of scattering.

The techniques described may be used in some cases for neutral atom–atom and atom–molecule collisions as well, provided that at least some of the collision products are charged or that the scattered neutrals are energetic enough to permit detection by secondary electron ejection at a surface. Energy loss of neutral scattered particles may be measured by a time-of-flight analysis of a pulsed scattered beam. Energy loss in ion–atom collisions is measured either by determining the "opening angle" between the scattered ion and the recoiling target in a delayed coincidence measurement, *or* by using an electrostatic energy analyzer to determine the energy of the scattered particles at some particular angle. Related spectroscopic techniques that reveal the dependence of deexcitation energies on impact parameter or scattering angle include the various scattered-ion–emitted-photon (or electron) delayed coincidence methods, which are limited by low counting rates but are inherently capable of very high resolution. The latter measurements can give very detailed information about collision dynamics especially when polarization or angular correlation measurements on the emitted particles are included.

The most widely used theoretical model for low-velocity inelastic collisions of ions heavier than H^+ is the molecular-orbital model. This is based on two main principles: (a) the Born–Oppenheimer separation of the nuclear from the electronic motion (valid for moderately low velocities), and (b) the approximate independence of the various electronic orbitals of the diatomic quasi-molecule formed during collisions (most nearly valid for inner-shell electrons and highly excited "Rydberg" electrons, but to some extent for valence electrons as well). The last assumption leads to the concept of "diabatic" molecular states, whose wave functions for many purposes may be considered as represented by antisymmetrized products of orthogonal H_2^+-like MOs. Diabatic state correlation diagrams, showing the continuous connection between molecular state electronic energies in the separated-atom and united-atom limits, allow at least qualitative inferences to be made about the distances of closest approach of the two nuclei at which inelastic processes are likely to occur. Inelastic transitions tend to occur in heavy ion collisions, by radial or rotational couplings, when there are crossings or near-crossings of two molecular potential curves. These transitions occur only to unoccupied orbitals, as specified by the Pauli exclusion principle.

Heavy particle collision studies are of practical importance in a number of applications. These include the physics and chemistry of upper-atmosphere and auroral processes, high-temperature gas and plasma properties, stellar atmospheres, gas laser excitation mechanisms, and the analysis of thin films and solid surfaces. Some of the techniques we have described (e.g., time-of-flight energy-loss spectroscopy) can probably be extended to studies of chemical reaction mechanisms at lower energy, using beams from chemical accelerators.

ACKNOWLEDGMENT

Work by Dr. Kessel was supported by the National Science Foundation. Work by Drs. Pollack and Smith was supported by the U.S. Army Research Office, Durham.

References

1. L. B. Loeb, *Science* **66**, 627 (1927).
2. O. Beeck and J. C. Mouzon, *Phys. Rev.* **38**, 969 (1931).
3. O. Beeck and J. C. Mouzon, *Ann. Phys. (Leipzig)* **11**, 737 (1931).
4. W. Weizel and O. Beeck, *Z. Phys.* **76**, 250 (1932).
5. M. Born and R. Oppenheimer, *Ann. Phys.* **84**, 457 (1927).
6. G. Herzberg, *Spectra of Diatomic Molecules*, van Nostrand–Reinhold, Princeton, New Jersey, 1950, Chapter 6.
7. W. Lichten, *Phys. Rev.* **164**, 131 (1967); see also U. Fano and W. Lichten, *Phys. Rev. Lett.* **14**, 627 (1965).
8. P. M. Morse and E. C. G. Stueckelberg, *Phys. Rev.* **33**, 932 (1929).
9. F. Hund, *Z. Phys.* **51**, 759 (1928).
10. J. Von Neumann and E. P. Wigner, *Phys. Z.* **30**, 467 (1929).
11. W. Lichten, *Phys. Rev.* **131**, 229 (1963).
12. M. Barat and W. Lichten, *Phys. Rev. A* **6**, 211 (1972).
13. Q. Kessel and B. Fastrup, *Case Studies in Atomic Physics* **3**, 137 (1972).
14. D. R. Bates, K. Ledsham, and A. L. Stewart, *Phil. Trans. R. Soc. (London)* **246**, 215 (1953).
15. F. P. Larkins, *J. Phys. B: Proc. Phys. Soc. (London)* **5**, 571 (1972).
16. R. S. Mulliken, *Chem. Phys. Lett.* **14**, 137 (1972).
17. J. S. Briggs and M. R. Hayns, *J. Phys. B: Proc. Phys. Soc. (London)* **6**, 514 (1972).
18. J. Eichler and U. Wille, *Phys. Rev. Lett.* **33**, 56 (1974); J. Eichler and U. Wille, *Phys. Rev. A* **11**, 1973 (1975).
19. J. N. Bardsley, *Phys. Rev. A* **3**, 1317 (1971).
20. E. W. Thulstrup and H. Johansen, *Phys. Rev. A* **6**, 206 (1972).
21. M. Krauss, NBS Technical Note 438 (Dec. 1967), U.S. Govt. Printing Office, Washington, D.C.
22. L. Landau, *Phys. Z. Sowjetunion* **2**, 46 (1932).
23. C. Zener, *Proc. R. Soc. A* **137**, 696 (1932).
24. H. S. W. Massey and E. H. S. Burhop, *Electronic and Ionic Impact Phenomena*: Vol. III, *Slow Collisions of Heavy Particles*, Oxford Univ. Press, London and New York, 1971.
25. E. W. McDaniel, *Collision Phenomena in Ionized Gases*, Wiley, New York, 1964, p. 241.
26. P. J. Martin and D. H. Jaecks, *Phys. Rev. A* **8**, 2429 (1973).
27. D. R. Bates, *Proc. R. Soc. (London) A* **240**, 437 (1957).
28. D. R. Bates, *Proc. R. Soc. (London) A* **243**, 15 (1957).
29. D. R. Bates, *Proc. R. Soc. (London) A* **245**. 299 (1958).
30. A. Russek, *Phys. Rev. A* **4**, 1918 (1971).
31. R. N. Olsen, *Phys. Rev. A* **6**, 1822 (1972).

32. Y. Demkov, *Sov. Phys. JETP* **18**, 138 (1964).
33. J. Perel and H. L. Daley, *Phys. Rev. A* **4**, 162 (1971).
34. V. Sidis and H. Lefebvre-Brion, *J. Phys. B* **4**, 1040 (1971); J. C. Brenot, D. Dhuicq, J. P. Gauyacq, J. Pommier, V. Sidis, M. Barat, and E. Pollack, *Phys. Rev. A* **11**, 1245 (1975); 1933 (1975).
35. B. Fastrup, G. Hermann, and K. J. Smith, *Phys. Rev. A* **3**, 1591 (1971); B. Fastrup and A. Crone, *Phys. Rev. Lett.* **29**, 825 (1972); B. Fastrup, G. Hermann, and Q. C. Kessel, *ibid.* **27**, 771 (1971).
36. S. Nagy, W. Savola, and E. Pollack, *Phys. Rev.* **177**, 71 (1969).
37. S. Nagy, S. Fernandez, and E. Pollack, *Phys. Rev. A* **3**, 280 (1971).
38. W. Savola, F. J. Erikson, and E. Pollack, *Phys. Rev. A* **7**, 932 (1973).
39. F. J. Eriksen, S. M. Fernandez, A. B. Bray, and E. Pollack, *Phys. Rev. A* **11**, 1239 (1975).
40. J. Perel, *J. Electrochem. Soc.* **115**, 343 (1968).
41. K. R. Spangenberg, *Vacuum Tubes*, McGraw-Hill, New York, 1948, Chapter 13.
42. D. Lorents, W. Aberth, and V. Hesterman, *Phys. Rev. Lett.* **17**, 849 (1966).
43. J. E. Jordan and I. Amdur, *J. Chem. Phys.* **46**, 165 (1967).
44. Y. N. Belyaev and V. B. Leonas, *Dokl. Akad. Nauk. SSSR* **173**, 306 (1967); Engl. transl.: *Sov. Phys. Dokl.* **12**, 233 (1967).
45. E. A. Mason and J. T. Vanderslice, in *Atomic and Molecular Processes* (D. R. Bates, ed.), Academic Press, New York, 1962, Chapter 17.
46. E. B. Jordon and R. B. Brode, *Phys. Rev.* **43**, 112 (1933).
47. V. V. Afrosimov, Yu. S. Gordeev, A. M. Polyanskii, and A. P. Shergin, Line widths of discrete energy losses in violent collisions of atomic particles, in *Sixth International Conference on the Physics of Electronic and Atomic Collisions*, Abstracts of Papers, MIT Press, Cambridge, Massachusetts, 1969, pp. 744–747; *Zh. Eksp. Teor. Fiz.* **57**, 808 (1969); English transl.: *Sov. Phys. JETP* **30**, 441 (1970).
48. E. N. Fuls, P. R. Jones, F. P. Ziemba, and E. Everhart, *Phys. Rev.* **107**, 704 (1957).
49. D. Lorents and W. Aberth, *Phys. Rev.* **139**, 1017 (1965).
50. D. H. Crandall, R. H. McKnight, and D. H. Jaecks, *Phys. Rev. A* **7**, 1261 (1973).
51. L. G. Filippenko, *Zh. Tehn. Fiz.* **30**, 57 (1960); Engl. transl.: *Sov. Phys. Tech. Phys.* **5**, 52 (1960).
52. A. V. Bray, Ph.D. Thesis, University of Connecticut, Storrs, Connecticut, 1975 (unpublished); see also A. V. Bray, J. D. Clark, and E. Pollack, *Bull. Am. Phys. Soc.* **18**, 1516 (1973); A. V. Bray, D. S. Newman, D. F. Drozd, and E. Pollack, in *Book of Abstracts, IX International Conference on the Physics of Electronic and Atomic Collisions* (J. S. Risley and R. Geballe, eds.), Univ. of Washington Press, Seattle, 1975, p. 627.
53. A. J. Dempster, *Phys. Rev.* **51**, 67 (1937).
54. M. Von Ardenne, *Tabellen zur Angewandten Physik*, Vol. I, Veb Deutscher Verlag der Wissenschaften, Berlin, 1962, Chapter 3, pp. 555–558.
55. G. D. Yarnold and H. C. Bolton, *J. Sci. Instr.* **26**, 38 (1949).
56. G. A. Harrower, *Rev. Sci. Instr.* **26**, 850 (1955).
57. D. J. Volz, Ph.D. Thesis, University of Nebraska, Lincoln, Nebraska, 1968 (unpublished).
58. S. M. Fernandez, F. J. Eriksen, and E. Pollack, *Phys. Rev. Lett.* **27**, 230 (1971).
59. M. Barat, J. Baudon, M. Abignoli, and J. C. Houver, *J. Phys. B* **3**, 230 (1970).
60. A. V. Bray, F. J. Eriksen, S. M. Fernandez, and E. Pollack, *Rev. Sci. Instr.* **45**, 429 (1974).
61. F. P. Ziemba, G. J. Lockwood, and E. Everhart, *Phys. Rev.* **118**, 1552 (1960).
62. B. L. Schram, H. J. H. Boerboom, W. Kleine, and J. Kistemaker, *Physica (The Hague)*, **32**, 749 (1965).
63. R. H. McKnight, D. H. Crandall, and D. H. Jaecks, *Rev. Sci. Instr.* **41**, 1282 (1970).
64. D. H. Crandall, R. H. McKnight, and D. H. Jaecks, *Phys. Rev. A* **7**, 1261 (1973).
65. H. D. Hagstrum, *Phys. Rev.* **89**, 244 (1953).

66. M. E. Rudd, T. Jorgensen, and P. J. Volz, *Phys. Rev.* **151**, 28 (1966).
67. G. Gerber, R. Morgenstern, and A. Neihaus, *J. Phys. B* **6**, 493 (1973).
68. For a comprehensive review of many electron emission experiments, together with an excellent discussion of the mechanisms of electron production in ion–atom collisions, see M. E. Rudd and J. H. Macek, *Case Studies in Atomic Physics* **3**, 47–136 (1972).
69. Q. C. Kessel and E. Everhart, *Phys. Rev.* **146**, 16 (1966).
70. For a review of Ref. 69 and a discussion of coincidence methods used prior to 1970, see Q. C. Kessel, Coincidence measurements, in *Case Studies in Atomic Collision Physics I* (E. W. McDaniel and M. R. C. McDowell, eds.), North-Holland Publ., Amsterdam, 1969, pp. 399–462.
71. M. Barat, J. C. Brenot, and J. Pommier, *J. Phys. B* **6**, L105 (1973).
72. R. Morgenstern, M. Barat, and D. C. Lorents, *J. Phys. B* **6**, L330 (1973).
73. W. Aberth and D. Lorents, *Phys. Rev.* **144**, 109 (1966).
74. F. T. Smith, R. P. Marchi, W. Aberth, D. C. Lorents, and O. Heinz, *Phys. Rev.* **161**, 31 (1967).
75. D. Coffey, Jr., D. C. Lorents, and F. T. Smith, *Phys. Rev.* **187**, 201 (1969).
76. F. T. Smith, H. H. Fleischmann, and R. A. Young, *Phys. Rev. A* **2**, 379 (1970).
77. J. Baudon, M. Barat, and M. Abignoli, *J. Phys. B* **3**, 207 (1970).
78. M. Barat, J. C. Brenot, D. Dhicq, J. Pommier, V. Sidis, R. E. Olson, E. J. Shipsey, and J. C. Browne, *J. Phys. B* **6**, 269 (1976).
79. S. W. Nagy, S. M. Fernandez, and E. Pollack, *Phys. Rev. A* **3**, 280 (1971).
80. S. W. Nagy, S. M. Fernandez, and E. Pollack, in *Proceedings of the Sixth International Conference on the Physics of Electronic and Atomic Collisions*, MIT Press, Cambridge, Massachusetts, 1969, Abstracts of Papers, p. 867.
81. R. Olson, private communication.
82. M. Barat, J. C. Brenot, and J. Pommier, *Book of Abstracts, VIII International Conference on the Physics of Electronic and Atomic Collisions* (B. C. Cobic and M. V. Kurepa, eds.), Graficko Preduzece "Buducnost," Zrenjanin, 1973, pp. 194–195.
83. F. J. Eriksen, Ph.D. Thesis, University of Connecticut, Storrs, Connecticut, 1975 (unpublished).
84. S. M. Fernandez, Ph.D. Thesis, University of Connecticut, Storrs, Connecticut, 1975 (unpublished).
85. S. M. Fernandez, F. J. Eriksen, A. V. Bray, and E. Pollack, *Phys. Rev. A* **12**, 1252 (1975).
86. A. Russek, to be published.
87. A. V. Bray, D. S. Newman, and E. Pollack, *Phys. Rev. A* **15**, 2261 (1977).
88. G. H. Morgan and E. Everhart, *Phys. Rev.* **128**, 667 (1962).
89. V. V. Afrosimov, Yu. S. Gordeev, M. V. Panov, and N. V. Fedorenko, *Zh. Tekhn. Fiz.* **34**, 1613, 1624, 1637 (1967); English transl.: *Sov. Phys. Tech. Phys.* **9**, 1248, 1256, 1265 (1965).
90. Q. Kessel, A. Russek, and E. Everhart, *Phys. Rev. Lett.* **14**, 484 (1965).
91. E. Everhart, G. Stone, and R. J. Carbone, *Phys. Rev.* **99**, 1287 (1955); F. W. Bingham, *J. Chem. Phys.* **46**, 2003 (1967). Extensive tables of R_0 and other classical parameters calculated by Felton Bingham are available as Document No. SC-RR-66-506 (unpublished) from Clearinghouse for Federal Scientific and Technical Information, National Bureau of Standards, U.S. Department of Commerce, Springfield, Virginia. See also M. T. Robinson, *Tables of Classical Scattering Integrals for the Bohr, Born–Mayer, and Thomas–Fermi Potentials*, Oak Ridge National Laboratory, ORNL-3493, UC-34-Physics, TID-4500, 21st ed.; ORNL-4556, UC-34-Physics, 1970.
92. J. D. Garcia, R. J. Fortner, and T. M. Kavanagh, *Rev. Mod. Phys.* **45**, 111 (1973).
93. W. W. Smith and Q. C. Kessel, Ion-induced x-rays from gas collisions, in *New Uses for Ion Accelerators* (J. Ziegler, ed.), Plenum Press, New York, 1975, p. 355.

94. J. S. Briggs and J. H. Macek, *J. Phys. B* **6**, 982 (1973); **5**, 597 (1972); **6**, 841 (1973).
95. U. Fano and J. H. Macek, *Rev. Mod. Phys.* **45**, 553 (1974); J. Macek and D. H. Jaecks, *Phys. Rev. A* **4**, 2288 (1971).
96. S. Dworetsky, R. Novick, W. W. Smith, and N. Tolk, *Phys. Rev. Lett.* **18**, 939 (1967).
97. R. L. Smick and W. W. Smith, in *Abstracts of Papers*, VII ICPEAC (I. Amdur, ed.), MIT Press, Cambridge, Massachusetts, 1969, p. 306; G. Rahmat, G. Vassilev, J. Baudon, and M. Barat, *Phys. Rev. Lett.* **26**, 1411 (1971); R. L. Smick and W. W. Smith, *Bull. Am. Phys. Soc.* II **16**, 563 (1971), and to be published.
98. S. Dworetsky, R. Novick, W. W. Smith, and N. Tolk, *Rev. Sci. Instr.* **39**, 1721 (1968).
99. See, for example, V. V. Zashkvara, M. I. Korsunskii, and D. S. Kosmacher in translation, *Sov. Phys. Tech. Phys.* **11**, 96 (1966); J. S. Risley, *Rev. Sci. Instr.* **43**, 95, (1972).
100. L. Wolterbeek-Muller and F. J. de Heer, in Ref. 110.
101. D. H. Jaecks, D. H. Crandall, and R. H. McKnight, *Phys. Rev. Lett.* **25**, 491 (1970); R. H. McKnight and D. H. Jaecks, *Phys. Rev. A* **4**, 2281 (1971).
102. L. T. Sin Fai Lam, *Proc. Phys. Soc. (London)* **92**, 67 (1967).
103. D. H. Jaecks, ICPEAC VIII, Invited Paper, 1973.
104. P. J. Martin and D. H. Jaecks, *Phys. Rev. A* **8**, 2429 (1973).
105. The theory of Stueckelberg oscillations is discussed in F. T. Smith, in *Lectures in Theoretical Physics: Atomic Collision Processes*, Vol. XIC (S. Geltman, K. T. Mahanthappa, and W. E. Brittin, eds.), Gordon & Breach, New York, 1969, p. 95.
106. L. D. Landau and E. M. Lifshitz, *Quantum Mechanics, Nonrelativistic Theory*, Pergamon, Oxford, 1958, pp. 160, 304 ff.; see also J. E. Bayfield, E. E. Nikitin, and A. I. Reznikov, *Chem. Phys. Lett.* **19**, 471 (1973).
107. The He^+ on Ne case and the general theory are discussed by R. E. Olson and F. T. Smith, in *Phys. Rev. A* **3**, 1607 (1971); **6**, 526 (1972); application of the theory to the Stueckelberg oscillations of the $2\,^3S$ state of He excited by He^+ collisions are given in R. E. Olson, *ibid.* **5**, 2094 (1972).
108. Indirect evidence of this coupling mechanism from low-resolution inelastic energy-loss measurements is discussed for the He^+ on He case in M. Barat, D. Dhuicq, R. Francois, R. McCarroll, R. D. Piacentini, and A. Salin, *J. Phys. B* **5**, 1343 (1972).
109. H. Rosenthal, *Phys. Rev. A* **4**, 1030 (1971); H. Rosenthal and H. M. Foley, *Phys. Rev. Lett.* **23**, 1480 (1969).
110. For examples of total cross section oscillations in other collision systems, see P. O. Haugsiaa, R. C. Amme, and N. G. Utterback, *Phys. Rev. Lett.* **22**, 322 (1969); S. V. Bobashov, *JETP Lett.* **11**, 260 (1970); L. Wolterbeek-Muller and F. J. de Heer, *Physica* **48**, 345 (1970); N. H. Tolk, C. W. White, S. H. Neff, and W. Lichten, *Phys. Rev. Lett.* **31**, 671 (1973); T. Anderson, A. M. Nielsen, and M. J. Olsen, *ibid.* **31**, 739 (1973).
111. G. Rahmat, G. Vassilev, and J. Baudon, *J. Phys. B* **8**, 1302 (1975).
112. W. Aberth, D. C. Lorents, R. P. Marchi, and F. T. Smith, *Phys. Rev. Lett.* **14**, 776 (1972).
113. D. A. Clark, J. Macek, and W. W. Smith, *Book of Abstracts, IV International Conference on the Physics of Electronic and Atomic Collisions* (J. S. Risley and R. Geballe, eds.), Univ. of Washington Press, Seattle, 1975, pp. 731–732; and to be published.
114. G. Vassilev, G. Rahmat, J. Slevin, and J. Baudon, *Phys. Rev. Lett.* **34**, 444 (1975).
115. D. H. Jaecks, F. J. Eriksen, W. de Rijk, and J. Macek, *Phys. Rev. Lett.* **35**, 723 (1975).
116. R. S. Thoe and W. W. Smith, *Phys. Rev. Lett.* **30**, 525 (1973); S. Sackmann, H. Lutz, and J. Briggs, *ibid.* **32**, 805 (1974); J. H. Stein, H. Lutz, P. Mokler, K. Sistemich, and P. Armbruster, *ibid.* **24**, 701 (1970); J. H. Stein et al., *Phys. Rev. A* **5**, 2126 (1972); E. Laegsgaard, J. U. Andersen, and L. C. Feldman, *Phys. Rev. Lett.* **29**, 1206 (1973).
117. G. Thomson, P. Landieri, and E. Everhart, *Phys. Rev.* **A1**, 1439 (1970); G. M. Thomson, W. W. Smith, and A. Russek, *Phys. Rev. A* **7**, 168 (1973).

118. D. H. Jaecks, W. D. Rijk, and P. J. Martin, *Abstracts of the VIIth International Conference on the Physics of Electronic and Atomic Collisions*, North-Holland Publ., Amsterdam, 1971, p. 424.
119. S. J. Young, J. S. Murray, and J. R. Sheridan, *Phys. Rev.* **178**, 40 (1969).
120. P. J. Wehrenberg and K. C. Clark, *Phys. Rev. A* **8**, 173 (1973).

4

Double Electron Transfer and Related Reactions

J. Appell

1. Introduction

When a beam of fast positive ions passes through a gaseous target, processes leading to the formation of negative ions are observed among many others. The main purpose of this chapter is to examine and to discuss what information can be obtained on the states of the ionized projectile or of the ionized target by measuring the translational energy spectra of the resulting fast negative ions. The cross section and mechanism of double electron transfer have been discussed in Chapter 2, and the use of this type of reaction as a source of negative ion mass spectra is presented in Chapter 5.

The main processes responsible for the formation of negative ions upon impact of a few kiloelectron-volt positive ions (A^+) on a gaseous target (M) are simple electron-transfer reactions, namely, the double-charge-transfer process in one collision:

$$A^+ + M \rightarrow A^- + M^{2+} \tag{I}$$

J. Appell • Laboratoire de Spectrometrie, Rayleigh-Brillouin, U.S.T.L., 34060 Montpellier, France. Formerly, Laboratoire des Collisions Ioniques, associated with the C.N.R.S., Université de Paris-Sud, 91405, Orsay, France.

and two successive single-charge transfers:

$$A^+ + M \rightarrow A + M^+ \tag{IIa}$$

$$A + M \rightarrow A^- + M^+ \tag{IIb}$$

Other processes, which can be generally represented as

$$A^+ + M \rightarrow A^- + M^{n+} + (n-2)e \quad \text{with} \quad n > 2$$

and

$$A^+ + M \rightarrow A + M^{m+} + (m-1)e$$
$$A + M \rightarrow A^- + M^{p+}(p-1)e \quad \text{with} \quad m \text{ and } p > 1$$

can lead to the formation of fast negative ions. However, the endothermicity of these processes is much larger than that for processes I and II, so that their contribution in the energy range considered here is negligible (see Section 2). Furthermore, A^- ions formed in such processes will have a continuous translational energy spectrum.

The energy necessary for process I or II can be supplied only by the incident relative translational energy of A^+ and M, namely $E_0\{m_M/(m_A + m_M)\}$ if the target is initially at rest, where m_A and m_M are the masses of A^+ and M, respectively. The outgoing negative ion has a translational energy E; the difference $E_0 - E$, hereafter denoted as the energy loss ΔE of the projectile, is given by the following relationships for processes I and II, respectively:

$$\Delta E_1 = I^{2+}(M) - I^+(A) - EA(A) - E_M \tag{1.1}$$

$$\Delta E_2 = I_a^+(M) + I_b^+(M) - I^+(A) - EA(A) - E_{aM} - E_{bM} \tag{1.2}$$

where $I^{2+}(M)$ is a double ionization potential of the target M (M is assumed to be initially in its ground state); $I_a^+(M)$ and $I_b^+(M)$ are ionization potentials of the target M which need not be ionized into the same state in both steps of process II; $I^+(A)$ is the ionization potential of A to the particular state of A^+ involved; EA(A) is the electron affinity of A; and E_M, E_{aM}, and E_{bM} are the recoil energies of the target M.

Under appropriate experimental conditions, namely a high incoming-ion energy E_0 and a geometry such that the secondary ions are observed at an angle close or equal to zero, the recoil energies of the target can be shown to be negligible compared with the other terms in (1.1) and (1.2) except when the projectile is significantly heavier than the target (see Appendix). Under these conditions the measurement of the energy losses ΔE of the product negative ion will be a direct measurement of the overall change in the internal energy of the colliding system. Each peak and thus each energy loss measured in a given spectrum can then be ascribed to a double-charge-transfer process (type I) or to a double collision process (type II) according to the manner in which its intensity varies with the target pressure (see Section 3).

As a rule we will, in the course of this chapter, refer to the (A^+, M) spectrum, meaning the energy-loss spectrum of the A^- ions formed upon impact of A^+ ions on a target M. It can easily be seen through relations (1.1) and (1.2) that the measurement of the energy losses ΔE in a given (A^+, M) spectrum can yield information on the target states if the projectile states are known, or conversely, on the projectile states if the target states are known.

The first experiments which were performed independently by Witteborn and Ali[1] and in our laboratory[2-6] were of the first kind, namely the study of the ionized states of various atomic and molecular targets by measuring the corresponding (H^+, M) spectra. We will discuss in Section 4 the results thus obtained. More recently measurements of some (A^+, M) spectra have been performed in our laboratory[7] with $A^+ = O_2^+, O^+, OH^+$ and $M = Ar$ and by Keough et al.[8] with A^+ a polyatomic ion and M the corresponding neutral molecule[8a] and with A^+ a halogen ion and M a rare-gas atom.[8b] We discuss these results and future possible applications in Section 5. However, before examining the results obtained by this method, which has been called double-charge-transfer spectroscopy (DCTS), we shall briefly examine in Section 2 the cross-section data given in the literature for the three collision events (I, IIa, and IIb) and discuss in Section 3 the experimental procedures used in double-charge-transfer spectroscopy.

2. Cross Sections

We do not intend to review the existing data on the cross sections for processes I, IIa, IIb and other related reactions; this would be far beyond the scope of the chapter. We will limit ourselves to a discussion of the general behavior of these cross sections as a function of energy and give their order of magnitude when the projectile ion is a proton in the energy range in which double-charge-transfer spectroscopy has been performed.

According to the adiabatic hypothesis formulated by Massey,[9] the cross section for a charge transfer process will be small as long as the "transition time" is small with respect to the "collision time". This defines an energy range where the collision is nearly adiabatic as the one where

$$\frac{aQ}{hv} \gg 1 \qquad (2.1)$$

where Q is the internal energy change, a is a parameter on the order of atomic dimensions, and v is the relative velocity of the two particles. As the energy increases, the cross section rises to a maximum and then decreases, the interaction time becoming too short for the transition to be likely.

Hasted has postulated that the velocity at which the maximum occurs is given by the relation $v \approx aQ/h$. This relation is known as the adiabatic maximum rule.[10-12] Its validity was verified by Hasted[12] on a number of typical single-charge transfers; a was shown then to be a constant equal to 7 Å. Furthermore, in the adiabatic region, the cross section should vary roughly as[10,13]

$$\sigma = Ke^{-aQ/(4\hbar v)} \tag{2.2}$$

where K is a constant that depends on the particular process considered.[13,14]

A charge-changing process can generally be represented by

$$A^k + B \rightarrow A^m + B^n + (m+n-k)e \quad \text{with} \quad m+n \geq k$$

The incident fast projectile with initial charge k collides with the neutral target B and undergoes a loss or capture of electrons thus going into charge state m while the target acquires a charge n so that $(m+n-k)$ electrons are released. In pure charge-transfer processes no electrons are released and $m+n=k$. The cross sections for such processes are denoted σ_{km}^{0n}, where the subscript refers to the initial and final states of the projectile and the superscript refers to those of the target (cf. Chapter 1, Section 1.1).

The first review, to our knowledge, of charge-changing cross sections was that by Allison and Garcia Munoz.[15] Since that time a considerable amount of data has accumulated on the cross sections of charge-changing collisions between H^+, H, or H^- and noble gases, alkali atoms, alkali earth atoms, or molecules, the energy of the collision ranging from several electron volts to several megaelectron volts. The experimental techniques used and the data have been reviewed by Fedorenko[16] and recently by Tawara and Russek.[17]

To date most experiments, where either the fast ions or the slow ions and electrons are detected after the collision, do not yield the individual cross sections σ_{km}^{0n} but the sum of some of these cross sections. For example, if the fast ions are detected, the experiment will yield the σ_{km} values which are the sums of the cross sections for all charge-changing processes where the charge of the projectile changes from k to m (see, e.g., Refs. 16 and 17). In order to measure the individual cross sections σ_{km}^{0n}, Afrosimov et al.[18-21] have developed an experiment in which the slow target ion and the fast projectile are detected in coincidence. The charge-changing collisions of H^+, H, and H^- on the rare-gas atoms have been studied in an energy range extending from 5 to 50 keV, except when the final state of the projectile was H^-, where the lowest energy was 13 keV.

As stated in Section 1, the contribution of processes other than pure charge transfer is negligible in double-charge-transfer spectroscopy. Not only do competing processes have different energy losses but the individual cross sections measured by the coincidence technique have quite different values; for example, at 5 keV σ_{10}^{02} is smaller than σ_{10}^{01} by an order of magnitude when the target is xenon, by two orders of magnitude when it is krypton, and by four orders of magnitude when it is neon or helium.[18]

This qualitative description of the variation of the cross sections with energy can be seen to apply reasonably well to the observed variations as they appear on the curves compiled by Fedorenko[16] and Tawara and Russek.[17] From the experimental data as well as from the adiabatic criterion,[3] it appears that most of the processes responsible for the measured (H^+, M) spectra take place in the adiabatic region. From these data compilations, further conclusions can be drawn concerning the relative magnitude of the cross sections for processes I, IIa, and IIb. For a given target at a given energy the cross sections decrease in the order $\sigma_{10}, \sigma_{0\bar{1}}, \sigma_{1\bar{1}}$ and this remains true over the whole energy range considered (several hundred electron volts to several megaelectron volts). At 5 keV and for most targets the cross section ratios are roughly

$$\sigma_{10}/\sigma_{0\bar{1}} \approx 1:10^{-2}, \qquad \sigma_{10}/\sigma_{1\bar{1}} \approx 1:10^{-3}$$

Finally we will quote two values of σ_{10}^{01} as measured by Afrosimov *et al.*[18] at the same energy (5 keV): $\sigma_{10}^{01} = 3 \times 10^{-17}$ cm^2 when the target is helium and $\sigma_{10}^{01} = 2 \times 10^{-15}$ cm^2 when the target is xenon. These two values are representative of the extreme values for the σ_{10} cross sections measured with most targets.[16,17]

The preceding discussion has been made without taking into account the internal states in which the projectile or target ions are formed. All cross sections are in fact a sum of the cross sections for the processes yielding the same charged projects in different electronic states. McNeal and Birely[22] have discussed the importance of the excitation of either the projectile or the target in a charge-changing collision and have reviewed the existing data which have been obtained by observing the light emission after the collision (cf. also Chapter 1). It should be noted that double-charge-transfer spectroscopy can provide some evidence pertaining to the relative cross sections of charge-transfer processes when the target is ionized in different electronic states as well as to the relative cross sections of single-charge-transfer processes leading to an excited state of the projectile (for this last point, see Ref. 23 and Section 4.6).

3. Experimental Techniques of Double-Charge-Transfer Spectroscopy

3.1. The Instrument

Most experiments reported in this section have been performed on a conventional 60° magnetic sector mass spectrometer to which a collision

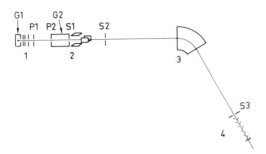

Fig. 1. Diagram of the mass spectrometer used for double-charge-transfer spectroscopy (sections 1 and 2 are shown larger than true scale). 1, Primary ion source; 2, collision chamber and deflecting plates; 3, energy analyzer (magnet); 4, ion detector (secondary electron multiplier). P1, P2, S1, S2, and S3 are collimating slits. G1 and G2 are gas inlets (see, e.g., Ref. 3).

chamber was simply added.[24] The experimental procedures used in DCTS are described herein and in more detail elsewhere.[3]

A schematic diagram of the apparatus is shown in Fig. 1. The primary ions are formed by 100-eV electron impact on an appropriate gas in a Nier-type ion source; they are extracted, focused, and accelerated to an energy E_0 before entering the collision chamber. The primary beam is collimated by the two slits P1 and P2 and under normal conditions, the energy spread of the beam is less than 0.5 eV. The secondary negative ions are formed in the collision chamber. The target gas pressure can be controlled over a range from $\sim 10^{-3}$ torr to less than 10^{-4} torr, and the pressure outside the chamber is typically equal to $\sim 10^{-6}$ torr. The secondary beam is collimated through slits S1, S2, S3; deflecting plates compensate for slight mechanical misalignments and for the deflection of the beam by the fringing field of the magnet before slit S2. The actual scattering angle of the collected ions is thus $0 \pm 0.5°$ while the angular width of the beam is equal to a few mrad. The negative ions are then momentum analyzed by the magnetic field and collected on the first dynode of a secondary electron multiplier; the resolving power of the analyzer was typically equal to 2000 (FWHM).

The pulses delivered by the electron multiplier are amplified, shaped, and sent to an integrator which delivers a voltage proportional to the number of counts per second; by this method counting rates as low as a few ions per second could easily be achieved. The resulting voltage is fed into the y channel of an $x-y$ recorder while the x channel is monitored by a voltage proportional to the variation of the magnetic field. The study of the (A^+, M) spectra requires, for the measurement of the energy losses ΔE, that the energy-loss scale be calibrated.

3.2. The Energy-Loss Scale

The unit of energy may be determined by two different methods: (a) The magnetic field variation (ΔH) across the spectrum is measured, and knowing the nominal translational energy E_0 of the primary ions and the mean value of the magnetic field H the energy variation (ΔE) across the spectrum is simply given by

$$\Delta E = \frac{2\,\Delta H}{H} \cdot E_0 \qquad (3.1)$$

(This relationship holds as long as ΔH and ΔE are small compared to H and E_0.) (b) Two spectra are recorded successively, the primary ion energy being E_0 for one spectrum and $E_0 + \Delta E_0$ for the other. The value of ΔE_0 is chosen to be of the same order of magnitude as Q, the energy defect for the reaction. The displacement of a given peak from one to the other spectrum readily gives the energy unit of the scale.

The zero of the energy-loss scale can in principle be determined by energy-analyzing negative ions formed in the ion source and accelerated to the same translational energy E_0 as the primary positive ions. This can be done by simply changing the signs of all potentials in the ion source; however, the equality of the absolute values of the potentials holds within a certain uncertainty, and furthermore the reversed extraction field in the ionization region will act differently on the ionizing electrons so that positive and negative ions will be formed at slightly different potentials. The zero of the energy losses can thus be determined only to within 2 or 3 eV. Although this precision is insufficient to provide an absolute calibration of the energy loss scale it was sufficient to secure, at least in the (H^+, M) spectra, a relative calibration method.[3]

In the double charge transfer spectra of (H^+, He) and (H^+, Ar) the lower energy-loss peak is due (see Section 4 and Fig. 6) to a double-collision process (type II). This peak could, by comparison with the energy of stable negative ions, be assigned without ambiguity to the process where the two He^+ or Ar^+ ions are formed in their ground state (the first excited state of He^+ is 41 eV above the ground state and the first excited state of Ar^+ is 13.4 eV above the ground state).[25] Although the ground state of Ar^+ is split into two components ($^2P_{3/2}$ and $^2P_{1/2}$ separated by 0.2 eV) the lowest energy-loss peak of the (H^+, Ar) spectrum has been used as a reference for the energy losses because of its higher intensity (see Refs. 2 and 3). From relation (1.2) this reference energy loss can be calculated and is equal to 17.4 ± 0.2 eV. The energy losses corresponding to the various peaks in a given (H^+, M) spectrum are taken at the maximum of each peak and measured with respect to the maximum of the lowest energy-loss peak in the (H^+, Ar) spectra recorded before and after the (H^+, M) spectrum under study.

3.3. An Apparatus with Higher Energy Resolution

The apparatus just described does not, because of its relatively poor energy resolution, allow for the separation of processes whose energy losses differ by less than ~ 2 eV; this is a severe drawback which can only be overcome by using more sophisticated instrumentation. The apparatus built by Fournier et al.[26] has been used to investigate the (H^+, noble gas) spectra[6,27] (see Section 4.1) with an energy resolution as high as 0.7 eV FWHM. The details of the apparatus are described by Fournier et al.[26] The primary ions are formed in a commercial unoplasmatron-type ion source (Colutron); they are mass analyzed by a Wien filter before entering the collision chamber, where the target gas is introduced. The resulting negative ions, scattered through an angle equal to 0 ± 3 mrad, can be decelerated before entering a 127° electrostatic analyzer; they are collected on a Channeltron multiplier. The double-charge-transfer spectra are accumulated in a multichannel analyzer.

For the study of the double-charge-transfer spectra of a molecular projectile ion, provision should be made for mass analysis of the primary ion beam. This is needed in order to avoid the fortuitous superposition of energy peaks. For example, an (A^+, Ar) spectrum and a dissociative double-charge-transfer spectrum (see Section 5.4) could overlap.

3.4. Target Pressure Dependences of the Negative Ion Intensities

We have stated in Section 1 that the peaks observed in a given (A^+, M) spectrum are ascribable either to double-charge-transfer processes (type I) or to double-collision processes (type II) according to the variation of their intensity with pressure (see Fig. 3). In the low-pressure range, where the double-charge-transfer spectra are measured, the intensity of the peak due to a double-charge-transfer process will be proportional to the pressure whereas the intensity of a peak due to a double-collision process will be proportional to the square of the pressure. The pressure range, in which these variation laws are valid, and which we denote as the "low-pressure" range, can in principle be determined for each (A^+, M) spectrum if the cross sections for processes I, IIa, IIb, and for the elastic scattering of the A^+ ions are known. As a rule the relative intensities of the various A^m components of a beam are determined as a function of gas target thickness by a set of coupled differential equations (see Refs. 10, 15–17). The low-pressure range will be defined as the range in which the rate of conversion of the positive projectile ions into the other components of the beam and the rate of loss of these projectiles by elastic scattering are sufficiently low so that the coupled differential equations can be simplified and lead to the above-mentioned variation laws.

At the relatively low energies where DCTS has been performed the loss of protons by elastic scattering is predominant whereas at higher energies (100 keV and above) the one-electron capture process will become predominant. We set, by definition, the low-pressure range to be the range in which the following inequality holds:

$$\sigma \times N \times l \leq 0.2 \tag{3.2}$$

where N is the number of target gas atoms (or molecules) per cubic centimeter, l is the collision length, and σ is the elastic scattering cross section of the proton at low energies and the one-electron capture cross section at high energies. Converting from number density to pressure, the upper limit of the low-pressure range is thus given by

$$P_{\max} \approx 1.6 \times 10^{-20} \frac{T}{\sigma \cdot l} \tag{3.3}$$

where the pressure $P_{\max}$ is in torrs, the temperature T is in degrees kelvin, and the cross section σ is in square centimeters.

Experimentally the upper limit of the low-pressure range can be readily ascertained: At higher pressures the intensity of each peak tends to grow more slowly with increasing pressure than in the low-pressure range and eventually levels off.

4. (H^+, M) Spectra and the Study of the States of the Target Species

We have already noted in Section 1 that (H^+, M) spectra should yield information on the states of the singly or doubly ionized target M. This comes about because of the existence of only one stable state of the H^- ion as well as of the proton. It may be asked whether H^- could not be formed in metastable states. Most excited states of H^-, which are known from resonance scattering of electrons by H and from theoretical calculations, are autoionizing states with very short lifetimes[28] and will not be observed in DCTS experiments, where the lifetime of a given state of H^- must be equal to or greater than 10^{-6} sec in order to be observed.[2,3,23] The first H^- excited state which could be long-lived enough is the $2p^2$, ^3P state; its autoionization is forbidden but its lifetime was calculated by Drake[29] to be about 10^{-7} sec, too short to allow for its observation in DCTS. This was confirmed by the (H^+, M) spectra measured when M is a noble-gas atom or the H_2 molecule (see Ref. 23).

Thus in (H^+, M) spectra the energy losses as derived from (1.1) and (1.2) are given for a double-charge-transfer process (type I) and for a double-collision process (type II), respectively, by

$$\Delta E_1 = I^{2+}(M) - 14.35 \tag{4.1}$$

and

$$\Delta E_2 = I_a^+(M) + I_b^+(M) - 14.35 \tag{4.2}$$

where the energy losses ΔE_1 and ΔE_2 and the ionization potentials $I^{2+}(M)$, $I_a^+(M)$, and $I_b^+(M)$ are in electronvolts. [$I^+(H) = 13.6$ eV and EA(H) = 0.754 eV.[30]]

The relationship between the energy losses and the ionization potentials as given by (4.1) and (4.2) can be used, together with the (H^+, M) spectra, to yield information on the selection rules which apply to the electronic transitions in processes I and II, when M is a target whose ionization potentials are well known. Alternatively, this approach allows the measurement of the double-ionization potentials of the target M when they are unknown. In the latter case, the assignment of the I^{2+} terms to given states of the doubly ionized target can then be discussed in terms of the selection rules established in the former case.

It must be stressed that double-charge-transfer spectroscopy yields double-ionization potentials of molecular targets regardless of the stability of the doubly ionized molecule with respect to dissociation. This is an advantage not only over the classical methods developed mostly for the study of the states of neutral or single ionized molecules, but also over recent methods developed to study the states of doubly ionized molecules (cf. Chapter 5); to our knowledge the only other method which does not have this drawback is Auger electron spectroscopy.[31-35]

4.1. The Spin Conservation Rule

The Wigner spin conservation rule requires that the total electron spin angular momentum of a pair of atoms or molecules does not change in the course of a collision.[9,36] We expect this rule to hold as long as the spin-orbit coupling remains small, that is, for collisions involving species composed of light atoms. However, in collisions in which the projectile is scattered through a nonzero angle and in which the spin momentum is coupled to the internuclear axis (e.g., in Hund's cases a or c), a coupling with rotation may lead to spin-changing collisions.

Moore[37] has investigated, using ion-impact energy-loss spectrometry, the validity of the spin conservation rule in inelastic and superelastic collisions of N^+ ions with energies of 1.5–3.5 keV on various targets; he has ascertained that spin-changing collisions are nearly three orders of magnitude less probable than spin-conserving collisions when the target is in a singlet state whereas in N^+–O_2 collisions there is some evidence that the rule does not hold as well.[38] The rule is also confirmed for double-charge-transfer processes, at least where the projectile ions are collected at zero or close to zero scattering angle, by the study of the $(H^+,$ noble gases) spectra. The high-resolution spectrum obtained by Fournier et al.[6] for (H^+, Xe) is displayed in Fig. 2. The appearance of the $^3P_2$ state of the Xe^{2+} (peak D of Fig. 2) shows that a transition with a change in the total spin of the colliding system is easily induced in a heavy target such as xenon whereas this

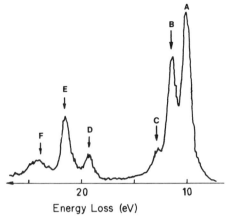

Fig. 2. (H$^+$, Xe) energy-loss spectrum.[6b] Collision energy $E_0 = 3$ keV. The peaks A, B, and C are due to double-collision processes (type II) in which the states of the two Xe ions are $^2P_{3/2}$, $^2P_{3/2}$ (peak A), $^2P_{3/2}$, $^2P_{1/2}$ (peak B), $^2P_{1/2}$, $^2P_{1/2}$ (peak C). The peaks D, E, and F are due to double-charge-transfer processes (type I) in which the state of the Xe^{2+} ion is $^3P_2$ (peak D), $^1D_2$ (peak E), and $^1S_0$ (peak F). The abscissa shows the measured energy loss of the fast particles, i.e., ΔE.

transition is not observed in argon and is very weak in krypton.[27] The observed double-charge-transfer processes in these noble gases are summarized in Table I, where the measured double-ionization potentials [as derived from Eq. (4.1)] are compared to the spectroscopic values. Further confirmation of the validity of the spin conservation rule for double-charge-transfer processes was found in the (H$^+$, O$_2$) spectra[4] (see Section 4.3 and Fig. 4).

A further observation, which can be important in the understanding of double-charge-transfer (I) and single-charge-transfer (IIa and IIb)

Table I. Double-Ionization Potentials (eV) of the Noble Gases Derived from the Double-Charge-Transfer Processes Observed in (H$^+$, Noble Gases) Spectra

Target M		Xenon		Krypton		Argon	
States of M^{2+}		a	b	a	b	a	b
...np^4	$^3P_2$	33.3 ± 0.05	33.327	38.7 ± 0.1	38.556	—	43.375
	$^3P_0$	—	34.335	—	39.215	—	43.513
	$^3P_1$	—	34.547	—	39.120	—	43.57
	$^1D_2$	35.5 ± 0.05	35.447	40.4 ± 0.05	40.370	45.1 ± 0.05	45.112
	$^1S_0$	38.0 ± 0.1	37.963	42.5 ± 0.1	42.656	47.6 ± 0.2	47.498

[a] Fournier et al. Xe: Ref. 6 (Kr) and Ref. 27 (Ar).
[b] Spectroscopic values from Moore.[25]

processes, has been made by Fournier et al.[6] on the (H^+, Xe) spectra; namely, that in the multiplet of states arising from the first electronic configuration of $Xe^+ (^2P_{3/2}$ and $^2P_{1/2})$ and of Xe^{2+} ($^3P_{2,1,0}$, $^1D_2$, and $^1S_0$) the states with higher J values were preferably formed in the charge-transfer processes. The experimental facts supporting this observation are discussed in Ref. 6, and the different models which might explain the predominance in double- and single-charge transfers are currently being worked out.

4.2. The Applicability of the Franck–Condon Principle

The potential curves of the ground state of H_2 and of the lower states of H_2^+ and the coulombic potential of H_2^{2+} are well known[39] so that the study of charge-transfer processes in collisions of H^+ on H_2 should provide evidence pertaining to the applicability of the Franck–Condon principle in atom–molecule collisions.

In their study of charge-changing processes in (H^+, H_2) using a coincidence method (Section 2) Afrosimov et al.[20] concluded from an analysis of the dissociation probability of H_2^+ formed in the $1s\sigma_g$ state, that the Franck–Condon principle applies to the electronic transitions in the target H_2 molecule over the whole energy range studied (5–50 keV).

Three peaks were observed in the (H^+, H_2) spectrum[2] shown in Fig. 3. The assignments of these peaks are summarized in Table II where the

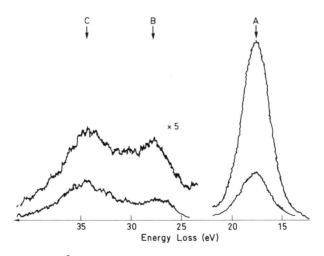

Fig. 3. (H^+, H_2) spectra.[2] Collision energy $E_0 = 4$ keV. The spectra are shown at two different target gas pressures: The peaks A and B can be seen to decrease much more rapidly than peak C, and they are ascribed to double-collision processes (types II and III) whereas peak C is ascribed to a double-charge-transfer process (cf. Table II and text).

Table II. Energy Losses and Assignments of the Peaks Observed in the (H^+, H_2) Spectrum (Measured at 4 keV)

Peaks	Order of target pressure dependence	Products	Calculated (Ref. 39)	Energy loss (eV) Measured Ref. 2	Ref. 1
A	2	$H_2^+(1s\sigma_g)$, $H_2^+(1s\sigma_g)$, H^-	17.7	17.6 ± 0.2	16.9 ± 1.0
B	2	$H_2^+(1s\sigma_g)$, $H_2^+(1s\sigma_g)$, H^- and $h\nu(2p \rightarrow 1s)$	27.9	28.1 ± 0.3	—
C	1	H_2^{2+}, H^-	36.8	34.6 ± 0.5	34.1 ± 1.0

measured energy losses at the peak maxima are compared to those measured by Witteborn and Ali[1] and to those calculated assuming vertical transitions to occur in H_2. For peak A, which is due to a double-collision process where both H_2^+ ions are formed in the ground ($1s\sigma_g$) state, as well as for peak B, due to a somewhat different double-collision process, discussed in Section 4.6, very good agreement is found between the measured and calculated values. However, the most probable transition in the double-charge-transfer process (peak C) requires somewhat less energy than the vertical transition occurring at the equilibrium distance of the initial state (0.74 Å). As discussed in Ref. 2 the observed maximum corresponds to a vertical transition occurring at 0.83 ± 0.02 Å in the classical form of the Franck–Condon principle. This internuclear distance is well within the Franck–Condon region. It was noted in Section 2 that the cross section for double-charge transfer is a decreasing function of the internal energy change [Eq. (2.2)] and it will thus be an increasing function of the internuclear distance. This led Fournier *et al.*[2] to the conclusion that the observed shift of the maximum reflects the variation of the cross section over the Franck–Condon region and does not imply that an extra momentum change takes place during the transition in violation of the Franck–Condon principle. Furthermore, the dependence of the relative cross section on the internal energy change was derived from the experimental data and from a calculation of the Franck–Condon factors. The cross section was found[2] to be an exponential function of the internal energy change in agreement with Eq. (2.2). A further check on the applicability of the Franck–Condon principle to electronic transitions taking place in the molecular partner in charge-transfer processes was provided by the (O_2^+, Ar) spectrum, which is discussed in Section 5.

The applicability of the Franck–Condon principle to electronic transitions occurring in molecules during charge-transfer processes has several implications.

1. The double-ionization potentials derived from the (H^+, M) spectra should be vertical ionization potentials, provided the doubly charged ion is not formed in a highly repulsive state; in the latter case the measured I^{2+} will be less than the vertical I^{2+}, as has been observed in H_2.
2. The energies measured for the formation of negative molecular ions in (AB^+, Ar) spectra (Section 5.3) are vertical energies which may well be slightly different from the electron affinities defined as adiabatic values.[40]
3. The determination of the geometrical structure of either positive or negative polyatomic ions by the study of (M^+, Ar) spectra as discussed in Section 5.2 is a direct consequence of the applicability of the Franck–Condon principle.

4.3. Selection Rules for the Symmetry of the Molecular States

In any collision process, if some symmetry elements of the colliding system are conserved throughout the collision, the symmetry properties (with respect to these elements) of the wave function describing the colliding system must evidently remain unchanged. This will, under given conditions, lead to selection rules for the symmetry of the electronic states of the species involved. Such rules have been discussed and tested by Goddard et al.[41] and Cartwright et al.[42] for the excitation of atoms and molecules by electron impact.

In a double-charge-transfer process between A^+ and M, the initial wave function is the product of (a) the electronic wave function* of M in its ground state, (b) the plane wave function associated with the incoming A^+ ions, and (c) the electronic wave function of A^+. Similarly, the final wave function is the product of (a) the electronic wave function of M^{2+}, (b) the divergent wave function associated with the scattered A^-, and (c) the electronic wave function of A^-. In most (A^+, M) cases no symmetry elements are conserved and thus no selection rules can be derived. However, for double-charge-transfer processes between H^+ and M as they are observed in (H^+, M) spectra, the H^- ion has a totally symmetric wave function and the H^- ions are collected at zero scattering angle so that, with the help of group theory, selection rules can in some cases be derived.

In the case in which M is an atom or a diatomic (or linear) molecule strict selection rules can be derived. In the former case the system has $C_{\infty v}$ symmetry whereas in the latter case it has C_s symmetry. The case in which M is a diatomic molecule has been investigated theoretically by Durup[43];

---

*We neglect here the influence of the rotation of the molecule. The molecule is described in the Born–Oppenheimer approximation.

qualitatively we can state that for zero-angle scattering events, a symmetry plane defined by the ion wave vector and the internuclear axis of the molecule will exist. This leads to a strict selection rule for nondegenerate states of the molecule, namely that transitions between Σ^+ and Σ^- states are forbidden. Furthermore, Durup[43] has shown that, at small scattering angles (θ), the cross sections increase as $\sin^2 \theta$ and usually reach a single maximum at small scattering angle.

Selection rules can further be derived for particular orientations of the molecule when this molecule is a homonuclear diatomic molecule[3,4] or a highly symmetrical polyatomic molecule.[3,5] Since the molecules have a random orientation with respect to the ion path, these rules will not hold strictly but they may provide us with a classification of the processes according to their probability of occurrence, taking into account the number of orientations in which they are forbidden.[5]

The (H^+, O_2) spectra shown in Fig. 4 provide experimental evidence[4] pertaining to the validity of the spin conservation rule already mentioned in Section 4.1, and to the validity of the $\Sigma^+ \leftrightarrow\!\!\!\!/\, \Sigma^-$ rule. The double-ionization potential (I^{2+}) values derived from the observed spectra are given in Table III together with experimental values obtained by Dorman and Morrison[44] and Moddeman et al.[33] and with values calculated by Hurley.[45] The ground

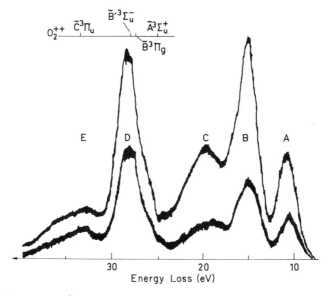

Fig. 4. (H^+, O_2) spectra.[4] Collision energy $E_0 = 4$ keV. The spectra are shown at two different target gas pressures. Peaks A, B, and C are due to double-collision processes (type II) where the two O_2^+ ions are formed in various electronic states. Peaks D and F are due to double-charge-transfer processes (type I) (see Table III and text).

Table III. Double-Ionization Potentials of O_2

		I^{2+} (eV)		
State of O_2^{2+}	DCTS Ref. 4	a	b	c
$\tilde{X}\,^1\Sigma_g^+$		35.5	36.3 ± 0.5	37.4
$\tilde{A}\,^3\Sigma_u^+$		39.6		
$\tilde{B}\,^3\Pi_g$	⎱ 43.0 ± 0.5 (peak D)	42.15		
$\tilde{B}'\,^3\Sigma_u^-$	⎰	42.5		
$\tilde{a}'\,^1\Sigma_u^-$		42.9		
$\tilde{a}\,^1\Pi_g$		43.7		
$\tilde{W}\,^1\Delta_u$		43.6		
$\tilde{C}\,^3\Pi_u$	48.0 ± 1.0 (peak E)	48.4		

[a] Calculated values.[45]
[b] By electron impact.[44]
[c] By Auger electron spectroscopy.[33]

state of O_2 is $\tilde{X}\,^3\Sigma_g^-$; the double-charge-transfer process leading to O_2^{2+} in its ground state $\tilde{X}\,^1\Sigma_g^+$ is thus spin-forbidden and the corresponding DCTS peak, if it exists, is buried in peak C (Fig. 4), which along with peaks A and B is due to a double-collision process.[4] From the spectra recorded at low pressure, it can be stated that the spin-changing process contributes an amount less than 5% of the height of the spin-conserving process (peak D). This finding supports the validity of the spin conservation rule in double-charge-transfer processes. By contrast, in both electron impact[44] and Auger electron spectroscopy[33] there is no hindrance to the formation of O_2^{2+} in a singlet state so that the ionization potential measured by both methods has been assigned to the $X\,^1\Sigma_g^+$ state of O_2^{2+} (cf. Table III).

As can be seen in Table III the lowest lying triplet state of O_2^{2+} is $\tilde{A}\,^3\Sigma_u^+$ whereas the next triplet states are $\tilde{B}\,^3\Pi_g$ and $\tilde{B}'\,^3\Sigma_u^-$; the comparison between the double-ionization potential derived from the maximum of peak D and the calculated potentials for these three states leads to the assignment of this peak to the processes in which O_2^{2+} is formed in the last two states.[4] No peak is observed at an energy loss corresponding to the process where O_2^{2+} would be formed in the $A\,^3\Sigma_u^+$ state (cf. Fig. 4) and in which the $\Sigma^+ \leftrightarrow\!\!\!/\, \Sigma^-$ rule would be violated.

4.4. Observation of Double-Charge-Transfer Processes at Different Scattering Angles

Up until now, double-charge-transfer processes have been studied by energy-analyzing the projectile ions scattered at an angle equal (or nearly so) to zero. The selection rules (see Sections 4.1 and 4.3) and the applicability of the Franck–Condon principle (Section 4.2) have thus only been tested for those processes in which the projectile ion is not deflected. It would clearly be of interest to examine the same points for processes in which the projectile ion is scattered through various angles. An angular study of the (H^+, Xe) spectra should also provide further information on the mechanism which favors the formation of Xe^+ or Xe^{2+} in those states of their multiplets having the highest J values.

For double-collision processes (type II), the observed scattering angle is the sum of the scattering angles in the two successive collisions. In order to define independently these two angles, an experimental setup could consist of two successive collision chambers, the ionic components of the projectile beam being swept out between those two chambers. This further provides a means to study another type of process, namely

$$A + M \rightarrow A^+ + M^-$$

leading to information on the states of the negative target ion regardless of its stability with respect to autodetachment or dissociation.[55]

4.5. Determination of the Double-Ionization Potentials of Some Molecules

We have already mentioned that among the results which can be derived from the study of (H^+, M) spectra are the double-ionization potentials of the target M. We have further stated that the method is particularly valuable when M is a molecule because this measurement can be made regardless of the stability of the doubly charged molecule with respect to dissociation.

The double-ionization potentials of some molecules, derived from the corresponding (H^+, M) spectra,[4,5,46] are listed in Table IV together with the values found in the literature.

Very little is generally known of the states of doubly ionized molecules; however, it is worthwhile to try to assign the experimental energies to given states of the doubly ionized molecules. For the diatomic molecules N_2 and NO, as well as for O_2, semiempirical calculations on the states of the doubly charged ions are available[45]; a comparison of experimental and calculated values together with the application of the selection rules already described have led to assignments which are discussed in Ref. 4.

For the polyatomic molecules, the assignments have been made using the same selection rules and a crude model for the states of the doubly charged ion. Moreover, double-charge-transfer processes, for which the energy losses are small, are predominant in the (H^+, M) spectra, so that M^{2+} will be formed in its lowest-lying accessible states. The lower-lying states of a doubly charged ion were derived from its first electronic configurations obtained by removing two electrons from the outer molecular orbitals in the

Table IV. Double-Ionization Potentials (eV)

Molecule	DCTS[a]	Electron impact[b]	Auger electron spectroscopy[c]
N_2	43.1 ± 0.5, 45.2 ± 0.5	42.7 ± 0.1, 43.8 ± 0.1	42.9,[d] 42.7,[e] 43.3, 46.2, 47.2[f]
NO	39.3 ± 0.5, 42.4 ± 1.0, 47.2 ± 0.5	39.8 ± 0.3	35.7, 40.1[N], 34.7[O][d]
NH_3	35.3 ± 0.7, 38.9 ± 0.7, 44.6 ± 0.7	33.7 ± 0.2, 36.8	
CH_4	38.9 ± 0.7		40.7 ± 0.8, 47.3 ± 1.7, 53.8 ± 1.2, 61.4 ± 0.8[g]
N_2O	37.3 ± 0.5, 39.7 ± 0.7, 43.3 ± 1.0		
C_2N_2	35.5 ± 0.5, 39.6 ± 0.5		
C_2H_2	33.7 ± 0.5, 38.5 ± 0.7, 45.5 ± 1.0		
C_2H_4	29.4 ± 0.5, 32.2 ± 0.7, 34.0 ± 0.7, 35.2 ± 0.7, 37.0 ± 0.7, 40 ± 1		
C_2H_6	33.4 ± 0.5, 37.7 ± 0.5		34.8 ± 0.8, 38.0 ± 0.8, 46.0 ± 0.8, 52.8 ± 0.8, 60.2 ± 1.2[g]
C_3H_8	31.5 ± 1.0		
C_6H_6[h]	26.0 ± 0.5, 30.3 ± 0.5, 33.1 ± 0.5	26.0 ± 0.2 (for C_6H_5D)	26.1 ± 0.8, 28.2 ± 0.8, 30.9 ± 0.8, 32.7 ± 0.9, 34.3 ± 0.8, 36.5 ± 0.9, 39.0 ± 0.8, 42.7 ± 1.1[g]
CH_3OH	33.2 ± 0.5, 39.5 ± 0.5		

[a] N_2 and NO from Ref. 4, NH_3 from Ref. 7, C_2H_4 from Ref. 54, other molecules from Ref. 5.
[b] Ref. 44.
[c] In Auger spectroscopy highly excited states of M^{2+} can be formed; the corresponding double-charge transfers have very low cross sections and are correspondingly not observed.
[d] Ref. 33. In NO the I^{2+} values are determined from the [N] and [O] spectra; Moddeman et al.[33] emphasize that the low I^{2+} are provided by lines which are possibly due to monopole excitation rather than normal Auger processes, in which case their lower I^{2+} would be 40.1 eV in agreement with our results.
[e] Ref. 34.
[f] Ref. 32.
[g] Ref. 35.
[h] Ast et al.[47] have found I^{2+} = 26.1 eV.

electronic configuration of the ground state of the neutral molecule.[3,5] These states were thus considered to be in the equilibrium geometry of the neutral molecule in its ground state. According to the Franck–Condon principle, which applies to double charge transfer as stated in Section 4.2, the doubly charged ions will be formed in this geometry although their equilibrium geometry, if any, may well be different.

As an illustration of a polyatomic case, the (H^+, C_2H_2) spectrum is shown in Fig. 5, and the assignments of the observed processes are given in Table V. Further discussion of these assignments along with that of the other

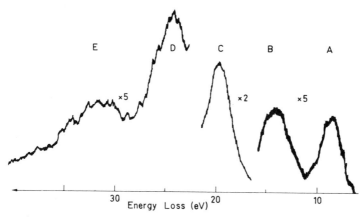

Fig. 5. (H^+, C_2H_2) spectrum.[5] Collision energy $E_0 = 4$ keV. A and B are double-collision processes; C, D, and E are double-charge-transfer processes. The assignments are given in Table V.

(H^+, M) spectra is to be found in Ref. 5. Very few theoretical calculations of the electronic energy levels of doubly ionized polyatomic molecules have so far been carried out. However, they can provide a model for the assignments of the measured I^{2+} values as has been demonstrated in the case of ammonia[46] and in the case of ethylene.[54]

Table V. Assignments of the Processes Observed in the (H^+, C_2H_2) Spectra[a]

States[b] of C_2H_2, $C_2H_2^+$, $C_2H_2^{2+}$

C_2H_2: $(1\sigma_g)^2(1\sigma_u)^2(2\sigma_g)^2(2\sigma_u)^2(3\sigma_g)^2(1\pi_u)^4 \tilde{X}\ ^1\Sigma_g^+$

$C_2H_2^+$: $\ldots (2\sigma_u)^2(3\sigma_g)^2(1\pi_u)^3 \tilde{X}\ ^2\Pi_u - \cdots - (2\sigma_u)^2(3\sigma_g)^1(1\pi_u)^4\ \tilde{A}\ ^2\Sigma_g^+ - \cdots - (2\sigma_u)^1(3\sigma_g)^2(1\pi_u)^4 - \tilde{B}\ ^2\Sigma_u^+$

$C_2H_2^{2+}$: $\cdots (2\sigma_u)^2(3\sigma_g)^2(1\pi_u)^2 - ^1\Delta_g,\ ^1\Sigma_g^+\ ^3\Sigma_g^- - \cdots - (2\sigma_u)^2(3\sigma_g)^1(1\pi_u)^3\ ^1\Pi_u,\ ^3\Pi_u$

	Double-collision processes			Double-charge-transfer processes		
	$I_a^+ + I_b^+$ (eV)				I^{2+} (eV)	
Peak	Measured	c	States of $C_2H_2^+$	Peak	Measured	State of $C_2H_2^{2+}$
A	22.8 ± 0.5	22.8	$\tilde{X}\ ^2\Pi_u - \tilde{X}\ ^2\Pi_u$	C	33.6 ± 0.5	$^1\Delta_g$ and/or $^1\Sigma_g^+$
B	29.2 ± 0.5	28.0	$\tilde{X}\ ^2\Pi_u - \tilde{A}\ ^2\Sigma_g^+$	D	38.5 ± 0.7	$^1\Pi_u$
		30.0	$\tilde{X}\ ^2\Pi_u - \tilde{B}\ ^2\Sigma_u^+$	E	45.5 ± 1	d

[a] Data from Ref. 5.
[b] C_2H_2 is linear in its ground state; its electronic configuration is given by Herzberg.[48]
[c] Calculated as the sum of the vertical I^+ values from Turner et al.[49]
[d] The assignment of this last value is uncertain owing to the unknown ordering of the higher excited states of $C_2H_2^{2+}$.

Table VI. Comparison of the Measured and Calculated Double-Ionization Potentials of NH_3^a

States[b] of NH_3 and NH_3^{2+}
NH_3: $(1a_1)^2(2a_1)^2(1e)^4(3a_1)^2 \tilde{X}\,^1A_1$
NH_3^{2+}: ...$(1e)^4(3a_1)^0 \tilde{X}\,^1A_1 - (1e)^3(3a_1)^1\,^1E, {}^3E - (1e)^2(3a_1)^2\,^1A_1, {}^1E, {}^3A_2$

States of NH_3^{2+}	Measured	From SCF calculations	From SCF–CI calculations
$\tilde{X}\,^1A_1$	35.3 ± 0.7	33.68	34.51
^1E	38.9 ± 0.7	37.04	38.96
^1E	44.6 ± 0.7	43.99	45.41
$^1A_1$	—	46.55	46.84

[a] Compare Ref. 46.
[b] NH_3 in its ground state is pyramidal; its electronic configuration is given by Herzberg.[48]

Ab initio calculations[46] on the lowest-lying states of NH_3^{2+} have been performed, at the equilibrium geometry of NH_3 in its ground electronic state, by two methods: (a) an SCF calculation, and (b) an SCF calculation followed by a large-scale CI. The double-ionization potentials obtained have been compared with the measured values (see Table VI and Ref. 46). The following observations can be drawn from this comparison. The SCF calculations do not provide good absolute I^{2+} values, which was to be expected because of the large difference in correlation energy between the ion and the neutral molecule; however, they do reproduce fairly well the energy differences between the first states of NH_3^{2+}. Similar SCF calculations on other molecules can thus be expected to prove useful in assigning the measured I^{2+} values to given states of the ion, provided these states are fairly well separated. On the other hand SCF–CI calculations are seen to yield accurate ionization potentials (the accuracy is estimated[46] to be better than 0.2 eV). Experimentally, the first excited state is observed closer to the ground state than to the second excited state. The quantity $2E_2 - (E_1 + E_3)$, where E_1, E_2, E_3 are the energies of the three states, is a measure of the displacement of the second state from a position exactly halfway between the two others. In the molecular orbital approximation this displacement is ascribed to differences in the Coulombic repulsion between electron pairs in each of these states and can be calculated from coulomb integrals[46]; the value obtained (-3.2 eV) is larger than both the experimental and the CI value (≈ -2 eV), showing that correlation energy differences reduce the effect of differences in the coulomb forces.

4.6. Observation of Radiative Transitions without Detection of the Photon

If in a double-collision process (IIa) the neutral projectile is formed in an excited state and emits a photon before taking part in the second collision (IIb), leading to the formation of the detected negative ion, this will show up in the final energy balance. Such a double-collision process is then

$$A^+ + M \to A^* + M^+$$
$$A^* \to A + h\nu \tag{III}$$
$$A + M \to A^- + M^+$$

and the energy loss of the A^- ion formed in such a process is

$$\Delta E_3 = I_a^+(M) + I_b^+(M) + h\nu - I^+(A) - EA(A) \tag{4.3}$$

Such a process can only be observed if at least two conditions are fulfilled: (a) The cross section for charge transfer with excitation of the projectile must be sufficiently large and (b) the radiative lifetime of the excited state must be small compared to the mean time between two collisions in the collision chamber.

The (H^+, M) spectra where M is a noble-gas atom or the H_2 molecule provide a good illustration of process III.[23] The (H^+, He) spectrum shown in Fig. 6 consists of two peaks which, because of their intensity variation with pressure, are ascribed to double-collision processes; in fact the double-charge-transfer (type I) process has not been detected because of its high endothermicity, which leads to a very small cross section (Section 2). We have already noted in Section 3.2 how the first peak (A in Fig. 6) corresponding to the lowest energy loss could be ascribed without any doubt to the

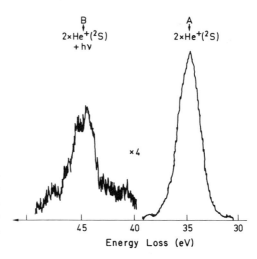

Fig. 6. (H^+, He) spectrum.[23] Collision energy $E_0 = 4$ keV. No double-charge-transfer process was observed. The cross section is very small because of the large endothermicity of such a process.

double-collision process (type II) where both He^+ ions are formed in their ground state. The energy loss corresponding to the second peak (B in Fig. 6) is larger by 10.1 ± 0.7 eV than the energy loss for peak A; this rules out the possibility of its being due to a double-collision process of type II, where one of the He^+ ions would be formed in an excited state (the first excited state of He^+ is 40.1 eV[37] above the ground state). Furthermore, an analogous peak due to a double collision and with an energy loss greater by roughly 10.2 eV than the one corresponding to the first double-collision peak, has been observed in other (H^+, M) spectra. In particular this is so when M is any noble-gas atom (see Ref. 27 for Ar, Kr, and Ne) or the H_2 molecule (cf. Fig. 3 and Ref. 2). (In xenon, as can be seen in Fig. 2, this peak unfortunately corresponds to an energy loss close to those of double-charge-transfer processes.) The conclusion that the occurrence of such peaks must be due to some excitation of the projectile is thus easily reached. A double-collision process (type II), where H^- would be formed in an excited state, is ruled out because of the absence of stable H^- states as already noted. It was ascertained that a double-collision process (type III) took place, where the H atom was formed in the excited $2p$ state and decayed to the $1s$ state by emitting a Lyman-α photon whose energy is well known (10.2 eV), before taking part in the second collision.*

These conclusions on the nature of the reaction have been confirmed by the high resolution $(H^+$, noble gases) spectra recently obtained by Fournier et al.[27] The possible excitation of the hydrogen atom formed in charge-transfer collisions has been extensively studied by measuring the light emission of the excited atom. A recent review of these studies has been given by McNeal and Birely.[22] Compare also Chapters 2 and 3 of this text.

Although the emission process III has so far only been observed when the projectile is a proton, its possible occurrence must be kept in mind when discussing any (A^+, M) spectrum.

5. (A^+, Ar) Spectra : An Insight into Future Possible Developments

As already noted, (A^+, M) spectra can provide information on the states of the projectile ion if the ionized states of the target M are known. This is the case if the target is a noble-gas atom the ionic states of which are well

---

*A contribution to the lowest energy-loss peak of a double-collision process where the intermediate H atom is formed and reacts in the $2s$ metastable state was incorrectly omitted in Ref. 23. In spite of the fact that the formation of $H^-(1s^2)$ through the capture of an electron by $H(2s)$ implies a two-electron transition, Dose and Gunz[50] have shown that the cross section for this process when the target is argon (for collision energies of some keV) is an order of magnitude greater than the cross section for the same process with $H(1s)$.

known.[25] In the experiments performed so far argon has been chosen as a target for two reasons: (a) It is light enough for spin–orbit coupling to be negligible and the spin conservation law can be applied when interpreting the spectra (Section 4.1); this is an advantage over heavier noble gases such as xenon. (b) Its single- and double-ionization potentials are lower than those of helium or neon so that the charge transfer cross sections are expected to be larger than in He or Ne (see Section 2) and the (A^+, Ar) spectra should thus be easier to obtain than (A^+, Ne) or (A^+, He) spectra. The observation of a given (A^+, Ar) spectrum is of course only possible when a sufficiently long-lived state of the negative A^- ion exists (the lifetime must be at least a few microseconds to allow for the energy analysis and for the detection of the ion). The information that can be obtained from such spectra is manifold and although the experiments are mostly in a preliminary state, in the following we describe briefly the different possible applications of DCTS of (A^+, Ar).

5.1. Determination of the State Composition of the A^+ Beam

The first (A^+, Ar) spectrum measured was the (O_2, Ar) spectrum obtained by Fehsenfeld et al.,[7] shown in Fig. 7. Its interpretation led to two conclusions.[7] First, as indicated in Section 4.2, the Franck–Condon principle applies to electronic transitions occurring in the molecular partner during a

Fig. 7. (O_2^+, Ar) spectrum.[7a] Peaks A and B are due to double-collision processes where the O_2^+ ion is initially in the $\tilde{a}\,^4\Pi_u$ and $\tilde{X}\,^2\Pi_g$ states, respectively. Peak C is due to the double charge transfer process where O_2^+ is initially in the $\tilde{a}\,^4\Pi_u$ state. No peak is observed due to a double-charge-transfer process where the O_2^+ ion would be initially in the $\tilde{X}\,^2\Pi_g$ state; because of the vertical transitions occurring in these processes, the O_2^- ions formed in this latter case are unstable with respect to autodetachment and are thus not detected by DCTS.

double-charge-transfer process. Second, the observed double-collision processes lead to a qualitative determination of the state composition of the incoming positive ion beam. In this particular case it could be ascertained that the O_2^+ beam was composed of ions in the ground state $X^2\Pi_g$ and in the metastable state $a\ ^4\Pi_u$ in agreement with the results of Turner et al.[51]

Preliminary results obtained by recording the (O^+, Ar) spectra when the O^+ ions are formed by electron impact on various molecules (O_2, CO_2, H_2O, CO, etc.) not only support the preceding statement that the qualitative state composition of the ion beam can be determined but further indicate that when A^+ is a fragment ion, the results should also provide evidence pertaining to the translational energy imparted to the fragment ion in a given electronic state during the dissociative ionization process leading to the ion in this particular electronic state.[7b]

5.2. Determination of the Geometric Structure of a Polyatomic A^+ or A^- Ion

In Section 4.2 it was noted that the determination of the geometric structure of an A^+ or A^- polyatomic ion by DCTS is a direct consequence of the applicability of the Franck–Condon principle to charge-transfer processes. This method of determination of the geometric structure of polyatomic ions was first proposed by Keough et al.[8a] in their study of the "$-E$ mass spectra" of polyatomic molecules (see Chapter 5).

The Franck–Condon principle simply states that the A^- ions will be formed with the same geometric structure as the incoming A^+ ion. Starting from an A^+ ion, whose geometric structure is known, the observation or the nonobservation of an (A^+, Ar) spectrum will provide evidence as to the stability of the A^- ion in this geometric structure. The stable geometric structure (if any) of A^- could be determined by using different molecular sources for producing the A^+ ions. It might also become possible to find the geometric structure of an ion A^+ formed by dissociative ionization of different molecules. Furthermore, it can be expected that the (A^+, M) spectrum should provide evidence pertaining to the geometric structure of an A^+ ion formed from a given molecule in a given electronic state. Work is currently in progress to ascertain the geometric structure of the $C_2H_2^+$ ion formed from different molecules.

5.3. The Electron Affinity of the Projectile Species A

A number of techniques have been developed to measure the electron affinities of atoms and molecules (see Refs. 10, 13, and 40). Clearly DCTS can be used to determine the electron affinity of the projectile species, as is apparent from Eqs. (1.1) and (1.2). However, this calls for a few remarks.

In the case of an atomic projectile, the electron affinity should be readily obtained from the (A^+, Ar) spectra. However, the accuracy achieved in DCTS cannot be expected, at the present time, to compete favorably with that of other techniques and particularly with that of photodetachment experiments.[52]

In the case of a molecular projectile, it must first be noted that, because of the validity of the Franck–Condon principle in charge-transfer processes (Section 4.2), the experimental value obtained from an (A^+, Ar) spectrum is not, strictly speaking, the electron affinity of A, which is defined as the adiabatic detachment energy.[40] Rather it represents a vertical detachment energy which may well be different. The experimental information on the electron affinities of molecules is rather meager and often contradictory, although very accurate EA values for some molecules have been measured recently by means of laser photodetachment and photoelectron spectrometry.[53] Therefore, it can be hoped that, by the use of double-charge-transfer spectroscopy (if the apparatus used provides for an absolute calibration of the energy loss scale), new information will be obtained concerning the heats of formation of negative molecular ions.

Although negative ions in excited states are expected to be unstable in most cases,[40] the (A^+, Ar) spectra should also provide data pertaining to these excited states when their lifetime is greater than a few microseconds.

5.4. Dissociative Double-Charge-Transfer Spectra

We have so far described the energy spectra of negative ions which were formed by double-charge-transfer or double-collision processes; in the negative-ion mass spectra recorded when a positive molecular ion beam collides with a target gas other peaks are observed, frequently at nonintegral apparent masses. This observation indicates that molecular negative ions (AB^-) formed in a double-charge-transfer (type I) or a double-collision (type II) process can dissociate. The resulting negative fragment ions (A^-) appear on the mass scale as if they had a mass close* to $(m_A)_{app} = m_A^2/m_{AB}$ as is well known from the similar dissociation of metastable positive ions in the field-free region of a mass spectrometer.

The observed peaks are broadened by the translational energy imparted to the fragment A^- ion during the dissociation of AB^-, and furthermore, this translational energy, which may be very small in the center-of-mass frame, is amplified in the laboratory frame exactly as in the

---

*The mass scale is determined for the nominal energy E_0 of the positive ions, and the apparent mass is calculated assuming AB^- has the same translational energy while in reality it has suffered an energy loss ΔE characteristic of its formation process. To a first approximation, because $\Delta E \ll E_0$, the apparent mass of A^- is given by the relation shown.

unimolecular or collision-induced dissociation processes described in Chapters 6 and 7. From these observations it can be foreseen that the dissociative double-charge-transfer spectra should become the counterpart, for molecular negative ions, of the translational energy spectra of positive fragment ions formed in metastable or collision-induced dissociation of positive molecular ions.

To obtain such a dissociative double-charge-transfer spectrum some experimental conditions have to be fulfilled, most of which are identical to those required for the translational energy spectra of positive fragment ions. A further condition is as follows: The AB^- ions should be formed either by only one process or by processes in which the energy losses are sufficiently close to each other to avoid a broadening of the peak simply due to the dissociation of AB^- ions having different initial translational energies; this should generally be possible by selecting the target pressure so that either double-charge-transfer processes or double-collision processes are predominant.

Appendix: Energy Loss of the Projectile in an Inelastic Collision

Initially the projectile A of mass m moves in a direction x with velocity $\mathbf{v}_0$ and a translational energy $E_0 = \frac{1}{2}mv_0^2$, the target B of mass M is at rest. During the collision the sum of the internal energy changes of A and B is Q. We want to calculate the translational energy E of the projectile A scattered in a direction whose angle to the incident direction x is θ (cf. introductory chapter, Fig. 4).

In the center-of-mass frame A and B move initially with a relative velocity $\mathbf{v}_0$ and the velocities $\mathbf{u}_A$ and $\mathbf{u}_B$ are given by

$$\mathbf{v}_0 = \mathbf{u}_A - \mathbf{u}_B$$

$$\mathbf{u}_A = \frac{M}{m+M}\mathbf{v}_0 \quad \text{and} \quad \mathbf{u}_B = -\frac{m}{m+M}\mathbf{v}_0 \quad \text{(A.1)}$$

After the collision, the relative velocity $\mathbf{v}'$, at an angle ϕ with respect to $\mathbf{v}_0$, and the velocities $\mathbf{u}'_A$ and $\mathbf{u}'_B$ will have the same relationship as (A.1) because of momentum conservation. Total energy conservation leads to

$$\frac{\frac{1}{2}mM}{m+M}v_0^2 = \frac{\frac{1}{2}mM}{m+M}v'^2 + Q$$

so that

$$v' = \left(v_0^2 - \frac{2(m+M)}{mM}Q\right)^{1/2} \quad \text{(A.2)}$$

In the laboratory frame, A will, after the collision, have a velocity **W**, the angle $\mathbf{v}_0\mathbf{W}$ being equal to θ, and the following relationships hold:

$$W_x = u'_A \cos\phi - u_B, \qquad W_y = u'_A \sin\phi \qquad (A.3)^*$$

$$\tan\theta = \frac{m \sin\phi}{M \cos\phi + m}$$

[this last relationship is valid as long as Q is small compared to $ME_0/(m+M)$] and we obtain the translational energy of A after the collision:

$$E = \tfrac{1}{2}mW^2 = \tfrac{1}{2}m[u'^2_A - 2u'_A u'_B \cos\phi + u^2_B] \qquad (A.4)$$

Replacing u'_A and u_B by their values as a function of v' and v_0, respectively [cf. Eq. (A.1)], (A.4) becomes

$$E = \tfrac{1}{2}m\left[\frac{m^2 v_0^2 + M^2 v'^2}{(m+M)^2} + \frac{2mM}{(m+M)^2} v_0 v' \cos\phi\right] \qquad (A.5)$$

Using the value of v' given by (A.2), we obtain

$$E = \tfrac{1}{2}m\left[\frac{m^2 + M^2}{(m+M)^2} v_0^2 - \frac{2M}{m(m+M)} Q \right.$$

$$\left. + \frac{2mM}{(m+M)^2} v_0^2 \cos\phi \left(1 - \frac{2(m+M)}{mM} \frac{Q}{v_0^2}\right)^{1/2}\right]$$

or as a function of E_0:

$$E = \frac{m^2 + M^2}{(m+M)^2} E_0 - \frac{M}{m+M} Q + \frac{2mM}{(m+M)^2} E_0 \cos\phi \left(1 - \frac{m+M}{M} \frac{Q}{E_0}\right)^{1/2} \qquad (A.6)$$

Expanding the square root in the last term and rearranging terms yields

$$E = E_0\left[\frac{m^2 + M^2 + 2mM \cos\phi}{(m+M)^2}\right] - Q\left[\frac{M + m \cos\phi}{m+M}\right] - \frac{1}{4}\frac{m}{M}\frac{(Q)^2}{E_0}\cos\phi \qquad (A.7)$$

The energy loss $\Delta E = E_0 - E$ can then be written as

$$\Delta E = Q + \frac{m}{m+M}(1 - \cos\phi)\left[\frac{2M}{m+M} E_0 - Q\right] + \frac{1}{4}\frac{m}{M}\frac{(Q)^2}{E_0}\cos\phi \qquad (A.8)$$

The two last terms on the right-hand side are the recoil energy of the target B:

$$E_B = \frac{m}{m+M}(1 - \cos\phi)\left[\frac{2M}{m+M} E_0 - Q\right] + \frac{1}{4}\frac{m}{M}\frac{(Q)^2}{E_0}\cos\phi \qquad (A.9)$$

*The center of mass has a velocity equal to $-\mathbf{u}_B$, the target being initially at rest.

We must now examine under what experimental conditions the energy loss ΔE can be set equal to the net change in the internal energy Q, thus neglecting the recoil energy of the target E_B.

The second term on the right-hand side of (A.9) can only become small if $E_0 \gg Q$, which is a condition met in DCTS where Q is equal to a few tens of electron volts while E_0 is equal to several kiloelectronvolts. Now the first term* in (A.9) will be maximized when the ratio $m/M = 1$, while the second term increases with increasing m/M. In DCTS experiments m is, as a rule, smaller than M so that by setting $m = M$ we obtain an upper value for E_B. E_B as a function of $t = \tan \phi/2$ is then given by

$$E_B = \frac{t^2}{1+t^2} E_0 + \frac{1}{4}\left(\frac{1-t^2}{1+t^2}\right)\frac{(Q)^2}{E_0} \qquad (A.10)$$

At zero scattering angle the first term on the right-hand side of (A.10) vanishes while the second reaches its maximum; taking an upper limit for $Q = 40$ eV we thus obtain a maximum value of E_B: $(E_B)_{max} = 0.1$ eV for $E_0 = 4$ keV (neglecting E_B in ΔE amounts to a maximum relative error of 0.25% in the determination of Q). The first term in (A.10) increases with the scattering angle; when $\phi = 1°$ [which corresponds, in this case, to an angle $\theta = 0.5°$ in the laboratory frame, cf. (A.3)] this term is equal to 0.3 eV. As already stated this is an upper limit of E_B; when m becomes smaller than M, E_B decreases while θ increases for a given ϕ [see relation (A.3)] so that it seems to be reasonable to neglect E_B as long as the collected projectiles are those scattered through small angles (θ less than some tenths of a degree).†

In the case of a double-collision process (type II) the observed scattering angle results from two individual scattering events so that the actual scattering angles could well be outside the limits set for the scattering angle in a double-charge-transfer process (type I) by the experimental setup. This could conceivably lead to a sum of individual recoil energies, computed as above, too large to be neglected. However, the differential scattering cross section for charge transfer is known to be strongly forward peaked above some hundred electron volts collision energy so that here again it seems reasonable to neglect the recoil energies of the two targets.

---

*We neglect in this term the contribution of Q. This is valid as long as $Q \ll [2M/(m+M)]E_0$. Q is at most equal to $10^{-2} E_0$ so this implies that $2M/(m+M) \gg 10^{-2}$, in other words, that $m \ll 200M$, a condition which is generally met.

†EDITOR'S NOTE: The above derivation also yields the disproportionation factor D, the ratio of the changes in energy Q of the projectile and target as a result of the inelastic collision. For the case of negligible laboratory scattering of the high-energy projectile we have from Eq. (A.8) that

$$D \approx Q \bigg/ \left\{\frac{1}{4}\left(\frac{m}{M}\right)\frac{(Q)^2}{E_0}\right\} = \left(\frac{4M}{m}\right)\left(\frac{E_0}{Q}\right) \qquad (A.11)$$

ACKNOWLEDGMENTS

I am indebted to Professor J. Durup for many enlightening discussions and for his helpful criticisms and comments on this chapter. I also wish to acknowledge the important contribution of Dr. P. Fournier to the development of DCTS.

References

1. F. C. Witteborn and D. E. Ali, *Bull. Am. Phys. Soc.* **16**, 208 (1971) and private communication.
2. P. Fournier, J. Appell, F. C. Fehsenfeld, and J. Durup, *J. Phys. B: At. Mol. Phys.* **5**, L58, 1810 (1972).
3. J. Appel, Thèse de doctorat d'Etat, Université de Paris-Sud, Orsay, 1972.
4. J. Appell, J. Durup, F. C. Fehsenfeld, and P. Fournier, *J. Phys. B: At. Mol. Phys.* **6**, 197 (1973).
5. J. Appell, J. Durup, F. C. Fehsenfeld, and P. Fournier, *J. Phys. B: At. Mol. Phys.* **7**, 406 (1974).
6. (a) P. Fournier, R. E. March, C. Benoit, T. R. Govers, J. Appell, F. C. Fehsenfeld, and J. Durup, VIII ICPEAC, Belgrade, 753 (1973); (b) P. Fournier, C. Benoit, J. Durup, and R. E. March, *C.R. Acad. Sci.* **278**, 1039 (1974).
7. (a) F. C. Fehsenfeld, J. Appell, P. Fournier, and J. Durup, *J. Phys. B: At. Mol. Phys.* **6**, L268 (1973); (b) Unpublished results of the present author.
8. (a) T. Keough, J. H. Beynon, and R. G. Cooks, *J. Am. Chem. Soc.* **95**, 1965 (1973); (b) *Int. J. Mass Spectrom. Ion Phys.* **16**, 417 (1975).
9. (a) H. S. W. Massey, *Rep. Progr. Phys.* **12**, 248 (1949); (b) H. S. W. Massey and E. H. S. Burhop, *Electronic and Ionic Impact Phenomena*, Oxford Univ. Press, London, 1952.
10. J. B. Hasted, *Physics of Atomic Collisions*, Butterworth, London, 1972.
11. J. B. Hasted, *Adv. At. Mol. Phys.* **4**, 237 (1968).
12. J. B. Hasted, *Proc. R. Soc.* A **205**, 421 (1951); A **212**, 235 (1952).
13. E. W. McDaniel, *Collision Phenomena in Ionized Gases*, Wiley, New York, 1964.
14. J. B. Hasted, *J. Appl. Phys.* **30**, 25 (1959).
15. S. K. Allison and M. Garcia-Munoz, in *Atomic and Molecular Processes* (D. R. Bates, ed.), Academic Press, New York, 1962, p. 722.
16. N. V. Fedorenko, *Sov. Phys. Tech. Phys.* **15**, 1947 (1971).
17. H. Tawara and A. Russek, *Rev. Mod. Phys.* **45**, 178 (1973).
18. V. V. Afrosimov, Yu. A. Mamaev, N. M. Panov, and N. V. Fedorenko, *Sov. Phys. Tech. Phys.* **14**, 109 (1969).
19. V. V. Afrosimov, Yu. A. Mamaev, M. N. Panov, and N. V. Fedorenko, *Sov. Phys. JETP* **28**, 52 (1969).
20. V. V. Afrosimov, G. A. Leiko, Yu. A. Mamaev, and M. N. Panov, *Sov. Phys. JETP* **29**, 648 (1969).
21. V. V. Afrosimov, *Zh. Eksper. Teo. Fiz SSSR* **62**, 2049 (1972).
22. R. J. McNeal and J. H. Bireley, *Rev. Geophys. Space Phys.* **11**, 633 (1973).
23. J. Durup, J. Appell, F. C. Fehsenfeld, and P. Fournier, *J. Phys. B: At. Mol. Phys.* **6**, L110, 1810 (1973).
24. J. Durup, P. Fournier, and D. Pham, *Int. J. Mass Spectrom. Ion Phys.* **2**, 311 (1969).
25. C. Moore, *Atomic Energy Levels*, Natl. Bur. Stand., Washington, D.C., 1949.
26. P. Fournier, C. Benoit, T. R. Govers, and R. E. March, to be published.

27. P. Fournier, J. Appell, C. Benoit, J. Durup, F. C. Fehsenfeld, and R. E. March, to be published.
28. P. G. Burke, *Adv. At. Mol. Phys.* **4**, 173 (1968).
29. G. W. F. Drake, *Phys. Rev. Lett.* **24**, 126 (1970).
30. C. L. Pekeris, *Phys. Rev.* **126**, 1470 (1962).
31. W. Mehlhorn, *Z. Phys.* **160**, 247 (1960).
32. D. Stalherm, B. Cleff, H. Hillig, and W. Mehlhorn, *Z. Naturforsch.* **24a**, 1728 (1969).
33. W. A. Moddeman, J. A. Carlson, M. O. Krause, B. P. Pullen, W. E. Bull, and G. K. Schweitzer, *J. Chem. Phys.* **55**, 2317 (1971).
34. J. Siegbahn, C. Nordling, G. Johansson, J. Heelman, P. F. Heden, K. Hamrin, U. Gelius, T. Bergmark, L. O. Werme, R. Manne, and Y. Baer, *ESCA Applied to Free Molecules*, North-Holland Publ., Amsterdam, 1969.
35. R. Spohr, T. Bergmark, N. Magnusson, L. O. Werme, C. Nordling, and J. Siegbahn, *Phys. Scripta* **2**, 31 (1970).
36. E. Wigner, *Nachr. Akad. Wiss. Goettingen, Math. Phys. Kl* **IIa**, 325 (1927).
37. J. H. Moore, Jr., *Phys. Rev.* A **8**, 2359 (1973).
38. For a discussion on the diradical-type triplet states, see, e.g., L. Salem and C. Rowland, *Angew. Chem.* **11**, 92 (1972).
39. T. E. Sharp, *LMSC 5. 10.69.9.*, Lockheed Palo Alto Research Lab., Palo Alto, California, 1969.
40. B. L. Moiseiwitsch, *Adv. At. Mol. Phys.* **1**, 61 (1965).
41. W. A. Goddard III, D. L. Huestis, D. C. Cartwright, and S. Trajmar, *Chem. Phys. Lett.* **11**, 329 (1971).
42. D. C. Cartwright, S. Trajmar, H. William, and D. L. Huestis, *Phys. Rev. Lett.* **27**, 704 (1971).
43. J. Durup, *Chem. Phys.* **2**, 226 (1973).
44. F. H. Dorman and J. D. Morrison, *J. Chem. Phys.* **35**, 575 (1961); **39**, 1906 (1963).
45. A. C. Hurley, *J. Mol. Spectrosc.* **9**, 18 (1962).
46. J. Appell and J. Horsley, *J. Chem. Phys.* **60**, 3445 (1974).
47. T. Ast, J. H. Beynon, and R. G. Cooks, *J. Am. Chem. Soc.* **94**, 6611 (1972).
48. G. Herzberg, *Molecular Spectra and Molecular Structure*, Vol. III, van Nostrand–Reinhold, Princeton, New Jersey, 1966.
49. D. W. Turner, C. Baker, A. D. Baker, and C. R. Brundle, *Molecular Photoelectron Spectroscopy*, Wiley-Interscience, New York, 1970.
50. V. Dose and R. Gunz, *J. Phys. B: At. Mol. Phys.* **5**, 1412 (1972).
51. B. R. Turner, J. A. Rutherford, and D. M. J. Compton, *J. Chem. Phys.* **48**, 1602 (1968).
52. L. M. Branscomb, in *Atomic and Molecular Processes* (D. R. Bates, ed.), Academic Press, New York, 1962, p. 100; B. Steiner, in *Case Studies in Atomic Physics II*, (E. W. McDaniel and M. R. C. McDowell, eds.), North-Holland Publ., Amsterdam, 1972, p. 483; W. C. Lineberger and B. J. Woodward, *Phys. Rev. Lett.* **25**, 424 (1970); H. Hotop, T. A. Patterson, and W. C. Lineberger, *Phys. Rev.* A **8**, 762 (1973); H. Hotop, R. A. Bennett, and W. C. Lineberger, *J. Chem. Phys.* **58**, 2373 (1973); H. Hotop and W. C. Lineberger, *ibid.* **58**, 2379 (1973).
53. M. W. Siegel, R. J. Celotta, J. L. Hall, J. Levine, and R. A. Benneth, *Phys. Rev.* A **6**, 607 (1972); R. J. Celotta, R. A. Bennett, J. L. Hall, M. W. Siegel, and J. Levine, *ibid.* **6**, 631 (1972).
54. C. Benoit and J. A. Horsley, *Mol. Phys.* **30**, 557 (1975).
55. M. Durup, G. Parlant, J. Appell, J. Durup, and J. B. Ozenne, *Chem. Phys.*, in press (1977).

5

Ionic Collisions as the Basis for New Types of Mass Spectra

D. L. Kemp and R. G. Cooks

1. Introduction

Recently developed techniques for ion structure determination employ ion–molecule reactions in which the ion of interest undergoes a change in charge. These charge permutation reactions can be classified into two categories: (a) charge stripping, in which electrons are removed (stripped) from the high-energy ion, or (b) charge exchange, in which electrons are transferred between the ion and a neutral collision gas. In this chapter we will be concerned with the following four specific charge permutation reactions of polyatomic organic ions.

$$m^{2+} + N \rightarrow m^+ + N^+ \tag{1.1}$$

$$m^+ + N \rightarrow m^{2+} + N + e^- \tag{1.2}$$

$$m^+ + 2N \rightarrow m^- + 2N^+ \quad \text{or} \quad m^+ + N \rightarrow m^- + N^{2+} \tag{1.3}$$

$$m^- + N \rightarrow m^+ + N + 2e^- \tag{1.4}$$

D. L. Kemp and R. G. Cooks • Department of Chemistry, Purdue University, West Lafayette, Indiana 47907.

We will also consider some fragmentation reactions which can be observed in studying these charge changes.

As will be suggested later, some of these reactions can be studied on single-focusing mass spectrometers. However, the methods we discuss were originally developed on double-focusing instruments which are superior. The versatility of conventional double-focusing mass spectrometers is such that their original application to exact mass determinations has been extended to such areas as the study of kinetics and mechanisms of gas phase reactions, energy partitioning in simple molecules, and ion structure determinations. This versatility stems in part from the fact that in these instruments, ions are transmitted on the basis of *two* fundamental properties: (a) their energies, and (b) their momenta. Ions produced in the source are accelerated through a potential V which is normally between 2 and 10 keV, and transmitted by a radial electric field of potential E. Mass analysis is then carried out by a magnetic field which is capable of discriminating between ions according to their momentum-to-charge ratios. Since all ions formed in the source possess identical energy-to-charge ratios, the energy analyzer permits all such ions to be transmitted simultaneously and as such is of no special use in the study of source ions beyond its contribution to improved resolution in the double-focusing mode. However, the energy analyzer is invaluable in the study of reactions occurring in the field-free regions of the mass spectrometer and its use in this regard has developed into the technique of ion kinetic energy spectrometry (IKES).[1]

Metastable ions produced in the source possess a small amount of internal energy which causes them to fragment between the source and the electric sector (first field-free region). The daughter ions thus produced in Eq. (1.5) possess only a fraction of the energy of the parent ion:

$$m_1^+ \rightarrow m_2^+ + m_3 \qquad (1.5)$$

specifically, this energy will be (m_2/m_1) eV. By setting the voltage on the electric sector plates to $(m_2/m_1)E$, the daughter ion m_2^+ can be detected by an electron multiplier placed after the β, or energy resolving slit of the mass spectrometer. Keeping the accelerating voltage V constant, and scanning the electric sector voltage from 0 to E, an ion kinetic energy spectrum is obtained which is a record of all metastable transitions which occur in the first field-free region. Figure 1 is an IKE spectrum of phthalic anhydride recorded on an RMH-2 mass spectrometer modified for this particular purpose. Although our interest here is not specifically in spectra of the type shown in Fig. 1, these are of interest from an historical point of view because of their contribution to the use of the energy analyzer as a useful analytical device in its own right.

Close examination of energy spectra such as that shown in Fig. 1 reveals reactions other than fragmentations. For example, the peak at $0.50E$ in Fig.

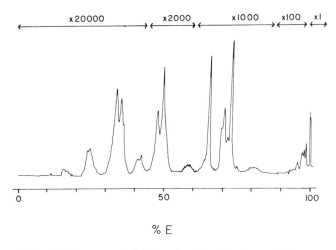

Fig. 1. IKE spectrum of phthalic anhydride using air as collision gas.

1 is due in part to the charge-stripping reactions exemplified by Eq. (1.2) as well as a special type of fragmentation reaction in which the daughter ion and neutral fragment are of equal mass. Keeping the electric sector constant at one-half the normal voltage while scanning the magnet has led to a new type of mass spectrum which includes the products of charge stripping as well as these particular fragment ions. Details of these spectra are discussed in Section 4.

Most work to date in mass spectrometry has been done on positive ions but there is a growing interest in negative ions. Conversion of the electric and magnetic sectors from their normal mode for transmission of positive ions to the mode for transmission of negative ions is readily accomplished and so the possibility exists for negative ion work on both single- and double-focusing instruments. High-energy ion–molecule reactions can be utilized to convert positive ions produced in the source to negative ions by the charge inversion process as demonstrated in Eq. (1.3). This form of reaction has been used to investigate the nature of negative polyatomic ions and has produced results which complement studies on negative ions produced by other methods. These results are analyzed in detail in Section 3.

General considerations of the factors which control high energy ion–molecule reactions are given in the Introduction to this volume and expanded upon in Chapter 7. Of particular note in connection with charge-changing collisions is the role of the collision gas, the nature of which affects the cross section and the thermochemistry of some reactions. For other reactions, such as charge stripping, the target acts merely as a cluster of electrons and its nature need not affect the overall thermochemistry. Other important considerations are the number of collisions (cf. Chapter 7, Section

2.4.1), the time-scale of the interaction (Chapter 7, Section 2.2.1), and the influence of resonance effects on reaction cross sections (Chapter 7, Section 2.3). In this chapter, we shall be dealing exclusively with new types of *mass spectra* which result when ions originally produced in the source undergo a charge-changing reaction in a field-free region of the mass spectrometer.

The ion abundances observed in charge-changing mass spectra are approximately 10^{-3} of those observed for normal mass spectra, so there is a need for high sensitivity in these types of measurements. This has led to the use of wide slits at the β position and at the final collector. Except for charge-stripping spectra, which require good energy resolution, the use of wide slits has proven to be beneficial in terms of providing high sensitivity for studying these charge-changing reactions. Another valuable procedure has been to employ a short collision region at the source focal point of a double-focusing mass spectrometer.[2] This instrumental modification permits the use of high-collision gas pressure to increase the number of ion–molecule collisions, thereby enhancing the abundances of the product ions. In addition, since all the collisions take place at or near the point of focus of the electric sector, this method (a) increases the energy resolution which can be obtained and (b) improves the collection efficiency of the instrument, thereby providing further enhancement of product ion abundances.

The technique of using the electric sector as an analyzer in its own right has facilitated the study of charge-changing reactions. As will be seen in the following sections, these new types of mass spectra are providing new insights into the structures of ionic species.

2. 2E Mass Spectra

The presence of multiply charged ions in mass spectra was first noted by the appearance of sharp peaks at nonintegral masses.[3] The most familiar examples are doubly charged ions (m_1^{2+}) which, having mass-to-charge ratios of $m_1/2$, appear at an apparent mass of one-half the actual mass. For odd-mass ions, the peaks appear at half-integral masses and so these ions are readily distinguishable from singly charged ions. But for doubly charged ions of even mass, the apparent mass is an integral number and is thus indistinguishable from that of the singly charged ion of one-half its mass. It is this fact which has reduced the usefulness of studies on doubly charged ions produced in the ion source. Although the abundances of these ions are great enough to make them easily detectable, the use of high-resolution techniques is required to determine the abundances of the even-mass ions.

These problems can be overcome by the use of a high-energy charge-exchange reaction in which the doubly charged ions gain an electron from a neutral collision gas introduced into the first field-free region of a conventional double-focusing mass spectrometer. This type of charge-transfer reaction is illustrated by

$$m^{2+} + N \rightarrow m^+ + N^+ \tag{2.1}$$

where N is the target species.

It is possible to separate these singly charged ions (m^+) from those produced in the source on the basis of their energy-to-charge ratios. All ions produced in the source are accelerated through the same potential, but because of the difference in charge the doubly charged ions gain twice the energy of the singly charged ions. If these ions now undergo a charge-exchange reaction in the first field-free region, they will have an energy-to-charge ratio twice that of the main beam of singly charged ions generated in the ion source. Increasing the electric sector to twice the normal value of E will transmit the charge-exchanged ions and no others. If the magnetic field is scanned with the electric sector in this configuration, a $2E$ mass spectrum results which is a record of all doubly charged ions which have undergone the charge exchange reaction.

A $2E$ spectrum is by no means an exact replica of the doubly charged spectrum which would result from plotting the abundances of doubly charged ions appearing in the normal mass spectrum. For this to be true, it would be necessary for all doubly charged ions to have the same cross section for charge exchange. Nevertheless, the available information, although limited, suggests a close relation between the abundances of ions in the $2E$ spectra and the abundances of doubly charged ions in the mass spectra.

Figure 2 shows the $2E$ spectrum of toluene using the compound itself as target gas. The spectrum displays several characteristics which are common to those of aromatic hydrocarbons. The ions at 86 and 90 have the

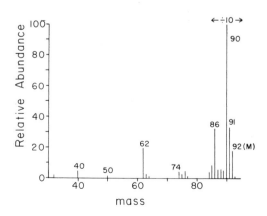

Fig. 2. $2E$ mass spectrum of toluene taken on the RMH-2 mass spectrometer using the compound itself as target gas in the short collision chamber.

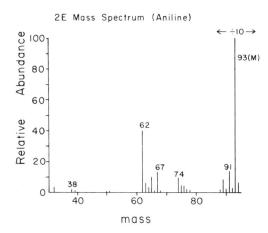

Fig. 3. 2E mass spectrum of aniline taken on the RMH-2 mass spectrometer using the compound itself as target gas in the short collision chamber.

compositions C_7H_2 and C_7H_6, respectively, and this is in agreement with the finding[4] that for this class of compounds, the ions of general formula $C_nH_2^{2+}$ and $C_nH_6^{2+}$ have high cross sections for the charge exchange to form the singly charged species. In contrast, Fig. 3 shows the 2E spectrum of aniline in which the molecular ion is the most intense peak. The presence of the heteroatom in aromatic amines lends stability to the molecular doubly charged ion and it is usually the base peak in the 2E spectrum.[5]

Before considering more of the chemistry of doubly charged ions, as represented in 2E mass spectra, it may be helpful to consider further the high-energy charge-exchange collision itself. This process involves electronic interaction between the ion and the target, with vibrational or rotational excitation apparently playing only a minor role. The collision has been termed quasi-Franck–Condon, since its time scale is rather less than that of one molecular vibration. An important consequence is that the configuration of the singly charged ion immediately after the collision is the same as that of the reactant doubly charged ion. This constraint is the same as that of the reactant doubly charged ion. This constraint in some cases means that the singly charged ion is necessarily formed in an excited electronic state. The other factor that determines the subsequent behavior of the singly charged ion formed by charge exchange is the energetics of the reaction. Unlike the charge-stripping process leading to $E/2$ spectra (see later) the nature of the target used *is* a factor in the thermochemistry of charge exchange. Thus, the exo- or endothermicity (Q) of the charge exchange reaction is given by the sum of the recombination energy of the doubly charged ion and the ionization potential of the target. As a general rule, the second ionization potential is greater than the first. This, therefore, represents a tendency for charge exchange reactions to be exothermic, to occur as superelastic collisions with the kinetic energy of the fast species carrying the bulk of the heat of reaction (see the Introduction). In polyatomic species, however, with large numbers of closely spaced levels there are numerous recombination energies and ionization potentials corresponding to different states. One would expect that those processes which come closest to resonance ($Q = 0$) will have the highest cross sections.[6,7] This would favor the formation of internally excited ions.

Thus, both the time-scale constraint and the energetics of the reaction will often favor the formation of charge-exchange products which are internally excited. Such ions may fragment, and be lost from the $2E$ spectra. That the charge-exchange products are internally excited is shown by the fact that spontaneous fragmentation is often observed. If this occurs after the charge-exchanged ion (m_1^+) has traversed the electric sector and before mass analysis, it gives rise to metastable peaks. These metastable peaks in $2E$ mass spectra are exactly analogous to those in normal mass spectra. For example, the $2E$ mass spectrum of pyridine shows broad low abundance ions at masses 32.9, 33.3, and 34.2. These are due to the further fragmentations of the singly charged ions produced by charge exchange of the doubly charged molecular ion ($79^{2+\cdot}$) and 78^{2+}:

$$79^+ \to 51^+ + 28 \qquad (2.2)$$

$$78^+ \to 51^+ + 27 \qquad (2.3)$$

$$79^+ \to 52^+ + 27 \qquad (2.4)$$

Figure 4 compares a region of the mass spectrum of pyridine with the same region of the $2E$ mass spectrum. The metastable peaks observed are quite different, in particular, the molecular ion $C_6H_5N^{+\cdot}$ generated by electron

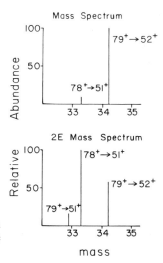

Fig. 4. Portion of the $2E$ mass spectrum compared with the mass spectrum of pyridine. All the peaks shown are metastable peaks.

impact fragments by loss of 27 mass units (HCN) but not detectably by loss of 28 mass units. However, this latter process is a significant metastable ion fragmentation of $C_6H_5N^{+\cdot}$ ions generated by charge exchange. It is likely that the two methods produce ions which have different structures. We shall see later how doubly charged ions tend to have linear structures with terminal hydrogens and that it is not unexpected for the charge-exchange products of such ions to be structurally distinct from ions generated directly by electron impact.

The energetics of the charge-exchange process leads one to expect that the nature of the target used to record $2E$ mass spectra can affect the cross sections for the electron transfer, the amount of internal energy of the ion, and hence the subsequent fragmentation of the product ion. This can be seen in Fig. 5 which shows the same region of the $2E$ mass spectrum of 1,4-dicyano-1-butene using five different target gases.

In the remainder of this discussion the chemistry of doubly charged ions is studied by considering their fragmentations in conjunction with their charge-exchange reactions ($2E$ spectra). The fragmentation reactions are followed by the usual methods of ion kinetic energy spectrometry. They are

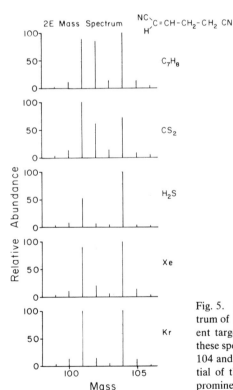

Fig. 5. Molecular ion region of the $2E$ mass spectrum of 1,4-dicyano-1-butene taken with the different target gases indicated. The prominent ions in these spectra are $(M-H_2)^{2+}$ and $(M-H_5)^{2+}$ at masses 104 and 101, respectively. As the ionization potential of the target decreases (from Kr to C_7H_8), a prominent $(M-H_4)^{2+}$ ion is also recorded.

both unimolecular and collision-induced and may [Eq. (2.5)] or may not [Eq. (2.6)] involve charge separation.[8,9] The types of reaction undergone by a particular ion and the kinetic energy release accompanying individual reactions can be used to characterize the structure of the doubly charged ion.

$$m_1^{2+} \to m_2^+ + m_3^+ \tag{2.5}$$

$$m_1^{2+} \to m_2^{2+} + m_3 \tag{2.6}$$

It was mentioned earlier in this section that the molecular ion in the $2E$ spectra of aromatic amines is usually the base peak in these compounds, this being due to charge stabilization by the heteroatoms. A study of the $2E$ spectra of the three geometrical isomers of phenylenediamine shows identical behavior in all cases.[5] In the cases of the molecular ions, the localization of charge on the nitrogen atoms should result in different cross sections for charge exchange if the ring remained intact and the amine groups remained meta, para, or ortho to each other in the respective isomers. The fact that the spectra are identical would indicate that isomerization occurs in the source and that the doubly charged ions formed in all three cases are identical. Further evidence for this is available. If these molecular ions were to fragment, with charge separation, the energy released would depend largely on the coulombic repulsion exerted by the two daughter ions against one another. This energy release can be measured by ion kinetic energy spectrometry and from this the intercharge distance r can be calculated. The value of r for the three isomers is found to be the same. Although these data are not conclusive as to the exact structure of the doubly charged molecular ion, it has been suggested that the para form, which allows maximum intercharge distance without disruption of the aromatic ring, is preferred in all three cases.

Another application of $2E$ spectra to the study of ion chemistry involves metastable transitions of the aromatic amines. In this class of compounds, the most favorable reaction of the singly charged ions is the loss of HCN from the molecular ion $M^{+\cdot}$ or, in a few cases, from the $(M-1)^+$ ion. A study of the behavior of the doubly charged molecular ions of amines and diamines has shown[5] that the favored decomposition is the loss of C_2H_2, and not HCN. The interpretation for this behavior is that the heteroatom(s) stabilizes the two charges on the molecular ion and that the loss of one of these nitrogen atoms would mean a decrease in the charge stabilization; hence loss of HCN is very unfavorable.

Metastable peaks observed in the $2E$ mass spectra result from the decomposition of singly charged ions in the field-free region between the electric and magnetic sectors. These are not, of course, singly charged ions produced in the source but rather those formed from doubly charged ions which have gained an electron via the charge-exchange reaction. As noted, the resulting singly charged ions may be internally excited, resulting in decomposition after being transmitted by the electric sector. The metastable peaks observed in the $2E$ spectra of the aromatic amines indicate that the loss of HCN, and not C_2H_2, is again the favored process. This lends support to the idea of charge stabilization by the nitrogen atoms but it also proves that the structure of the doubly charged ion is such that it can gain an electron and resume the normal behavior expected for singly charged ions.

Structural studies on doubly charged ions are in some respect more tractable than those on singly charged ions, the kinetic energy release in charge separation reactions providing data which relate directly to the geometry of the fragmenting ion. This point, among others, has been explored in a study of the $2E$ mass spectra of hydrocarbons.[10]

Since even-electron species are generally more stable than odd-electron species,[11] it is expected that most doubly charged ions which reach the field-free region and are available for charge exchange will be even-electron ions. This is an important factor in determining the nature of $2E$ mass spectra and it means that the abundant peaks in these spectra fall at even mass (for C-, H-, and O-containing ions). The spectrum of 1,7-octadiene (Fig. 6) illustrates this feature as well as showing the unusual stability of ions in the $C_nH_2^{2+}$ and $C_nH_6^{2+}$ series, namely, 86 ($C_7H_2^{2+}$), 90 ($C_7H_6^{2+}$), and 102 ($C_8H_6^{2+}$). The ions $C_7H_7^{2+\cdot}$ and, in several other hydrocarbons, $C_9H_7^{2+\cdot}$ are unexpectedly stable for odd-electron ions. Structural information on some of these ions has been obtained. The metastable ion reactions of $C_8H_6^{2+}$

$$C_8H_6^{2+} \rightarrow C_7H_3^+ + CH_3^+ \qquad (2.7)$$

$$C_8H_6^{2+} \rightarrow C_5H_3^+ + C_3H_3^+ \qquad (2.8)$$

imply an extended cumulene type of structure.

$$H_3C-\overset{+}{C}=C=C=C=C=\overset{+}{C}-CH_3$$

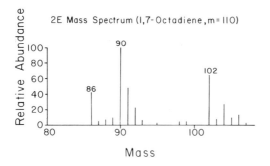

Fig. 6. $2E$ mass spectrum of 1,7-octadiene using benzene as target gas in the first field-free region of the RMH-2 mass spectrometer.

This observation applies to other ions of the $C_nH_2^{2+}$ and $C_nH_6^{2+}$ classes. Kinetic energy releases accompanying these reactions yield intercharge distances consistent with these linear structures. For example, 7.3 Å for the CH_3^+ loss in reaction (2.7). It must be noted, however, that these ion structural results refer to spontaneously fragmenting ions, and it is always possible that nonfragmenting ions will have different structures. Nevertheless, the importance of the carbon-wire type of structure in doubly charged ion chemistry seems to be established. Further evidence comes from a consideration of fragmentations which do not involve charge separation. In the hydrocarbon series, such reactions are predominantly H_2 and C_2H_2 losses so that the doubly charged molecular ion of 2-phenylnaphthalene ($C_{16}H_{12}^{2+}$), for example, fragments as follows to give ions which are abundant in the $2E$ spectrum:

$$C_{16}H_{12}^{2+} \xrightarrow{-H_2} C_{16}H_{10}^{2+} \xrightarrow{-C_2H_2} C_{16}H_8^{2+}$$
$$\downarrow {-C_2H_2} \qquad \downarrow {-C_2H_2}$$
$$C_{14}H_{10}^{2+} \xrightarrow{-H_2} C_{14}H_8^{2+}$$
$$\downarrow {-C_2H_2} \qquad \downarrow {-C_2H_2}$$
$$C_{12}H_8^{2+} \xrightarrow{-H_2} C_{12}H_6^{2+}$$

A most striking and not atypical result in the $2E$ spectra of hydrocarbons is the presence of an ion at mass 126 in the n-decane spectrum. To generate this species, the doubly charged molecular ion of n-decane must lose in succession eight molecules of H_2 in approximately 1 μsec (source residence time). The stability of the species $C_{10}H_6^{2+}$ clearly provides a powerful driving force.

It seems from the results of this section that fragmentation reactions are often in competition with charge-changing collisions. By studying the $2E$ mass spectra, valuable insights are obtained regarding the relationship between the structures of singly and doubly charged ions. On the other hand, by a study of the competition between fragmentations, much is learned about the chemistry of doubly charged ions generated by electron impact. In Section 4, we will see how some of these relationships extend to doubly charged ions generated by collision.

3. $-E$ Mass Spectra

In the preceding section, we discussed those ion–molecule reactions in which an electron is transferred from a neutral collision gas molecule to a doubly charged ion, thereby forming a singly charged positive ion. The resulting $2E$ spectra have been shown to be useful in structural studies of *positive* ions, and in this section, we will discuss the application of another type of charge exchange reaction to the investigation of *negative* ion structures.

A beam of high-energy singly charged positive ions passing through a region of high collision-gas pressure can potentially undergo many types of reactions, one of these being the capture of two electrons from the neutral molecules as shown by reaction (3.1) forming singly charged anions. (The term charge inversion is used to describe this type of process, which is also covered in Chapter 4.)

$$m^+ + N \rightarrow m^- + N^{2+} \quad (3.1)$$

Some of the anions (m^-) may be formed in excited states and will fragment or eject an electron, but others may be stable for at least several microseconds. If the charge-exchange reaction takes place in the field-free region prior to the electric sector of a conventional double-focusing mass spectrometer, the negative ions can be separated from the positive ions simply by reversing the polarity of the electric sector plates. Subsequent mass analysis, with the polarity of the magnetic sector reversed so as to transmit negative ions, yields a mass spectrum of negative ions which have undergone this charge inversion reaction. Spectra taken in this manner have been termed $-E$ mass spectra[12] to represent the type of reaction being studied and the technique by which they are investigated.

Figure 7 shows the normal mass spectrum and the $-E$ spectrum of the dinitrile **1** using methyl iodide as the collision gas. A comparison of the two spectra shows many interesting differences. The $-E$ spectrum bears little resemblance to the mass spectrum, which indicates that for this compound at least, the cross sections for charge inversion are very much dependent on the ion structure and composition. As a simple case which points out how different structures are indicated by different abundances of the negative ions, consider the ions of mass 104^+ and 105^+. Loss of ethylene from the molecular cation accounts for the 104 peak so that this ion retains the dinitrile moiety, whereas 105 represents the loss of HCN from the molecular ion, yielding a nitrile. The difference in functionality between these two ions is borne out by the large cross section for charge inversion of 104^+ (dinitrile) as opposed to the low cross section for 105^+ (nitrile structure). This supports the findings from negative ion mass spectra obtained by low energy electron capture[13,14] that dinitrile compounds (dicyanodiacetylene in particular) give intense molecular anions (cf. 104), whereas aliphatic nitriles do not give molecular anions (cf. 105).

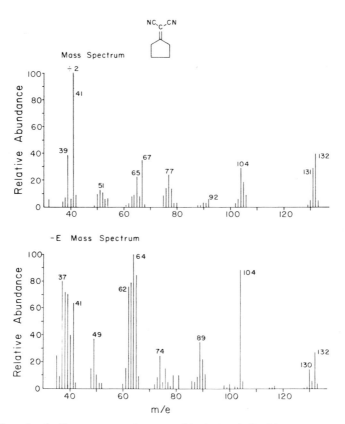

Fig. 7. Normal and $-E$ mass spectra of compound **1** using methyl iodide as collision gas in the short collision region of the RMH-2 mass spectrometer.

A similar situation is observed with ions of mass 89 and 92. The low cross section for charge inversion of 92^+ indicates that the dinitrile moiety is not present so that the loss of C_2H_2N seems more reasonable than loss of C_3H_4. The higher cross section for charge inversion of 89^+, on the other hand, indicates that the correct fragmentation is loss of C_3H_7, forming the dinitrile, rather than loss of C_2H_5N. These two predictions as regards the compositions of the ions of mass 89 and 92, based on their stabilities as negative ions, have been shown to be correct from exact mass measurements of the positive ions.

$-E$ spectra taken on a series of aromatic compounds have shown that the series of ions $C_n^{-\cdot}$ and C_nH^- formed by the charge-inversion reaction are very stable.[15] This has been rationalized on the basis that, for the C_2 case, the molecular orbital configuration is $(\sigma_1)^2(\sigma_1^*)^2(\pi)^4(\sigma_2)^0$ so that up to two more electrons can be accommodated by the (σ_2) orbital to form either $C_2^{-\cdot}$ or

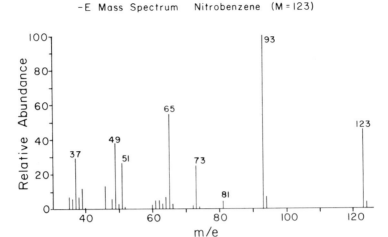

Fig. 8. $-E$ mass spectrum of nitrobenzene using methyl iodide as target gas in the short collision region of the RMH-2 mass spectrometer.

HC_2^-. The $C_5H_5^-$ ion was also found to be very abundant in the $-E$ mass spectra of nitrobenzene, toluene, anisole, phenol, phenetole, and aniline, and it has been suggested that this is due to a cyclopentadienyl structure of the C_5H_5 cation as opposed to, say, a linear structure. These $-E$ spectra are relatively insensitive to the particular aromatic compound and thus are of limited value analytically. Figure 8 shows the $-E$ spectrum of nitrobenzene and is typical of the aromatic compounds studied.

Charge-inversion reactions can also be effected in the second field-free region of a double-focusing mass spectrometer if the positive ions formed in the source are transmitted through the electric sector and then allowed to collide with a neutral gas. A spectrum similar to the $-E$ spectrum is produced by scanning the magnet with the polarity reversed. This type of spectrum has been termed a charge-inversion mass spectrum so as to indicate the method of analysis (momentum analysis) as well as the nature of the reaction (viz., charge inversion). Figure 9 shows the charge-inversion mass spectrum of compound **1** using methyl iodide as collision gas.

The most significant difference between this spectrum and the $-E$ mass spectrum is the appearance of additional broad peaks in the charge-inversion spectrum. These include the peaks at masses 46, 47, 59, and 60 which are due to collision-induced fragmentations occurring before the magnet. These fragmentations cannot be observed in the $-E$ spectra because the fragment daughter ions possess only a fraction (m_2/m_1) of the energy which is required for transmission by the electric sector.

Loss of H˙ is a favored fragmentation in the charge-inversion mass spectrum of **1**, as seen by broad peaks corresponding to formation of C_5^- from C_5H^-, and C_5H^- from $C_5H_2^-$. The broad peak at m/e 47, due to the transition $49^{-\cdot} \rightarrow 48^- + H^\cdot$, has an abundance of 30% of that of the parent ion 49^-. By contrast conversion of even 1 to 2% of a positive ion to fragments by collision-induced dissociation is unusually high.[16]

To interpret these results, we must return to previous discussions concerning the mechanism of electron transfer from the collision gas to the high-energy ion. In considering the charge-exchange reactions of doubly charged positive ions to form single charged positive ions, it was noted that the initial and final structures of the ion were necessarily the same since electron transfer occurs on a time scale which is less than a molecular vibration. The same is true for the charge-inversion reaction, and the nascent negative ion will have the same structure as the positive precursor ion. During the transfer of two electrons from the collision gas to the ion, a certain amount of kinetic energy is being converted into electronic excitational energy of the ion. The dramatic difference in appearance of the $-E$ or charge-inversion spectra as compared with the normal mass spectra and the large abundance of fragment ions observed in the charge inversion spectra lend evidence to the claim that in general the stable structures of positive and negative ions are very different. A negative ion whose stable structure is

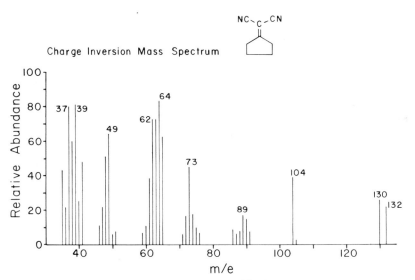

Fig. 9. Charge-inversion mass spectrum of compound **1** using methyl iodide as collision gas in the second field-free region of the RMH-2 mass spectrometer. Positive ions from the source undergo charge inversion after transmission through the electric sector so that metastable and CID transitions are recorded also.

different from the corresponding cation is more likely to fragment if formed by the charge inversion process than it is to rearrange to a more stable form. This is certainly true if the anion is in an excited state as well as an unfavorable conformation.

With these facts in mind we can conclude with a brief discussion of the merits of $-E$ spectra, which are basically twofold. The first is analytical in nature, with the $-E$ spectra being used in a complementary fashion to the normal mass spectra to aid in identification of molecules. The second is the inference of ion structures from the cross sections observed for formation of negative ions by charge inversion. Knowing the structure of either the cation or anion automatically defines the other when the cross section is large, whereas a small cross section implies an unstable form of the anion.

4. E/2 Mass Spectra

The inelastic collision of a fast-moving ion and an atomic or molecular target may occur with the transfer of one or more electrons from target to ion. Such processes form the basis for $2E$ and $-E$ mass spectra as already discussed. Another general type of inelastic collision process results in the *stripping* of electrons from the ion. This type of process is an analog of electron-impact ionization: A direct electron–electron interaction is responsible for removal of the electron from the ion. In this type of process a singly charged ion (m_1^+) can be converted into a doubly charged ion (m_1^{2+}) as illustrated by

$$m_1^+ + N \rightarrow m_1^{2+} + N + e^- \tag{4.1}$$

Ignoring the relatively small changes in kinetic energy of the ions associated with the endo- or exothermicity of the stripping reaction, the products of the transformation $m_1^+ \rightarrow m_1^{2+}$ will have half the kinetic energy-to-charge ratio of the reactants.

Since the electric sector of a double-focusing mass spectrometer analyzes for kinetic energy-to-charge ratio, it will, when set to $E/2$ (one-half its normal value), transmit all ions which undergo charge stripping. Mass analysis can then be used to identify the individual product ions, and the resultant spectrum has therefore been termed an $E/2$ mass spectrum.[17,18] The $E/2$ mass spectrum of 3-hexanone obtained in this manner is shown in Fig. 10. The peak at mass 39, for example, is due to the ion $C_3H_3^+$ being stripped of an electron to give $C_3H_3^{2+}$. When the peak due to this single process is examined with good energy resolution, it is found to be sharp and structureless, and broadened relative to the main beam of noncolliding ions by only a few volts (in 8000 eV).

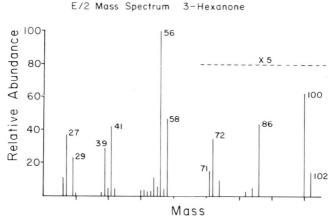

Fig. 10. $E/2$ mass spectrum of 3-hexanone. Peaks are plotted in terms of the mass of the reactant ion.

This example of an $E/2$ spectrum suggests that they be applied in characterizing singly charged ions via the extent to which they can be converted into stable doubly charged ions. The observed doubly charged ion abundance represents the resultant of the cross section for stripping and that for fragmentation of the stripped ions. This novel approach to characterizing ions is considered in more detail later.

Before pursuing this, however, it is instructive to return to the original point made in this section regarding the types of inelastic collisions which can occur in the kilovolt energy range. In addition to charge transfer and electron stripping another type of inelastic collision can be recognized. In this case collisions lead to electronic excitation or deexcitation of the collision partners. Because of the experimental methods available, our focus here is on events in which the ion is excited and the consequences of this excitation. Four such consequences can be recognized: (a) radiation may be emitted; (b) the ion, if diatomic or polyatomic, may fragment; (c) the ion may be stable in its excited state on the time scale of the experiment; and (d) the ion may eject one or more electrons. Process d is of concern here but it may be noted that processes a and c form the basis for energy loss spectroscopy as discussed in Chapter 1. Process b is the subject of Chapters 6 and 7. The process of interest here, d, corresponds essentially to a delayed electron-stripping reaction, excitation being to a state of the singly charged ion which lies above the energy of the doubly charged ion. Autoionization is the term used to describe the corresponding process in which a neutral molecule is excited to a state lying higher in energy than the ionization potential. Although ejection of an electron may occur within one molecular vibration of the collision, the process can in principle be distinguished from a direct electron-stripping mechanism in which some particular electron acquires by electron–electron impact such a kinetic energy that it is stripped from the ion *during* the collision. The point which deserves emphasis here is that the overall stripping reaction is, at least in one form, closely related to collision-induced dissociation. Thus, we can view the collision as yielding an excited ion m_1^{+*} which from certain states may competitively fragment [Eq. (4.2)] or lose an electron [Eq. (4.3)]:

$$m_1^{+*} \longrightarrow m_1^{2+} + e^- \quad (4.2)$$
$$\phantom{m_1^{+*}} \longrightarrow m_2^+ + m_3 \quad (4.3)$$

Furthermore, if the doubly charged ion is not stable, it may fragment by charge separation:

$$m_1^{2+} \rightarrow m_2^+ + m_3^+ \quad (4.4)$$

thus providing another route to the fragment ion. This further emphasizes the close connection between charge stripping and fragmentation.

Not only are there these fundamental relationships between stripping and collision-induced dissociation, but there is also a more practical relationship evident in $E/2$ mass spectra. This arises because product ions having half the kinetic energy-to-charge ratio of the reactant ions may be generated *either* by charge stripping, as described above, or by fragmentation into products of equal mass. Hence the reaction

$$m_1^+ + N \rightarrow m_1^+/2 + m_1/2 + N \quad (4.5)$$

will give product ions which *also* appear in the $E/2$ mass spectrum. The apparent mass of the ion $m^+/2$ produced by this fragmentation is identical to that of the doubly charged ion m_1^{2+} produced by charge stripping. Thus, a given reactant ion m_1^+ will give rise to only one peak in the $E/2$ spectrum regardless of whether it undergoes reaction (4.1) or (4.5), or both.

Fortunately, these processes can be distinguished by the characteristic peak broadening associated with fragmentation reactions. As detailed elsewhere (Chapters 6 and 7), the fragmentation of a fast-moving ion is accompanied by release of kinetic energy which is amplified in the laboratory frame of reference. This broadening of a fragmentation peak will be evident in a kinetic energy scan but it can also be recognized in the $E/2$ mass spectrum. Figure 11 illustrates this latter situation, showing how narrow peaks due to charge stripping are sometimes associated with broader peaks due to fragmentation at the same mass. It is possible, although tedious, to energy analyze each peak in an $E/2$ spectrum and so to separate the contributions from stripping and fragmentation. Pure charge-stripping spectra have been obtained in this way[19] and representative features are

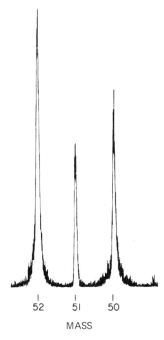

Fig. 11. Region of $E/2$ mass spectrum of aniline using the compound itself as target gas in the first field-free region. Broad bases on 50 and 52 result from kinetic energy released upon fragmentation. Sharp components are due to charge stripping.

discussed later. Experiments of this type have also revealed that fragmentations in which the ion and neutral products are not of exactly equal mass may also contribute to $E/2$ mass spectra. This is not a consequence of poor resolution but a fundamental consequence of the kinetic energy release just discussed. Consider an ion $(m_1 + \Delta m)^+$, where Δm may be positive or negative, fragmenting to give an ion $m_1/2^+$ and a neutral $(m_1/2 + \Delta m)$ [reaction (4.6)].

$$(m_1 + \Delta m)^+ \rightarrow \left(\frac{m_1}{2}\right)^+ + \left(\frac{m_1}{2} + \Delta m\right) \qquad (4.6)$$

Because of the isotropic release of kinetic energy in the center of mass, some of the fragment ions $(m_1/2)^+$ may have resultant kinetic energies which exactly correspond to transmission by the electric sector set at the value $E/2$. This is so even though the average kinetic energy of these product ions will be $[(m_1/2)/(m_1/2 + \Delta m)]eV$, where eV is the kinetic energy of the reactant ion $(m_1 + \Delta m)^+$. These reactions will give rise to peaks in the $E/2$ spectrum at the same apparent mass as those due to the symmetrical fragmentation. Since the daughter ions produced by this reaction appear to be produced from parent ions of twice their mass, this can give rise to peaks which appear to arise from parent ions whose masses are greater than the molecular ion. For example, the $E/2$ spectrum of 3-hexanone already shown in Fig. 10 shows peaks at masses 100 and 102. The peak plotted at mass 100 could be due to charge stripping of the molecular ion

$$100^+ + N \rightarrow 100^{2+} + N + e$$

or to its symmetrical fragmentation.

$$100^+ + N \rightarrow 100^{+*} \rightarrow 50^+ + 50$$

A kinetic energy analysis of the peak shape [FWHM = 200 eV (lab)] revealed that it was entirely due to the fragmentation process with no contribution from the narrow peak which would be associated with charge stripping. Similarly, the plotted peak at 102 is due to fragmentation, not the symmetrical fragmentation of 102^+ which of course is not formed from this compound, but the fragmentation

$$100^+ + N \rightarrow 100^{+*} \rightarrow 51^+ + 49$$

Figure 12 shows the results of kinetic energy analysis of the peak at mass 56 in the 3-hexanone spectrum. Two broad peaks are evident; the minor one corresponds to the symmetrical collision-induced dissociation

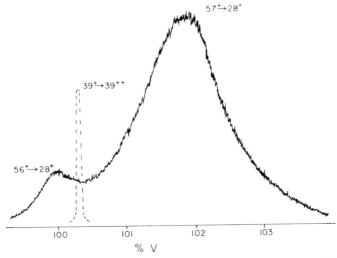

Fig. 12. Kinetic energy profiles of the peaks at 56 (full line) and 39 (dashed line) in the $E/2$ mass spectrum (Fig. 10) of 3-hexanone. The figure illustrates the three processes which can give rise to peaks in $E/2$ mass spectra, stripping as for mass 39 and symmetrical and unsymmetrical fragmentations ($56^+ \rightarrow 28^+$ and $57^+ \rightarrow 28^+$, respectively), both contributing to mass 56 in the $E/2$ spectrum.

$56^+ \to 28^+$ and the major to fragmentation of 57^+ to give 28^+. As just discussed, this ion is transmitted under $E/2$ conditions (with the accelerating voltage set at V) only because of the large energy release resulting in some ions having the correct laboratory kinetic energy. The figure also shows, for contrast, the energy profile of the peak at mass 39 which is due only to charge stripping.

The $E/2$ spectrum of 3-hexanone is noteworthy in that it consists chiefly of ions at even mass. As a result of the mass scale in $E/2$ spectra, ions at odd mass must correspond to stripping (see Ref. 19 for a detailed explanation), whereas those at even mass may be due to stripping or to fragmentation. Apparently excitation of the ions formed from an aliphatic compound such as 3-hexanone does not readily lead to formation of stable doubly charged ions, consistent with the absence of doubly charged ions in the normal mass spectrum of 3-hexanone.

In general, stripping reactions leading to stable doubly charged ions are expected to be much more prominent in aromatic compounds, and much of the remaining discussion will center on these compounds. Many interesting features have been discerned in the $E/2$ spectra of simple aromatic compounds, with ions at mass 39, 51, and 61–65 usually being the predominant doubly charged ions. Fragmentations as exemplified by reactions (4.5) and (4.6) are not observed in the C_3 and C_5 regions since it is impossible for these ions to form a neutral and a daughter ion of the same, or nearly the same, mass. Significant peaks are often observed for masses 50, 52, 72, 74, 76, and 78 which are due primarily to the formation of the singly charged fragment ions 25, 26, 36, 37, 38, and 39, respectively. The parent ions giving rise to these fragment ions vary of course from compound to compound, but it has generally been found that both reactions (4.5) and (4.6) are significant in contributing to the peaks observed in the $E/2$ spectrum. In many cases, it is possible to deduce the reaction leading to a particular fragment ion. Consider, for example, Fig. 13, which shows the normal, the charge-stripping, and $E/2$ mass spectra of p-anisidine. It has been found that the relative cross sections for the formation of doubly charged ions from stable singly charged ions are dependent on ion composition but that for a given compound, the ratio of the highest observed cross section to the lowest is usually less than 10. This appears to be due to the fact that the cross section for charge stripping is dependent on the energy defect for the reaction, which for organic ions lies between 8 and 13 V. This relatively narrow range means that the cross section for charge stripping is nearly independent of ion composition, at least for the rather similar ions encountered among a representative set of simple aromatic molecules. The presence of an intense peak in the $E/2$ spectrum corresponding to an ion which is much weaker in the mass spectrum indicates that a more intense ion of nearly the same mass is fragmenting via reaction (4.6). Thus, the appearance of abundant ions at

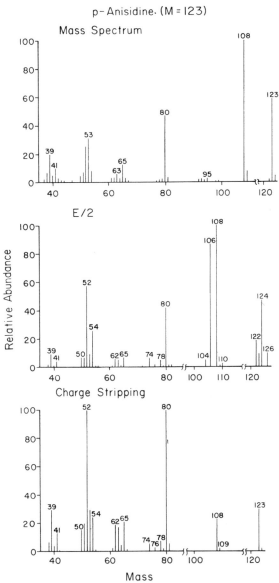

Fig. 13. Normal, $E/2$, and charge-stripping mass spectra of p-anisidine taken on the RMH-2 mass spectrometer using N_2 as target gas in the short collision chamber.

masses 104, 106, and 110 in Fig. 13 would indicate that these are not doubly charged ions but rather singly charged fragment ions of masses 52, 53, and 55. Since 108^+ is abundant in the mass spectrum, and since it has almost twice the mass of the observed fragment ions, it could be inferred that the reactions observed in this spectrum are $108^+ \rightarrow 52^+ + 56$, $108^+ \rightarrow 53^+ + 55$, $108^+ \rightarrow 54^+ + 54$, and $108^+ \rightarrow 55^+ + 53$. Experiments have confirmed that these are indeed the reactions which give rise to these ions, and this demonstrates that it is not always necessary to carry out an accelerating voltage scan to elucidate the reactions leading to certain fragment ions. After correcting for fragmentation reactions the charge-stripping spectra of aromatic compounds tend to reflect very closely the abundance of ions in the normal mass spectra.

In addition to providing data on some potentially interesting dissociative processes, $E/2$ spectra in conjunction with charge-stripping spectra occasionally provide insight into the competition between the tendency of an excited ion to fragment versus the tendency to form a stable doubly charged ion by ejection of an electron. The molecular ion region of the p-anisidine spectrum presented in Fig. 13 is an excellent example of this. The excited 123^+ ion can eject an electron to form 123^{2+}, or it can fragment to give daughter ions of mass 61^+, 62^+, or 63^+ (those ions appearing as 122^{2+}, 124^{2+}, or 126^{2+}). The total abundance of these fragment ions is much greater than the abundance of 123^{2+}, which indicates that the probability for formation of the stable doubly charged ion is far less than the probability for fragmentation. Considering the fact that only a small fraction of the total ion current for, say, $123^+ \rightarrow 61^+ + 62$ is actually collected at the accelerating voltage used and in view of the fact that only two specific types of fragmentation reactions are observed by this method [reactions (4.5) and (4.6)] we are led to the conclusion that the formation of a doubly charged molecular ion of p-anisidine via an ion–molecule reaction of the stable singly charged ion is an unfavorable process.

There are several features worthy of special mention in the charge-stripping spectra (as distinct from the uncorrected $E/2$ spectra) of simple aromatic compounds. The special stability shown from $2E$ spectra to be associated with ions of the structure $C_6H_2^{2+}$ is also reflected in the charge-stripping and hence the $E/2$ spectra. Thus, the ion 74^{2+} in the charge-stripping spectrum typically has a much greater abundance than the very weak 74^+ ion in the mass spectrum would otherwise lead one to expect. The tendency for stable doubly charged ions to be those which are highly unsaturated is further seen in the 3-hexanone spectrum (Fig. 10). Masses 41^+ and 39^+ are very minor peaks in the mass spectrum, compared with 43 ($C_3H_7^+$) for example. However, these ions dominate the stripping spectrum.

We conclude this section by discussing $E/2$ spectra determined using instrumentation other than a conventional double-focusing mass spectrometer. A reversed-sector mass spectrometer, or mass-analyzed ion kinetic energy spectrometer (MIKES), can serve equally well to record such spectra. If the collision occurs prior to both sectors, then the order in which the sectors are traversed by the product ion is immaterial and the spectrum is exactly analogous to that recorded on conventional instrumentation. Alternatively, the collision may occur in the region between the sectors, i.e., after mass analysis but before energy analysis, so that these spectra differ in that the reactant rather than the product ion is mass-analyzed. This has the advantage that the mass scale for $E/2$ spectra is identical to that for mass spectra so that calibration is greatly simplified. Figure 14a shows the $E/2$ mass spectrum of toluene taken by this method. This spectrum closely resembles the spectrum taken using the conventional geometry RMH-2 instrument (Fig. 14b) but one significant difference between the two spectra concerns the unsymmetrical fragmentation. These processes appear on the MIKES at the *same* mass as do the symmetrical fragmentations since mass analysis precedes the collision. Thus, some important fragmentations such

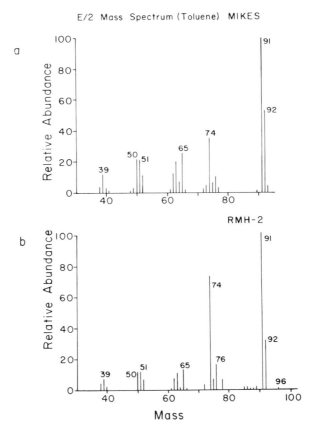

Fig. 14. (a) $E/2$ mass spectrum of toluene taken on the reversed-sector instrument using air as collision gas. (b) $E/2$ mass spectrum of toluene taken on the RMH-2 mass spectrometer (conventional geometry) using N_2 as collision gas.

as $91^+ \rightarrow 48^+ + 43$ in toluene will be observed on the conventional instrument but not on the MIKE spectrometer in this type of scan.

So far our entire discussion has concerned ions produced by electron impact. Some data are becoming available on charge-stripping and $E/2$ spectra produced by chemical ionization. A noteworthy fact is that the cross sections for charge stripping of quasi-molecular ions $(M+H)^+$ formed by chemical ionization is sometimes markedly greater than that for stripping of the corresponding molecular ions $(M^{+\cdot})$ formed by electron impact.[20] This has been found to be true for a number of aliphatic ketones among other compounds. The lower internal energies of ions formed by CI may be responsible. The doubly charged ions formed in the CI experiment are expected to have a lower internal energy than those doubly charged ions formed from vibrationally hot singly charged ions generated by electron impact. Hence, the doubly charged ions formed by stripping of electron-impact-generated species would be expected to show preferential fragmentation over those formed via the CI route.

In conclusion, $E/2$ spectra provide a convenient summary of a great deal of interesting gas phase chemistry which results from high-energy collisions. The cooccurrence of fragmentation and stripping reactions and competition between these reaction channels is of particular note.

5. +E Mass Spectra

$E/2$ mass spectra contain peaks due to the charge stripping of singly charged positive ions to form doubly charged ions. Another important charge-stripping reaction which is amenable to study by the methods of ion kinetic energy spectrometry involves the loss of two electrons from singly charged negative ions to form singly charged positive ions:

$$m^- + N \rightarrow m^+ + N + 2e^- \qquad (5.1)$$

The positive ions possess essentially the same energy as the negative precursors although a small fraction of the kinetic energy of the ion is converted to internal energy upon colliding with the neutral target gas. With only a slight correction for this kinetic energy loss, the positive daughter ions can be transmitted by the electric sector if its polarity is reversed from that employed in transmitting the main beam of negatively charged ions. Subsequent mass analysis of the beam of positive ions will record all ions which have undergone this type of charge stripping reaction, and following the system of nomenclature used for $-E$, $E/2$, and $2E$ spectra, this type of spectrum is termed a $+E$ mass spectrum.[21]

The production of negative ions in the ion source occurs primarily via low-energy electron capture by neutral reagent molecules,[22] and thus the primary and sometimes the only component of the negative ion current is the negative molecular ion.[23] Consequently, $+E$ mass spectra in such cases would consist only of one peak due to the molecular ion ($M^- \rightarrow M^+$) with perhaps a few weak metastable peaks due to delayed fragmentations which occur in the second field-free region. These spectra are, therefore, not of great interest except as regards the cross section of the process (5.1). However, several closely related techniques in which charge inversion of negative ions is studied in conjunction with associated fragmentation reactions do summarize a great deal of negative ion chemistry. Thus, Bowie and Blumenthal[21] have shown that there is a significant abundance of positive fragment ions produced as a result of the high-energy collisions between a molecular anion and the target gas. These fragment ions have been recorded by two methods. The first employs ion kinetic energy spectrometry without mass analysis. In this method the polarity of the electric sector is reversed to transmit positive ions which are recorded on a multiplier placed immediately after the electric sector. Such spectra have been termed, inappropriately in our view, $+E$ IKE spectra. A longer, but more appropriate term is charge-inversion ion kinetic energy spectra. In the second type of spectrum, mass analysis of the positive ions is performed; the electric sector is set to transmit the negative ions formed in the ion source and the magnet is scanned to record all positive ions generated by fragmentation in the second field-free region. This type of spectrum may be referred to as a charge-inversion mass spectrum.

Figure 15 shows the charge-inversion ion kinetic energy spectrum produced from the flavanone anion and Fig. 16 shows the charge-inversion mass spectrum of the same anion. These figures are essentially identical except that the peaks in Fig. 16 are better resolved. Figure 17 is the normal

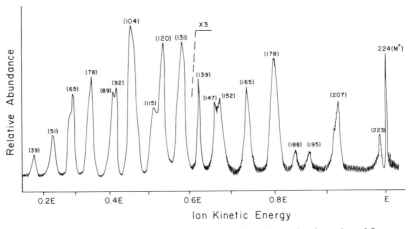

Fig. 15. Energy spectrum of positive ions generated from the molecular anion of flavanone. Spectrum obtained by scanning the electric sector without mass analysis. Retraced.

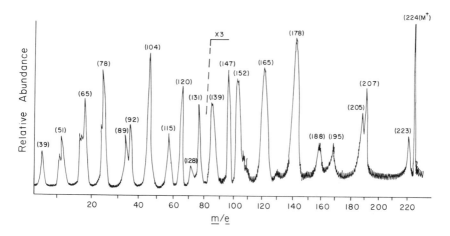

Fig. 16. Momentum spectrum of positive ions generated in the second field-free region from the molecular anion of flavanone. Spectrum obtained by scanning the magnetic sector. Retraced.

mass spectrum of flavanone and has been included for comparison. The peak centered at E in Fig. 15 results only from the molecular ion of flavanone, but in other compounds, this peak may contain fragment ions formed in the source which undergo the charge-inversion reaction. The charge-inversion mass spectrum would resolve these ions so this type of spectrum has the advantage over the IKE spectrum in that it records charge inversion of all ions formed in the source and of all their fragments.

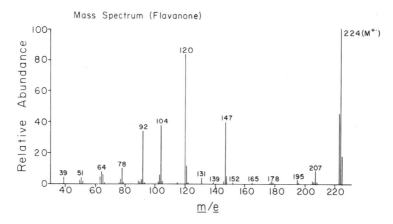

Fig. 17. Mass spectrum (positive ions) of flavanone.

The fragment ions recorded in Figs. 15 and 16 may be formed either before or after charge inversion. These reactions are given by

$$m_1^- \rightarrow m_2^- + m_3 \tag{5.2a}$$

$$m_2^- + N \rightarrow m_2^+ + N + 2e^- \tag{5.2b}$$

$$m_1^- + N \rightarrow m_1^+ + N + 2e \tag{5.3a}$$

$$m_1^+ \rightarrow m_2^+ + m_3 \tag{5.3b}$$

The close similarity between the positive ion mass spectrum of flavanone and the two spectra resulting from the charge-stripping processes has important implications as regards the structures of the molecular anion and cation and the reaction mechanisms associated with these spectra. The fragmentations observed in the mass spectrum are the same as those observed in the charge-inversion spectra, which is a good indication that the structures of the reactant ion in both cases are similar if not the same. Thus it is concluded that the fragment ions observed in Figs. 15 and 16 are generated from the molecular cation, and that the fragmentation sequence is, in the general case,

$$m_1^- + N \rightarrow m_1^{+*} + N + 2e^- \rightarrow m_2^+ + m_3 \tag{5.4}$$

The alternative possibility, given by

$$m_1^- + N \rightarrow m_2^- + m_3 + N \rightarrow m_2^+ + m_3 + 2e^- \tag{5.5}$$

is less probable in this case because we would expect the negative ion to undergo different reactions, the structural features which stabilize positive and negative ions being different.

6. *Other Types of Spectra*

Other types of charge-changing reactions resulting from high-energy ion–molecule collisions can be studied on double-focusing mass spectrometers with little or no instrumental modifications. On the Hitachi RMH-2, for example, it is possible to study the charge-exchange reaction occurring between positively charged ions of kilovolt energies with a neutral target gas, as illustrated by

$$M^+ + N \rightarrow M + N^+ \tag{6.1}$$

This reaction occurs in a collision region held at a potential which is less than the full accelerating voltage but greater than ground potential. The product ions, represented in the general case by N^+, are accelerated to several kilovolts and subsequently mass analyzed. A schematic representation of

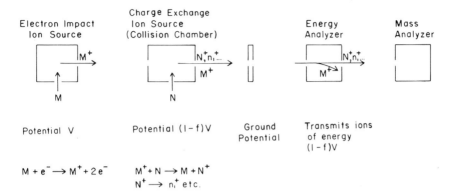

Fig. 18. Schematic representation of RMH-2 mass spectrometer illustrating the production of high-energy ions (N^+, n_1^+, etc.) through a charge-exchange reaction.

instrumentation which allows the study of this type of charge exchange reaction is shown in Fig. 18. Ions M^+ produced by electron impact in the source are accelerated to a fraction f V of their ultimate kinetic energy. In a region of constant potential, the collision chamber, these ions collide with the target gas N and some of these collisions result in charge exchange, producing N^+. If N^+ is a polyatomic ion, it may fragment to produce product ions n_1^+, n_2^+, etc. The ions produced in this region will then be accelerated by a potential $(1-f)$ V. The energy analyzer is set to transmit these ions of energy $(1-f)$ eV and consequently removes ions of all other energies including the reagent ions (M^+) of energy eV. Subsequent mass analysis may then be carried out to determine the charge-exchange mass spectrum.

Charge-exchange reactions at kilovolt energies have also been studied using tandem mass spectrometers.[24] These much more complex instruments have advantages over the method just given in terms of reagent ion selection. The charge exchange process is important because it results in the deposition of a well-defined internal energy in the polyatomic ion N^+. By varying the charge-exchange reagent ion M^+, it is possible to follow the effects of internal energy upon the fragmentation pattern. Studies of this nature can be performed on conventional double-focusing mass spectrometers using the method just described; one set of results is illustrated in Fig. 19. This shows charge-exchange mass spectra of cyclohexene taken with argon, hydrogen, and benzene as the source of the reagent ions. As the recombination energy of the ion M^+ decreases and less energy is deposited on the cyclohexene, one observes less fragmentation and a dramatic increase in the molecular ion (m/e 82) abundance.

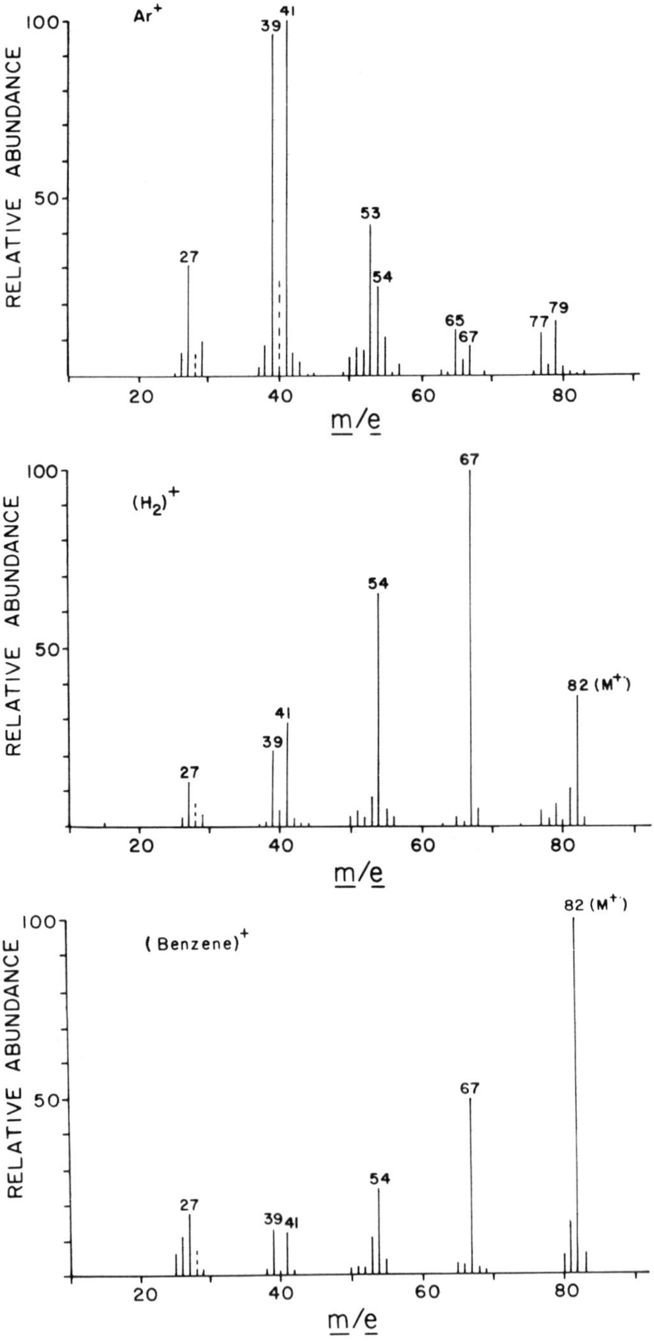

Fig. 19. Charge-exchange mass spectrum of cyclohexene produced using the technique illustrated in Fig. 18. The effect of reagent ion (M^+) on ion abundances (N^+, n^+, etc.) is illustrated.

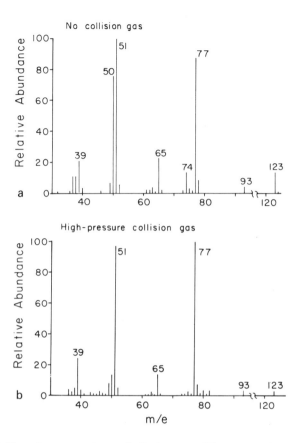

Fig. 20. (a) Normal mass spectrum of nitrobenzene; (b) mass spectrum obtained with high-collision gas pressure in first field-free region of RMH-2 mass spectrometer. Ions must be stable relative to both fragmentation and charge changing in order to pass through the collision gas.

An indirect method of examining charge-changing collisions is by examining the distribution of ions which survive passage through a collision chamber containing a collision gas at high pressure without fragmentation or change in charge. The relative extent to which different polyatomic ions are transmitted through gas cells varies greatly, as the comparison of the initial and final ion intensities for nitrobenzene given in Fig. 20 shows.

ACKNOWLEDGMENT

The Purdue work described in this chapter was supported by the National Science Foundation.

References

1. J. H. Beynon, R. M. Caprioli, W. E. Baitinger, and J. W. Amy, *Int. J. Mass Spectrom. Ion Phys.* **3**, 313 (1969).
2. J. H. Beynon, R. G. Cooks, and T. Keough, *Int. J. Mass Spectrom. Ion Phys.* **14**, 437 (1974).
3. (a) R. Conrad, *Phys. J.* **31**, 888 (1930); (b) J. J. Thomson, *Phil. Mag.* **24**, 668 (1912); (c) S. Meyerson and R. W. Vander Haar, *J. Chem. Phys.* **37**, 2458 (1962).
4. T. Ast, J. H. Beynon, and R. G. Cooks, *Org. Mass Spectro.* **6**, 749 (1972).
5. T. Ast and J. H. Beynon, *Org. Mass Spectro.* **7**, 503 (1973).
6. D. L. Smith and L. Kevan, *J. Am. Chem. Soc.* **93**, 2113 (1971).
7. D. L. Smith and L. Kevan, in *Recent Developments in Mass Spectrometry* (K. Ogata and T. Hayakawa, eds.), Univ. of Tokyo Press, 1970, p. 883.
8. F. L. Mohler, V. H. Dibeler, and R. M. Reese, *J. Chem. Phys.* **22**, 394 (1954).
9. J. H. Beynon, G. R. Lester, and A. E. Williams, *J. Phys. Chem.* **63**, 1861 (1959).
10. T. Ast, J. H. Beynon, and R. G. Cooks, *Org. Mass Spectrom.* **6**, 741 (1972).
11. F. W. McLafferty, in *Mass Spectrometry of Organic Ions* (F. W. McLafferty, ed.), Academic Press, New York, 1963, p. 318.
12. T. Keough, J. H. Beynon, and R. G. Cooks, *J. Am. Chem. Soc.* **95**, 1695 (1973).
13. V. H. Dibeler, R. M. Reese, and J. L. Franklin, *J. Am. Chem. Soc.* **83**, 1813 (1961).
14. C. E. Brion and L. A. R. Olsen, *Int. J. Mass Spectrom. Ion Phys.* **9**, 413 (1972).
15. T. Keough, Ph.D. Thesis, Purdue University, 1972, p. 127.
16. E. G. Bloom, F. L. Mohler, J. H. Lengel, and C. E. Wise, *J. Res. Natl. Bur. Std.* **40**, 437 (1948).
17. T. Ast, Ph.D. Thesis, Purdue University, 1972, p. 89.
18. R. G. Cooks, J. H. Beynon, R. M. Caprioli, and G. R. Lester, *Metastable Ions*, Elsevier, Amsterdam, 1973, p. 139.
19. D. L. Kemp, J. H. Beynon, and R. G. Cooks, *Org. Mass Spectrom.* **11**, 857 (1976).
20. J. F. Litton, unpublished results.
21. J. H. Bowie and J. Blumenthal, *J. Am. Chem. Soc.* **97**, 2959 (1975).
22. J. Marriott and J. D. Craggs, *Applied Mass Spectrometry*, Institute of Petroleum, London, 1954.
23. M. Von Ardenne, K. Steinfelder, and R. Tummler, *Angew. Chem.* **73**, 136 (1961).
24. C. F. Giese and W. B. Maier, III, *J. Chem. Phys.* **39**, 739 (1963).

6

Collision-Induced Dissociation of Diatomic Ions

J. Los and T. R. Govers

1. Introduction

It is the aim of this chapter to show how, in recent years, relatively simple mass spectrometry experiments have contributed considerably to the general understanding of dissociation in heavy-particle collisions. When a well-defined primary beam of (diatomic) molecular ions impinges on a gas target at the object point of a mass spectrometer, this instrument can be used to measure the laboratory momentum distribution of the charged fragments which originate from dissociative collisions. This laboratory distribution can be converted into a momentum distribution in the center-of-mass coordinate system of the parent molecule, provided some requirements are met. In a dissociative collision of a diatomic molecule and an atom, taking into account conservation of energy and momentum, five and in many cases six velocity components of the separating particles should be measured simultaneously in order to describe the collision completely. Since only the momentum distribution of the charged fragment is determined, the interpretation of the measurements requires a number of basic concepts and

---

J. Los • F.O.M.-Instituut voor Atoom- en Molecuulfysica, Kruislaan 407, Amsterdam/Wgm, The Netherlands. *T. R. Govers* • Laboratoire de Résonnance Électronique et Ionique (part of the Laboratoire de Physico-chimie des Rayonnements, associated with the C.N.R.S.), Université de Paris-Sud, 91405-Orsay, France.

models in the first place. Moreover, certain rather stringent requirements should be met in order to extract meaningful conclusions from the data. The most important of these requirements is that the laboratory deflection of the molecular center of mass may be neglected with respect to the laboratory deflection of the fragment ions which results from their dissociation velocity component in the molecular center-of-mass system.

It is this requirement, the necessity of converting laboratory to center-of-mass distributions, which limits the scope of this chapter. It entails that we will restrict ourselves to considering only those experiments in which the primary beam energy is of the order of at least a few keV. In this case, with the exception of dissociations due to violent collisions, the momentum transfer between molecule and target atom is so small that deflection of the molecular center of mass can be neglected. This leaves out a large number of experiments carried out in the low-energy range (~100 eV) which are of extreme interest, but which cannot be transformed in a unique way from the laboratory to the center-of-mass system. For an extensive bibliography of all experimental measurements on collision-induced dissociation up until 1970, however, we refer the reader to the treatise by McClure and Peek[1] entitled *Dissociation in Heavy Particle Collisions*. This chapter is not intended to be such a comprehensive review. We will stress the information which can be obtained from measuring momentum distributions of fragment ions about the mechanisms operative in collision-induced dissociation and about properties of molecular ions especially concerning dissociative and predissociative states. The experimental evidence obtained so far allows a classification of the different mechanisms of collision-induced dissociation. Such a classification was first given by Durup[2] in 1969, and is still almost up to date. We also refer the reader to a brief review on collision-induced dissociation by Los.[3]

The early investigations on collision-induced dissociation by means of mass spectrometric techniques, especially of H_2^+, H_3^+, and several other small hydrogenic molecular ions, were performed in view of possible applications in plasma physics. Collision-induced dissociation as well as field dissociation might offer the possibility of injecting beams into confined plasmas. In recent years, however, there has been a growing interest in dissociative collisions from a more general point of view. Cross sections for dissociation of molecules are of importance for such widely different fields as aeronomy, astrophysics, laser physics, and discharge chemistry. Although essentially concerned with dissociation due to electron impact or photon absorption, the proceedings of a workshop on the dissociative excitation of simple molecules[4] gives several reviews on the applications of molecular dissociation.

Since this chapter deals mainly with the second major point of interest in the study of dissociative heavy-particle collisions, namely, dissociation mechanisms and dissociative states, we will emphasize those experimental findings which give the most clear and unique information about these subjects. We will therefore refrain from discussing total dissociation cross-section measurements, as in general several dissociation mechanisms will simultaneously contribute to the formation of fragment ions. In total cross-section data these contributions cannot be distinguished.

In the next section we will first pay attention to the different mechanisms of collision-induced dissociation for which a theoretical model has been formulated. For energies of the primary beam molecules in the keV range, a very convenient approximation can be introduced in the theoretical description, the two-step model. In this approximation the dissociative collision is thought to proceed in two distinct steps. First the molecule is excited to the continuum of the initial bound state, or to an excited dissociative state. The next step is then the dissociation of the molecule. The different excitation mechanisms are briefly described.

The third section gives an elementary review of the different experimental designs which have been used to measure the momentum distribution of fragment ions from collisional dissociations. All these methods are essentially based on the Aston band technique, well known in mass spectrometry. Special attention is paid to those features which are important for the uniqueness of the data obtained. The Aston band method has to be refined in order to avoid convolution effects in the fragment-ion spectra stemming from poor angular resolution or from an ill-defined location of the collision region. Methods which have been applied to specify the internal state of the molecule, and to determine the final internal state of the fragments are discussed briefly.

The experimental data discussed in Section 4 are selected in order to illustrate the different dissociation mechanisms. Therefore, most attention is paid to collision-induced dissociation of H_2^+ and HeH^+ at collision energies of ~ 10 keV. Moreover this has the advantage that the potential energy curves, including those of a number of excited states, are quite well known.

In Section 5 we pay attention to those dissociative processes in which a discrete spectrum of fragment ions is formed. A discrete spectrum can be due either to predissociation or to photodissociation. In these cases the spectrum reveals the energy levels of bound or quasi-bound states and the transition probabilities of these states to the continuum of a dissociative state.

2. Dynamics of Dissociation

Even in studying the simplest case of collision-induced dissociation, namely the dissociation of a diatomic molecule due to collision with an atom, the number of variables to be controlled or to be measured for a complete description of the collision process is almost prohibitive. Considering the dissociative collision

$$AB + M \rightarrow A + B + M$$

one first needs to specify the initial state of the system. Even when the collision energy is well defined and when AB and M are both in the electronic ground state, one still has to account for the vibration and rotation of the molecule. If the AB molecules are thermalized, they will nearly all be in the vibrational ground state, and their rotational population will be described by a Boltzmann distribution. In many experiments the situation will be quite different, in particular when a beam of diatomic ions is extracted from the source of a mass spectrometer. In the case of H_2^+, for instance, direct ionization by high-energy electrons yields a Franck–Condon distribution over all vibrational levels v, the maximum of which is at $v = 3$. The distribution over the rotational levels will not differ much from the Boltzmann distribution of the parent molecule. With sources where ion–molecule reactions take place, however, the resulting beam may contain molecular ions with a high degree of both rotational and vibrational excitation.

Given an exact description of the internal states and velocities of the colliding particles, the kinematics of the collision are completely determined by the nine velocity components of the three separating atoms. The requirement of conservation of energy and momentum reduces the number of variables to be measured to five, provided the electronic states of the dissociation fragments and the target atom are known. When, as has been done almost exclusively until now, only the momentum vector of one particle is measured, this means that one integrates over the two remaining unknown velocity components. If the collision also leads to electronic excitation, this integration extends over three variables.

As pointed out by Schöttler and Toennies[5] it is possible to investigate collision-induced dissociation by either of two distinct methods. In both cases the momentum distribution of one of the three separating atomic particles is determined. In the first method one measures the energy loss of the incident atom together with its scattering angle in the detector plane; in the second method one measures the momentum distribution of one of the fragments of the dissociated molecule. The two methods yield complementary information.

When the first method is employed, as is possible in a crossed-beam or beam-gas configuration, the energy-loss distribution of the incident atoms (or, most often, atomic ions) is measured at a fixed laboratory scattering angle. Scattering angle together with energy loss determines the recoil angle and energy of the molecule, from which the inelastic energy transferred to the molecule can be found. Dissociation takes place if the inelastic energy exceeds the binding energy of the molecule, provided no electronic excitation of the separating atoms has occurred. These measurements yield doubly differential dissociation cross sections: differential with respect to scattering angle and differential with respect to inelastic energy loss. It is clear that in this way one has integrated over all possible directions in which the fragments of the molecule fly apart. In other words, no information is obtained about the probability for dissociation as a function of the orientation of the molecule at the moment of collision.

Most of these experiments have been done with alkali ions scattered on hydrogen or deuterium.[6-9] Since the alkalis are heavier than the target molecules, the laboratory scattering angles are confined to a small cone, allowing a study of large deflection angles in the center-of-mass system, where large inelastic energy losses are expected. Actually Dittner and Datz[6,7] as well as van Dop et al.[8] confined themselves to the determination of the energy-loss distribution in the primary-beam direction. Schöttler and Toennies[9] also measured energy loss distributions for laboratory scattering angles different from zero.

Experiments of the second type, which are the subject of this chapter, employ a primary beam of molecular ions. These ions are made to collide with either a crossed-beam or a random-gas target, and measurements are made of the distribution of momentum vectors (angle and energy) of the resulting ionic fragments. For the information thus obtained to be physically significant, two conditions have to be satisfied. In the first place, the separating particles should not be electronically excited, or, alternatively, additional information should be available about electronic excitation of either the two fragments or the target atom. In some cases such information can be inferred from measurements of the fragment momentum distribution. The second, more stringent, requirement concerns the contribution of the scattering of the center of mass of the incident molecule to the scattering of the fragments in the laboratory frame. This contribution should either be negligible or should be deducible from the measured momentum distribution. If these two conditions are met, the laboratory momentum distribution of the fragments can be transformed to a momentum distribution in the center of mass of the dissociated molecule. In this way dissociation cross sections are obtained which are differential with respect to the orientation of the molecule and with respect to the translational energy of the separating fragments, but averaged over all scattering angles of the center of mass of the incident molecule.

It may be remarked here that for photodissociation, which is discussed in the section on translational spectroscopy, the conditions formulated herein are rigorously fulfilled. Also, in the process of dissociation by electron impact, the center of mass of the molecule will hardly be deflected.

2.1. The Two-Step Model

In the two-step model, the process of collision-induced dissociation is considered to occur in two successive steps. The first step is the collision between the atom and the molecule; the second is the dissociation of the molecule. The model is essentially based on a comparison of the collision time and the molecular vibration time. When the collision takes place in a time short with respect to the vibrational period, it is indeed likely that these two aspects of the process can be treated separately. However, the way in which the two-step model is applied in different theoretical approaches varies widely and depends on the mechanism of the collisional dissociation process.

If the dissociation is due to *electronic excitation* of the molecule to a repulsive state, the application of the two-step model to relatively light molecules is quite justified when the collision energy is in the kiloelectron volt range. Consider, for example, the dissociation of 10-keV H_2^+ due to collision with a helium atom. Assuming that the dissociation is due to electronic excitation from the ground $1s\sigma_g$ state to the first excited $2p\sigma_u$ state, the characteristic times can be classified as follows:

$$\tau_{coll} \cong 10^{-16} \text{ sec}$$
$$\tau_{vibr} \cong \tau_{diss} \cong 10^{-14} \text{ sec}$$
$$\tau_{rot} \cong 10^{-13} \text{ sec}$$

It is obvious that in a theoretical treatment of the electronic excitation, the vibrational and rotational motion can be considered as "frozen." If in addition no significant momentum transfer takes place in the collision,[2] the two-step approach is equivalent to the Franck–Condon principle applied in photoabsorption or -emission. Moreover, since the characteristic time for dissociation is one order of magnitude smaller than that for rotation, a second approximation can be made for the dissociation step: the dissociation may be considered to take place along the momentary direction of the molecular axis at the time of collision. This "axial recoil" assumption can also be formulated in terms of energy: if the energy associated with the separation of the molecular fragments is large with respect to the initial rotational energy, the axial recoil model will be justified. This approximation has been discussed by Zare[10] who showed that the phase shift of the radial motion of the dissociating particles is independent of the rotational quantum number, provided the energy of separation is large with respect to the rotational energy. When this approximation is no longer applicable, such as close to the dissociation threshold or in case of very high rotational levels, the rotational motion tends to smear out the distribution of dissociation fragments. In the extreme limit, the fragments are ejected perpendicular to the initial direction of the molecular axis, and we could speak of tangential recoil.

The approximation of axial recoil is part of a more general method, the reflection method. What it actually comes to is that the radial wave function of the separating fragments is approximated by a three-dimensional Dirac delta function, $\delta(\mathbf{R}-\mathbf{R}^*)$. The vector $\mathbf{R}$ denotes the relative position of the molecular atoms before dissociation; the vector $\mathbf{R}^*$ is colinear with the direction of dissociation, which represents axial recoil, and the magnitude R^* is that of the classical turning point on the repulsive potential curve where the kinetic energy is equal to the energy of separation being considered. This method and its limitations have been discussed very thoroughly by McClure and Peek.[1]

The reflection method is essential for the appraisal of the information contained in the momentum distribution of the dissociation fragments. Direct electronic excitation of keV-ions generally involves inelastic collisions where the momentum transfer is small enough for the deviation of the molecular center of mass to be negligible with respect to the recoil of the fragments. In that case the laboratory momentum distribution of the charged dissociation fragments can be directly converted to the momentum distribution in the centers of mass of the dissociated molecules. The distribution with respect to angle then reflects the probability of excitation as a function of the orientation of the molecule at the moment of impact. The distribution with respect to the final velocity (or energy) of the fragments reflects the variation of excitation probability with internuclear distance, since the energy with which the fragments separate is determined by the position of the classical turning point on the repulsive potential-energy curve.

In *vibrational/rotational dissociation*, the two-step model is applied in a quite different way. Again this method is based on the fact that the collision time is short with respect to the dissociation time, permitting one to treat the process of excitation separately from the process of dissociation. In this case, dissociation results from vibrational (and/or rotational) excitation of the molecule into the continuum of its initial electronic state. Excitation occurs by momentum transfer to one or both atoms of the molecule. Roughly speaking, the model assumes that dissociation takes place when the energy transferred to the molecule exceeds its binding energy. Theoretical treatments of this model usually suppose that only binary encounters are important; i.e., the collision of atom A of the molecule AB with atom M is considered to be independent of the presence of B.

Collision-induced dissociation in the 1-10-keV range usually takes place via one of the two excitation mechanisms mentioned in this section. In these cases, excitation into a continuum of either another or the initial state, gives rise to what one may call *direct dissociation*. Alternatively, the collision may populate an intermediate state wich fragments through *predissociation*, as discussed further in Section 2.4. First, we will consider mainly the excitation step, and focus successively on electronic excitation (Section 2.2) and vibrational/rotational excitation (Section 2.3).

2.2. Electronic Excitation

Figure 1 illustrates the direct dissociation of H_2^+ resulting from a vertical electronic excitation from the vibrational level $v = 4$ of the $1s\sigma_g$ state to the first excited $2p\sigma_u$ state. This process gives rise to fragments having a separation energy e_d which lies between two limits. According to Franck's classical principle, an electronic transition in a molecule requires conservation of both the relative position and momentum of the nuclei. Referring to Fig. 1, this means that transitions are possible whose endpoints lie on the dotted curve; this curve is a reflection of the $1s\sigma_g$ potential below $v = 4$ onto the $2p\sigma_u$ curve. Thus the classical Franck–Condon principle predicts fragments with a variable e_d, its extremes being the differences between the asymptote of the potential curves and the limits labeled A and B in Fig. 1. Photodissociation experiments, which are considered in more detail in Section 5.3, can be quite well interpreted in terms of this picture. When the excitation occurs by collision with a heavy particle, the possibility of

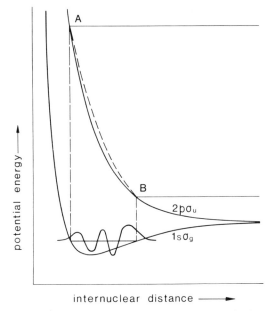

Fig. 1. Dissociation of $H_2^+(1s\sigma_g, v = 4)$ by electronic excitation to the $2p\sigma_u$ state. Illustration of the Franck–Condon principle and the reflection model.

transferring a variable amount of energy and the variation of excitation probability with internuclear distance strongly favor the formation of fragments with separation energies which fall between the same limits A and B in Fig. 1. This is the range of e_d values predicted by the *reflection model*, in which the continuum wave function is represented by a three-dimensional Dirac delta function. This model restricts the vertical electron jump to points which fall on the potential-energy curve of the excited state and neglects the conservation of momentum required by the Franck–Condon principle. Although the reflection model can certainly not be used to explain the details of photodissociation experiments, it is probably a very reasonable model for the treatment of dissociation resulting from electronic excitation in heavy-particle collisions. It is only applicable to dissociation of diatomic species; in polyatomics energy partitioning is much more complex.[34]

Dissociation via the mechanism of vertical electronic excitation to a dissociative state generally becomes important in the energy range between 1 and 10 keV. From the theoretical point of view, this process has received relatively much attention. The first Born approximation in particular is often used to describe dissociation events around 10 keV or above. In this energy range classical theory fails to predict the proper behavior of the dissociation cross section as a function of collision speed. Recently a very comprehensive review on the theoretical methods which have been used to discuss dissociative collisions has been given by McClure and Peek,[1] and the reader is referred to this work for further details. Particular attention is paid to the Born approximation and the additional approximations needed to evaluate the transition matrix elements.

In the Born approximation the initial and final motions of the colliding particles are approximated by plane waves, which combine to a plane wave along the momentum-transfer vector **K**, where **K** is the difference between initial and final wave-propagation vectors $\mathbf{K} = \mathbf{k}_i - \mathbf{k}_f$. It has been pointed out by Dunn[11] that in this case the entire system has a symmetry axis parallel to **K**, and that symmetry of the initial state should be preserved in the final state, leading to angular selection rules. In H_2^+, for example, transition from the $1s\sigma_g$ ground state to the first excited $2p\sigma_u$ state is allowed if the molecular axis is aligned with **K**, but forbidden if it is perpendicular to **K**. A $1s\sigma_g \to 2p\pi_u$ transition, however, is forbidden for a parallel orientation and allowed for a perpendicular one. The $1s\sigma_g \to 2s\sigma_g$ transition is allowed in both directions. The angular selection rules are given by Dunn in tabular form, for both homonuclear and heteronuclear molecules. When the collision energy is not high enough for the Born approximation to be valid, the outgoing wave is spherical "and no definite symmetries for the entire system are evident."[11]

The qualitative arguments of Dunn imply that the excitation is often strongly anisotropic with respect to **K**. If the axial recoil assumption is valid, i.e., when the dissociation takes place along the instantaneous orientation of the molecular axis, the resulting fragments will thus be distributed anisotropically with respect to **K**. As **K** will in general not be distributed isotropically with respect to the axis of the incident beam, the momentum distribution of the dissociation fragments will be anisotropic with respect to the initial beam direction. As a consequence, the measurement of the angular distribution of the dissociation fragments can be a useful tool in establishing with which molecular state the observed fragments are correlated.

The qualitative arguments of Dunn are fully confirmed by Born calculations. A very beautiful example showing how the angular dependence of the dissociative excitation varies with the specific transition involved, has been given by the Born calculations of Green and Peek[12] for H_2^+ colliding on He and Ar targets. They calculated explicitly the dissociation cross section $\sigma(R, \phi)$ as a function of ϕ, the angle between $\mathbf{k}_i$ and **R**, with the internuclear distance R of the H_2^+ as a parameter. The transitions considered are $1s\sigma_g \to 2p\sigma_u$, $1s\sigma_g \to 2p\pi_u$, and $1s\sigma_g \to 2s\sigma_g$. The cross section $\sigma(R, \phi)$ is given by

$$\sigma(R, \phi) = 4v_0^{-2} \int_{k_i-k_f}^{k_i+k_f} dK\, K^{-3} |\varepsilon_T|^2 \int_0^{2\pi} d\xi\, |\varepsilon(K, \delta, R)|^2 \quad (2.1)$$

The matrix element $|\varepsilon_T|$ describes the target excitation; it reduces to unity when no such excitation takes place. $|\varepsilon(K, \delta, R)|$ is the matrix element giving the H_2^+ transition moment as a function of the momentum transfer K, the angle δ between **K** and **R**, and R. v_0 is the speed of the incident beam and ξ is the azimuthal angle of **K** in the laboratory frame. Figure 2 is reproduced

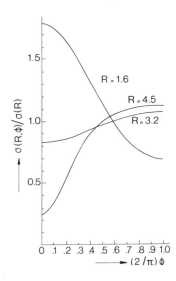

Fig. 2. Born approximation calculation[12] for dissociative $1s\sigma_g \to 2p\sigma_u$ excitation of 10-keV H_2^+ incident on Ar. The quite different dependences of the cross section on the angle ϕ between the molecular axis and the beam direction is shown for three internuclear distances (in a_0).

from the paper by Green and Peek[12] and shows the ratio $\sigma(R, \phi)/\sigma(R)$ as a function of $(2/\pi)\phi$ for the transition $1s\sigma_g \rightarrow 2p\sigma_u$ produced by impact of 10-keV H_2^+ on Ar; the Ar atom is not excited by the collision. $\sigma(R)$ is given by

$$\sigma(R) = \tfrac{1}{2}\int_0^\pi \sigma(R, \phi) \sin \phi \, d\phi \tag{2.2}$$

The figure shows clearly how strongly the angular variation of the dissociative excitation probability depends on the internuclear distance R. For small R values the dissociation is primarily forward–backward with respect to the axis of the incident beam. Intermediate values of R yield an almost isotropic distribution, and at large R values the dissociation is peaked perpendicular to the primary beam. Green and Peek[12] explain this behavior from the properties of the transition element $\varepsilon(K, \delta, R)$. For the $1s\sigma_g \rightarrow 2p\sigma_u$ transition the lowest-order term in the multipole expansion of this matrix element is proportional to $\cos \delta$. The magnitude of the integrand in Eq. (2.1), on the other hand, is zero for $K = 0$, rises to a maximum around $K = 1$ atomic unit (a.u.), after which it rapidly decreases. So when $K_{\min}$, the minimum momentum transfer needed for dissociative excitation, is already close to 1 a.u., as is the case for $R = 1.6$ a.u., the integrand is heavily weighted for K values close to $K_{\min}$. Since the angle χ between $\mathbf{K}$ and $\mathbf{k}_i$ is approximately given by

$$\sin \chi = \left(1 - \frac{K_{\min}}{K}\right)^{1/2} \tag{2.3}$$

the most probable value of χ is close to zero, which means that the most probable direction of $\mathbf{K}$ lies approximately along the beam axis. This results in a transition probability which is approximately proportional to $\cos^2 \phi$. For large R values, on the other hand, only a relatively small amount of inelastic energy transfer is required for dissociative excitation (see Fig. 1), so that $K_{\min}$ tends to zero. The integrand in Eq. (2.1), however, is maximal for a value of K around 1 a.u., resulting in an average value of χ close to $\pi/2$ [see Eq. (2.3)]. That is, the favored direction of $\mathbf{K}$ is perpendicular to the incident beam and dissociation is most probable for $\phi \approx \pi/2$.

The $1s\sigma_g \rightarrow 2p\pi_u$ transition requires a $K_{\min}$ value between 1 and 2 a.u. while $|\varepsilon(K, \delta, R)|$ varies nearly as $\sin \delta$. This results in a pronounced peaking of the dissociative excitation probability for orientations perpendicular to the incident beam, almost independent of R.

The Born approximation treats collisional dissociation as resulting from an excitation within the diatomic molecule. However, dissociation may also occur as the result of a diabatic passage through an avoided crossing between two potential surfaces for the triatomic system: molecule + atomic collision partner (see Chapter 3, Section 2). In this case a Landau–Zener description is more appropriate. From scattering experiments in the case of $H^+ + H_2$ there is definite evidence[13] that such an avoided crossing exists between two surfaces of the $H_3^!$ molecule, and that it plays an important role in the inelastic scattering and in charge transfer when H^+ collides with H_2. So far, there are no theoretical calculations for this type of transition in connection with collisional dissociation in the 1–10 keV range.

2.3. Vibrational/Rotational Dissociation

There are two completely different mechanisms which can give rise to vibrational/rotational dissociation. The first one, already mentioned in Section 2.1, is the process in which the momentum change of one or both nuclei of the molecule by the target atom is sufficiently large to cause dissociation. A theoretical analysis of this process in the case of 10 keV H_2^+ colliding with He has been given by Green[14] on the basis of the classical impulse binary encounter model. In this model it is assumed that the momentum transfer takes place to one or the other of the molecular nuclei, and that the probabilities for dissociation can be added. In order to achieve dissociation by a momentum transfer K to atom one in the molecule, energy conservation requires

$$\frac{K^2}{2m_1} \geq D_v$$

where D_v is the dissociation energy of the initial vibrational state v of the molecule and m_1 is the mass of atom one. For the H_2^+ molecule $D_v \approx 0.1$ a.u. (2.7 eV) for $v = 0$, so that $K \geq 20$ a.u. Still following the arguments of Green this leads to a scattering angle of the H_2^+ center of mass of order $K/(2m_1 v_0)$, which for 10-keV H_2^+ on He is about 0.5°. It is clear that for these relatively violent collisions the deflection of the molecule cannot be neglected. The velocity component of the molecule perpendicular to the primary beam becomes of the same order of magnitude as the fragment velocities in the H_2^+ center-of-mass frame. The information obtained by measuring the laboratory distribution of proton velocities is blurred by the unknown deflection of the H_2^+ center of mass.

The results obtained by Green[14] are analogous to those of Meierjohann,[15] who used a purely classical impulse approximation, but who did not assume that the molecular atoms behave independently. From classical trajectory calculations Meierjohann obtained the momentum transfer to each of the molecular nuclei $\mathbf{K}_1$ and $\mathbf{K}_2$. The vectorial sum of these two momentum changes gives the deflection of the center of mass of the molecule, the vectorial difference gives the extent of vibrational/rotational excitation. Some of the results of this calculation, which was performed for

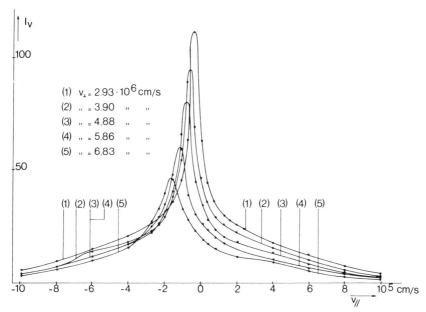

Fig. 3. Probability distribution I_v of proton velocity with respect to the H_2^+ center of mass, calculated by a classical impulse approximation.[15]

20.4 keV H_2^+ on Ar, are shown in Fig. 3. The parallel and perpendicular velocity components of the H^+ fragments, $v_\|$ and $v_\perp$, are given with respect to the initial H_2^+ center-of-mass velocity. It is seen that for values of $v_\perp$ other than zero the most probable value of $v_\|$ is smaller than $v_\perp$. Together with the preponderance of relatively small $v_\perp$ values, this results in a large probability for the dissociation fragments to have velocities relative to that of the incoming H_2^+, which are perpendicular to the axis of the incident beam. In other words, if one neglects the deflection of the H_2^+ center of mass and transforms the laboratory distribution to the center-of-mass system of the incident H_2^+, there results an apparent distribution which peaks sharply around 90°. The same conclusion was reached by Green.[14]

In these analyses the screening of the nuclei by the surrounding electrons is very critical. This especially influences the probability for dissociations due to momentum transfer just above threshold.

A new mechanism for vibrational/rotational excitation and dissociation of molecules has been predicted by Russek.[16] This dissociation mechanism is of a much more subtle character than the violent nuclear vibrational dissociation just discussed. Russek made the observation that in the energy range under consideration (1–100 keV), the Born approximation, although reasonable for the description of the motion of the heavy nuclei, does not take into account that the electrons are able to adjust themselves considerably during the collision. He therefore developed a modified Born approximation which allows for adiabatic behavior of the electrons. Electronic excitation, however, is incorporated in this treatment. This adiabatic behavior leads to polarization-induced forces which can give rise to vibrational/rotational excitation and dissociation. Although the theory of Russek is a self-consistent scattering formulation of the collision process, it can be understood in terms of the well-known induction and dispersion forces. The molecular ion induces a dipole in the target atom. The potential on account of this induction is $U \propto -\alpha/r^4$, where α is the polarizability of the atom and r is the distance between atom and molecule. This potential only contributes to elastic scattering. For homonuclear molecules, the induced dipole in turn induces a dipole in the molecule giving a potential term proportional to α/r^6. For heteronuclear molecules the induced dipole interacts with the permanent dipole of the heteronuclear molecule giving rise to a potential term, proportional to α/r^5. These two terms in the expansion of induction and dispersion forces give rise to rotational/vibrational excitation and, for high initial vibrational/rotational states, to dissociation. The proportionality to r^{-6} and r^{-5}, respectively, is only valid for large distances; at smaller distances screening occurs and for this reason a cutoff parameter a is introduced. As the perturbations are now proportional to $\alpha/(r^6+a^6)$ or $\alpha/(r^5+a^5)$ the cross sections for excitation or dissociation are proportional to α^2/a^6 or α^2/a^5, respectively. Since these interactions act at large distances they can, even though they are weak, contribute considerably to collision-induced dissociation, especially when the initial state is vibrationally highly excited. In the case of electronic excitation, the modified Born approximation yields results identical to those of the Born approximation.

2.4. Predissociation

As indicated earlier, collision-induced dissociation can proceed not only in a direct manner, i.e., by excitation into a continuum, but also by excitation to a bound state which subsequently decays by predissociation. This type of process is referred to a number of times later on in this chapter, and for this reason it may be useful to consider briefly a few aspects of predissociation.

Predissociation is a radiationless transition from a metastable "discrete" energy state to an energetically degenerate level in a dissociation continuum. In a diatomic molecule, it results from either of two phenomena:

1. In the first case a series of bound vibration–rotation levels of one electronic state overlaps the dissociation continuum of another

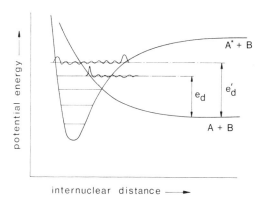

Fig. 4. Illustration of predissociation by electronic transition.[33]

electronic state; this gives rise to *predissociation by an electronic transition*, also called *predissociation by curve crossing* or *type-I predissociation*[17] and illustrated by Fig. 4; a variant of this is *accidental predissociation* which involves the additional interaction with an intermediate bound state.[17,18]

2. In the second case, a series of quasi-bound rotation–vibration levels of an electronic state overlaps the dissociation continuum of the same state. These levels lie above the dissociation limit of the electronic state in question, but they are separated from the continuum by the rotational barrier in the effective potential; they can predissociate by tunneling through the effective potential barrier. This *rotational predissociation*, or *case-III predissociation*[17] is illustrated in Fig. 5. Levels very near the top of the barrier may appear as shape- or orbiting resonances in scattering experiments.[19,20]

In polyatomic molecules one in addition encounters vibrational (type II[17]) predissociation, as discussed in Chapter 7.

In optical spectroscopy, predissociation was first observed through the broadening of rotational absorption lines.[21] It can also be recognized by the widening of emission lines and by the weakening or breaking off of rotational structure in emission spectra.[17] Such observations have been used extensively in the determination of dissociation energies of neutral diatomics.[17,22] In the last few years, there has been a revival of interest in the study of rotational predissociation,[23,24] largely as a result of a paper by Bernstein[19] who drew attention to the possibility of using rotational predissociation data as a probe of the long-range behavior of the interatomic potential. Similarly, predissociation by curve crossing has recently received considerable attention as a source of information on repulsive energy curves.[25] This information is contained in the variation in predissociation

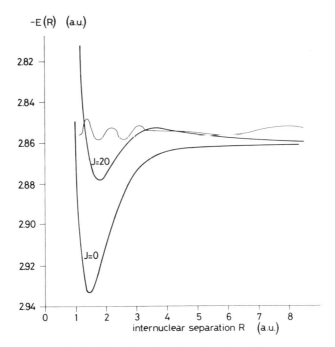

Fig. 5. Illustration of rotational predissociation.

probability from one rotational/vibrational level to another—a variation which reflects the applicability of the Franck–Condon principle.

Optical *absorption* spectra, the main source of information on predissociation, are relatively rare in the case of ions—mainly because of the experimental difficulties associated with the production of sufficiently high ion concentrations. In addition, predissociation results in observable line broadening only when the line width becomes larger than the Doppler width, so that only rapid predissociations (predissociation rates $\geq 10^9$ sec^{-1}) can be studied in this way.[17] Weakening in *emission* of course implies that predissociation competes effectively with radiative decay. To the extent that the relevant emission spectra often correspond to dipole-allowed transitions (emission rates $\geq 10^6$ sec^{-1}), this method too is usually limited to rather rapid predissociations, although precise lifetime measurements can detect predissociation rates down to $\sim 10^6$ sec^{-1}.[54]

On the other hand, mass spectrometric techniques such as used in the study of collision-induced dissociation, allow us to observe predissociation of molecular ions in the most direct way possible, namely by observing the resulting charged fragments—and the accessible range of lifetimes is much wider than in optical spectroscopy. If the predissociation competes effectively with radiative decay, predissociation rates from a low of, say, 3×10^3 sec^{-1} to a high of 10^{13} sec^{-1} can be studied.

The lower limit can be reached by studying the unimolecular fragmentation of metastable ions as they traverse the field-free region of the mass spectrometer. In the case of metastable N_2^+ mentioned in Section 6, flight-time considerations show that these metastable ions should have lifetimes τ between 6×10^{-8} and 3×10^{-4} sec, the optimum value being $\tau \approx 10^{-6}$ sec.

The fastest predissociation accessible by mass spectrometry can be observed by detecting fragments formed immediately after collisional excitation of the parent ions. In that case the lower limit to the excited state's lifetime is $\approx 10^{-13}$ sec, i.e., a vibrational period: when shorter times are involved the concept of fragmentation via predissociation is replaced by that of direct dissociation. For $\tau = 10^{-12}$ sec, for example, the state would have an energy width, $\Gamma = \hbar/\tau$, of 2.4×10^{-5} a.u. ($\approx 6.5 \times 10^{-4}$ eV or ≈ 5.3 cm^{-1}), and it would be difficult to observe spectroscopically. A number of examples involving predissociation are discussed further in the following.

3. Experimental Techniques

The usefulness of a mass spectrometer in studying the dissociation of molecular ions was recognized by Aston[26] as early as 1920. Apart from well-defined primary peaks representing singly or multiply charged ions extracted from the ion source, magnetically scanned mass spectra show diffuse peaks or bands which are due either to collisional phenomena or to spontaneous fragmentation of metastable ions. These bands appear as a consequence of any process, collisional or unimolecular, in which ions change their charge-to-mass ratio as they travel from the ion source to the detector. They are generally known as "Aston bands" or "metastable peaks." In the early days of mass spectrometry they most often arose from collisions with the background gas in the apparatus, but even in 1932 Friedländer et al.[27] recognized the occurrence of unimolecular decay in the case of CO^{2+}. In order to distinguish between these two possibilities, we will designate collision-induced fragmentation peaks as "Aston bands," and those resulting from the unimolecular decay of long-lived states formed in the ion source as "metastable peaks."

The study of Aston bands in connection with collision-induced dissociation and charge transfer has been discussed very thoroughly by McGowan and Kerwin.[28] More recent applications of this method have taken advantage of a number of important refinements, but the basis of the technique remains the same. "Aston banding" can also be used to study ion–molecule reactions, and many aspects of the technique are also applicable to the investigation of truly metastable peaks. For a comprehensive review which focuses mainly on applications to organic chemistry, we refer to a recent book by Cooks and co-workers at Purdue.[29]

A second mass spectrometric technique which is useful in the investigation of collision-induced and, especially, unimolecular dissociations of ions, is the appearance-potential method. This technique is also discussed in the paper by McGowan and Kerwin quoted earlier. Conversely, Chupka has drawn attention to the effect of unimolecular decay on the interpretation of appearance potentials.[30] In recent years, the appearance-potential method has, for instance, been used by Durup and co-workers at Orsay to unravel the contribution of different excitation processes to collision-induced band spectra in H_2^+ and N_2^+.[31,32]

3.1. The Aston Band Method

In this section we will restrict ourselves to the application of the Aston band method to dissociation processes in diatomic ions. It is supposed that the instrument used is a simple magnetic sector mass spectrometer which comprises the elements illustrated in Fig. 6. A useful modification is the addition of a selection magnet immediately after the ion source. Another well-suited arrangement is that based on the sequence ion source, magnetic sector, electrostatic sector, and detector, which is used, e.g., by Fournier and co-workers at Orsay[33] and by the group at Purdue.[34]

The fragments of fully accelerated fast molecular ions which dissociate in the field-free drift region before the magnetic analyzer have approximately the same velocity as the parent ion. It can easily be shown[28] that they appear on the mass scale with an apparent mass-to-charge ratio given approximately by

$$\left(\frac{m}{q}\right)_{\text{app}} = \left(\frac{m}{q_f}\right)^2 \left(\frac{q_p}{M}\right) \quad (3.1)$$

where m and M denote the mass of the fragment and parent ion, respectively. Their charges are denoted by q_f and q_p. In the case of singly charged ions this relation reduces to

$$m_{\text{app}} = \frac{m^2}{M} \quad (3.2)$$

Now Aston bands as they appear in mass spectrometry do not only come from dissociations in the drift region before the analyzer. Dissociation also occurs in the acceleration region, in the magnetic field, and in the drift region between analyzer and detector. As a result, the complete Aston band extends from the position given by Eq. (3.1) to the location of the parent-ion peak. This occurs quite apart from the broadening due to the excess energy

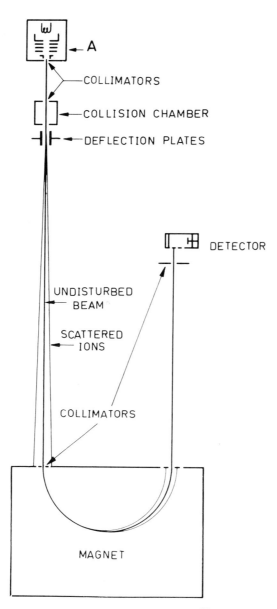

Fig. 6. Schematic drawing of the Amsterdam apparatus[35] for studying collision-induced dissociation.

carried off by the dissociation fragments (see later). The occurrence of dissociation at various locations in the mass spectrometer detracts from the uniqueness of the information contained in the Aston band. It is therefore of great value to mount a collision chamber at the object point of the analyzing magnet. Its length should be short with respect to the focal length of the magnet. Fragment ions formed in the collision chamber are then focused on the collector slit, ensuring maximum resolution. The background pressure should be very low with respect to pressure in the collision chamber, in order to minimize the relative contribution to the Aston band from fragments formed outside the collision chamber. In the measurement of truly metastable peaks, one can ensure that the fragments originate from inside the collision chamber by using a technique which is often employed in merging-beam experiments. In this case the collision chamber only serves as a localization device: for example, it is put at a positive potential U with respect to the field-free region. For molecules dissociating inside the collision chamber there is a net gain in fragment momentum because the velocity increase which the fragments receive as they leave the collision chamber is larger than the velocity decrease experienced by the parent ion as it enters this region. The difference in momentum between fragments of mass m_1 formed inside and outside the collision chamber is given approximately by

$$\Delta(m_1 n) \approx \frac{1}{2}\left(\frac{m_2}{m_1}\right)\left(\frac{q_f U}{E_0}\right) \times m_1 v \qquad (3.3)$$

where E_0 is the kinetic energy of the parent molecule and $m_2 = M - m_1$. A potential U of a few hundred volts is often sufficient to separate these two groups of fragments in the mass spectrum.

An example of Aston bands in a mass spectrum up to mass number 4 is shown in Fig. 7. The source gas consists of a mixture of H_2, D_2, and He, giving rise to primary peaks at the integral mass numbers $1(H^+)$, $2(H_2^+, D^+, He^{2+})$, $3(HD^+, H_3^+)$, and $4(He^+, D_2^+)$. Even on this scale the broad Aston bands due to collision-induced dissociation can be clearly distinguished from the primary peaks. The position on the mass scale indicates which process is responsible for the Aston band, although the assignment is not always unique. For example,

$\left(\dfrac{m}{q}\right)_{\text{app}} = \dfrac{1}{2}, \qquad H_2^+ \rightarrow H^+ + H$

$\left(\dfrac{m}{q}\right)_{\text{app}} = \dfrac{4}{3}, \qquad H_3^+ \rightarrow H_2^+ + H, \quad HD^+ \rightarrow D^+ + H$

$\left(\dfrac{m}{q}\right)_{\text{app}} = \dfrac{16}{5}, \qquad HeH^+ \rightarrow He^+ + H$

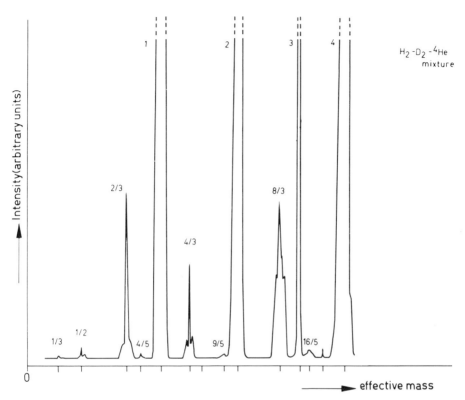

Fig. 7. Mass spectrum of H_2, D_2, and He. The broadened peaks at nonintegral mass numbers are due to collision-induced dissociations of molecular ions.[3]

In the derivation of Eq. (3.1) it has been assumed that the dissociation fragments have zero velocity with respect to their center of mass. As a result, this formula locates only the center of the Aston band (or metastable peak). In the case of collision-induced dissociation, Eq. (3.1) implies one more approximation since it neglects the inelastic energy transfer in the collision. This transfer results in a laboratory energy loss of the molecular center of mass which we shall call ΔE. The determination of ΔE is often valuable in the interpretation of Aston bands, and this can be done by comparing the position of the band with that of a primary ion peak nearby. In some cases this is possible because of a near-coincidence between the Aston band and such a "marker" peak, as in the case of N_2^+ illustrated in Fig. 8. The center of the Aston band produced by collisional dissociation of N_2^+ lies at an apparent mass slightly smaller than 7, the latter position being marked by primary N_2^{2+}. McGowan and Kerwin showed that ΔE can be determined quite accurately when use is made of isotopic molecules. If there is no marker peak close to the Aston band under consideration, a situation similar to the

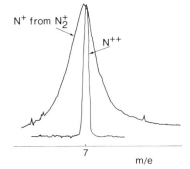

Fig. 8. Qualitative drawing of part of the Aston band which is due to the dissociation of N_2^+ on collision with N_2. At mass 7 the sharp marker peak N^{2+} is shown.[28]

example just mentioned can be achieved by an appropriate change in acceleration potential. For the Aston band resulting from the dissociation of 10-keV H_2^+ ions, one can for instance use the peak corresponding to 5-keV H^+ as marker.

The relatively broad appearance of Aston bands and metastable peaks is due to the conversion of internal energy of the excited parent molecule into translational energy of the dissociation fragments with respect to their center of mass. The amount of internal energy thus transferred is designated by e_d; for diatomic ions it is the excess energy of the excited parent molecule with respect to the dissociation limit corresponding to the resulting fragments. The value of e_d is generally very small with respect to the laboratory energy of the incident beam, typical values being 1 eV and 5 keV, respectively. Nevertheless, because a magnetic mass spectrometer is a momentum analyzer, e_d can be determined with relatively high accuracy. If we consider only fragments scattered forward and backward with respect to the incident beam, we find k_f, the final laboratory momentum of the fragment with mass m_1, to be

$$k_f = m_1 v_1 = k_i \pm \kappa = k_i \left\{ 1 \pm \left[\frac{m_2 e_d}{m_1(E_0 - \Delta E)} \right]^{1/2} \right\} \quad (3.4)$$

where κ is the magnitude of the momentum with respect to the center of mass of the excited parent ion: $\kappa = (2\mu e_d)^{1/2}$; k_i is the momentum of fragment 1 traveling at the speed of the excited parent ion, v^*:

$$k_i = m_1 \left[\frac{2(E_0 - \Delta E)}{m_1 + m_2} \right]^{1/2}$$

E_0 is the laboratory energy of the incident ion and ΔE its laboratory energy loss due to excitation; and $\mu = m_1 m_2/(m_1 + m_2)$ is the reduced mass of the molecular fragments. Note that under the experimental conditions described in this chapter the measured energy loss ΔE is usually, to a good approximation, equal to the energy defect or inelastic energy loss Q (see the introductory chapter). If the analyzing magnet has a resolution $\delta k_f/k_f$, the resolution in e_d which can in principle be obtained by measuring the momentum of fragment 1 is

$$\frac{\delta e_d}{e_d} \approx 2\left(\frac{m_1 E_0}{m_2 e_d}\right)^{1/2} \frac{\delta k_f}{k_f} \tag{3.5}$$

One sees that the best resolution is achieved when $m_1 < m_2$ and when E_0 is small and e_d large. For $m_1 = m_2$, $E_0 = 10$ keV, and $\delta k_f/k_f = 10^{-4}$, it is for instance possible to distinguish between e_d values of 1.00 and 1.01 eV. The lowest measurable e_d can be estimated by setting $\delta e_d = e_d$; in the conditions just mentioned this yields a lower limit of $e_d \approx 10^{-4}$ eV. Similar considerations apply to experiments in which an electrostatic energy analyzer is used, as outlined by Fournier[33] (cf. Chapter 7, Section 3.2).

In order to take full advantage of the possibility of measuring e_d with high resolution, precautions have to be taken concerning the angular width of the primary beam and the angular aperture of the analyzer. Stringent angular collimation of primary and scattered beam is a necessity: The effective angular aperture, expressed in radians, should not exceed the relative momentum spread, $\delta k/k$, resulting from the finite resolution of the analyzer and the energy spread of the beam. This requirement is discussed in more detail in Section 3.3.

Stringent collimation at the same time offers the possibility of measuring the angular distribution of the dissociation fragments in the laboratory system. This can be performed either by placing a set of deflecting plates behind the collision chamber by means of which the scattered fragments can be deflected, or by rotating the primary beam around the center of the interaction chamber by mechanical or electrostatic means. As is apparent from the vector diagram of Fig. 9 the laboratory velocity of fragment 1 given by v_1 and θ can be uniquely transformed to the center-of-mass velocity given by u and ϕ, provided the inelastic energy loss is known.

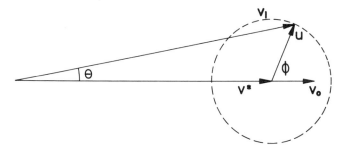

Fig. 9. Velocity vector diagram illustrating the relation between laboratory and center-of-mass velocities. The difference between the initial velocity of the molecule v_0 and the final velocity v^* is due to inelastic energy loss.

This brings us to the proper choice of the accelerating voltage. As already mentioned in connection with Eqs. (3.4) and (3.5), the highest resolution in κ, and thus in energy release e_d, is obtained with the lowest acceleration voltage. However, there are other considerations which favor a high primary-beam energy. In the first place, in collision-induced dissociation the inelastic energy loss is a function of the energy release e_d. Even if the Aston band has a well-defined central peak, its location only allows the determination of the energy loss for fragments with $e_d \approx 0$. For the other fragments the inelastic energy loss can generally not be determined in a unique way. In the case of H_2^+, for instance, the energy loss for the $1s\sigma_g \rightarrow 2p\sigma_u$ transition varies from almost zero to ~ 15 eV. A high energy of the primary-beam particles diminishes the effect of this variable energy loss on the shape of the fragmentation band. The transformation from the laboratory momentum distribution to that in the center of mass therefore is relatively more accurate. Naturally, this also means that the determination of the inelastic energy loss Q becomes more difficult at higher beam energies. This is a disadvantage because Q can be an important indication of the electronic state by which the dissociation took place.

A second point concerns the effect of the energy spread in the primary beam. In most cases this spread, which finds its origin largely in the ion source, is nearly independent of energy, so that its relative importance decreases with increasing beam energy. For this reason one often increases the beam energy until the spread in the measured momentum due to the finite resolution of the analyzer becomes comparable to that resulting from the energy spread in the primary beam. Plasma ion sources in general have a larger energy spread than do Nier sources. Instruments using plasma ion sources are therefore often used at ~ 10 keV or higher, whereas mass spectrometers with Nier sources typically operate in the 2–3 keV range.

Finally, it is useful to work at rather high energies from a physical point of view, because the Born approximation has yielded the most reliable and most detailed predictions for collisional dissociation by diatom–atom collisions. Even for H_2^+, energies as high as 40 keV are needed to get into the range in which the Born approximation is fully justified.*

Figure 6 shows an early version of the apparatus used at the F.O.M. Institute in Amsterdam.[35] The central part of the instrument is a 180°

Note Added in Proof: Prof. J. Durup has kindly drawn our attention to the fact that working at high acceleration voltage also leads to an absolute decrease of the velocity spread due to thermal motion in the source. This follows from the expression

$$\tfrac{1}{2}Mv_0^2 = E_0 + \tfrac{1}{2}Mv_t^2$$

so that the spread δv_0 in initial beam speed due to a change δv_t in the axial thermal velocity component can be estimated from

$$\delta v_0 \approx \frac{v_t}{v_0} \delta v_t$$

inhomogeneous magnet, the dispersive power of which is about 10 times that of a comparable homogeneous magnet. Its momentum resolution, $\delta k/k$, is $\sim 2 \times 10^{-4}$. The object and focus are located at about 150 cm from the entrance and exit points of the magnet, respectively. The center of the collision chamber has been placed at the object point; its length is ~ 3.5 cm. The ion source is of the unoplasmatron type. The acceleration optics are adjusted to provide optimum transmission of the primary beam through the two collimators ($\varnothing = 0.4$ mm, separation 35 cm). The half-width of the beam is $\approx 10^{-3}$ rad, and the beam current at the entrance of the magnet is typically 10^{-8} Å at 10 keV; the energy spread of the beam is ≈ 2 eV. The exit slit of the collision chamber measures 0.7×5 mm, and collimation of the scattered fragments is performed at the entrance of the magnet by means of a rotating disk having holes with diameters between 0.3 and 5 mm. Deflecting plates, mounted just behind the collision chamber, are used to measure the laboratory angle distribution of fragment ions.

In a more recent version of the apparatus,[36] a magnet has been added for mass selection of the primary beam. Differential pumping ensures operating pressures of $\sim 10^{-8}$ torr in the collimation region and of $\sim 10^{-9}$ torr in the region of the deflecting plates and the analyzing magnet.

The experimental procedure depends on the process that is the subject of study. When one expects quasi-discrete values of the separation energy of the fragments, e_d, as in the case of predissociation mechanisms or photodissociation experiments, the determination of these values is often the main goal of the measurements. One then aims for the best resolution possible, and this is achieved by measuring fragments recoiling forward or backward along the incident beam axis, using stringent collimation. When the dissociation process is relatively slow, so that fragments are formed in significant numbers all along the free-flight region, one can meet the requirement of high angular resolution by putting a potential on the collision chamber. This makes it possible to study fragments originating there separately from the others, as outlined in Section 3.1. On the other hand, it is often of great importance to know the angular distribution of the dissociation fragments, particularly in the case of "direct" collision-induced dissociation. The procedure followed in Amsterdam is aimed at obtaining the angular distribution of fragments in the center of mass at a fixed energy release, e_d. With the assumption that the deflection of the center of mass is negligible the vector diagram of Fig. 9 can be used to calculate the laboratory speed and deflection angle of the chosen fragment, v and θ, for each given e_d (i.e., u) as a function of the center-of-mass angle ϕ. Thus the mass spectrometer can be programmed with the appropriate magnetic-field and deflection-plate settings to measure the intensity of fragments having a fixed value of e_d at chosen intervals of ϕ. A reasonable estimate is made for the inelastic energy loss Q.

Similar instruments and techniques are used by other groups. Vogler and Seibt[37] use a parabola spectrometer to analyze mass and energy of the fragment ions. Angular scans are performed by mechanical rotation of the primary beam around the center of the collision chamber. The energy distribution is measured at a series of chosen laboratory angles θ to yield the velocity distribution in the center of mass. Sector-type mass spectrometers are used, for example, by Durup's group[31] Newton and Sciamanna,[38] Moran and co-workers,[39] and Wakenne and Momigny.[40] Mass spectrometers which incorporate two magnetic analyses or a magnetic and an electrostatic analyzer are also often used.[34] This short list is by no means complete; for an exhaustive bibliography on instruments and techniques, we refer the reader to the monograph by McClure and Peek.[1]

3.2. Initial-State Preparation

The preparation of the parent molecular ions in a definite quantum state has received relatively little attention in mass-spectrometric studies of collision-induced or spontaneous dissociation. Although the ultimate ideal is to start with molecules in a well-defined quantum state, it is not an absolute requirement for every investigation. We have already pointed out that inelastic energy loss obscures the interpretation of the measurements less seriously at higher collision energies. In those cases the spread in energy loss caused by the population of the incident molecules over a range of energy levels is of less consequence. As far as angular distributions is concerned, the Born approximation indicates that the initial rotational/vibrational state is of minor importance. To what extent the validity of the reflection approximation which is used to obtain $\sigma(R, \phi)$ (cf. Section 2.2) depends on the initial vibrational state is still open to question. In photodissociation experiments, the majority of the fragments arise from transitions where the photon energy fits the energy difference between classical turning points on the upper and lower potentials. Such experiments, as well as theoretical predictions, indicate that the reflection approximation does not depend strongly on the initial vibrational state. Finally, we may mention that in studies of rotational predissociation such as those discussed in Section 5.2, the strong variation in lifetime from one vibration/rotation level to the next strongly limits the number of levels that can be observed. In those cases, too, precise knowledge of the vibration–rotation population is not required.

The simplest step in initial state prepreparation concerns the selection of the parent ion. In many instances the mass of the parent ion can be deduced directly from the apparent mass of the fragments. In other cases, a preselecting magnet is required; even then the assignment is not always unambiguous, as the example of H_3^+ and HD^+ in Section 3.1 shows.

The ion source is the most critical part of the apparatus as far as initial state preparation is concerned. Plasma sources yield the worst defined primary beams, both with respect to rotational and vibrational excitation and with respect to electronic excitation. Often the spread in translational energies is wider than for controlled electron-impact sources, but this effect can be minimized by raising the acceleration potential (Section 3.1). However, plasma sources do offer a number of advantages. Because of the many secondary reactions taking place, even molecular ions which have no stable neutral parent can be formed in substantial amounts. These reactions can also give rise to ions with a high degree of rotational and/or vibrational excitation, allowing the study of rotational predissociation. Finally the yield of plasma sources is relatively high: Even with fairly low extraction potentials (required to reduce the energy spread), a unoplasmatron typically delivers total currents of 10^{-6} to 10^{-5} A.

An alternative to plasma sources is the classical Nier source, where most of the ions are created by direct electron impact. At electron energies of ~ 100 eV, most of the ions are populated by Franck–Condon-type transitions, so that the vibrational and rotational populations are fairly well known. In many cases the ions are mostly in the electronic ground state, but there is ample evidence for the presence of metastable excited states in ion beams, either because they are present in substantial amounts (e.g., for NO^+)[41-43] or because the experiment detects even small fractions of metastables (e.g., for N_2^+).[32,44] An important advantage of Nier-type sources is that the population of the energy states of the parent ions can be controlled by variations in the energy of the ionizing electrons. The measurement of appearance potentials, for example, is very helpful in establishing the role of excited electronic states.[32] One may also work at sufficiently low energies to populate only the lower vibrational levels of the electronic ground state.[31] Such experiments are of course limited by the severe drop in ionization efficiency near the threshold. A less precise method is to work at energies between threshold and, say, 30 eV, in order to promote vibrational populations which differ from those expected for Franck–Condon ionization.[45]

A much more refined method, which, to our knowledge, has not be used in high-velocity dissociation experiments, is that of photoionization. The ideal would be to detect fragments in delayed coincidence with the photoelectron ejected in the ionization of the parent molecule. In fact, Berkowitz and Chupka[46] have shown that autoionization in H_2 shows a preferential production of photoelectrons with minimum kinetic energies. As a result, narrow control of the H_2^+ vibrational population can be achieved by choosing the photon wavelength to coincide with one of the autoionizing transitions.

Well-established techniques which can be used to control the internal energy of the parent ions are those which rely on charge transfer or chemical ionization. Such reactions have been exploited by Cermák and Herman,[47]

Field,[48] and Lindholm,[49] among others. An attractive possibility seems to be the use of ion-cyclotron sources: In these sources the ions can be stored during times which are sufficiently long to ensure a thermal population of energy levels. Richardson *et al.*[50] have used such a source to measure alignment of H_2^+ by means of selective photodissociation.

When it cannot be controlled, collisional alignment may be a problem, although probably not a serious one, in dissociation experiments as well as in other experiments with neutral or charged molecular beams. This alignment arises from the fact that the cross section for dissociation, e.g., with the source gas at the exit of the ion source, generally varies with the orientation of the molecule with respect to the beam axis.[12] Van Asselt *et al.*[36] found evidence for such alignment in studying the angular distribution of protons resulting from the photodissociation of an H_2^+ beam (cf. Section 5.3).

A unique possibility for vibrational state selection is offered by photodissociation. Ozenne[51] has observed that in the photodissociation of H_2^+ with a pulsed ruby laser, one vibrational level was nearly completely depopulated. This makes it possible to study the collision-induced dissociation of that particular vibrational level by synchronizing the detection of fragments with the laser pulse.

3.3. Final-State Analysis

As mentioned earlier, a complete final-state analysis generally requires the measurement of three velocity components in addition to the three which define the velocity of the charged fragment. They can relate to either the neutral fragment or to the target atom. If none of the separating atoms is electronically excited, two rather than three additional velocity components are needed. If one detects only fragments scattered exactly along the primary beam axis (which excludes any significant deviation of the molecular center of mass) and if there are no electronically excited products, it is sufficient to measure just the speed of the charged fragment.

Coincidence measurements which meet the requirements of the most general case are at present practically impossible. A less ambitious but useful experiment would, for example, consist in detecting the neutral fragment in delayed coincidence with the mass-analyzed fragment, and this for pairs of scattering angles which are chosen to discriminate between nuclear vibrational dissociation and dissociation by electronic excitation. In the former case the two fragments will be scattered in nearly the same laboratory angles, since their deflection results mainly from the scattering of the molecular center of mass. In the latter case the momentum transfer will usually be small so that the corresponding scattering angles can be deduced from the vector diagram of Fig. 9. Experiments aimed at distinguishing between these dissociation processes are being undertaken at Giessen.[52]

A relatively simple way to perform a partial final-state analysis is to measure the charged fragments in coincidence with photon emission, either from the target atom or from one of the fragments. A first attempt in this direction has been made by Jaecks et al.[53] These authors studied the collision-induced dissociation of keV H_2^+ ions and measured Lyman-α photons in delayed coincidence with H atoms or protons, scattered over a certain laboratory angle. By applying quenching techniques they could measure not only L_α photons emitted by H atoms produced in the $2p$ state but also those originating from H($2s$). Because of the low coincidence rate, however, the angular resolution was rather poor, and velocity analysis of charged fragments was not performed. This technique is very attractive since among the large number of fragments due to vibrational or to electronic excitation it selects only those which are the result of a specific electronic excitation of the molecule. In the case investigated by Jaecks et al. these were the $2p\pi_u$ and $2s\sigma_g$ states of H_2^+. For further discussions of coincidence methods the reader is referred to Chapter 3.

3.4. The Apparatus Function

One of the most tricky problems in the analysis of the results of the present type of experiments is the transformation of the laboratory data to the molecular center of mass. The problems arise from the fact that the fragments are measured with finite resolution, both in momentum (or energy) and scattering angle. The momentum resolution is discussed in many textbooks. Mass spectrometers and energy analyzers have a constant resolution, $R = \delta k/k$, and $R = \delta\varepsilon/\varepsilon$, respectively. The resolution depends on the geometry of the apparatus, the radius of the analyzer, slit widths, and so on. Of importance is, of course, the stability of magnetic and electrostatic fields, the stability of the accelerating voltage, and the energy spread of the primary beam. However, angular resolution is a requirement which has not often been discussed in mass spectrometry. For intensity reasons one often tries to make the acceptance angle as large as possible. The aberration effects are then corrected for by higher-order focusing. In dissociation measurements, however, the angular resolution is just as important as the momentum resolution. This is the reason for several of the features of the Amsterdam apparatus mentioned in Section 3.1: the location of the scattering chamber at the object point of the mass spectrometer, the stringent collimation of the primary beam and the scattered fragments, and the application of a potential to the collision chamber in order to obtain a spectrum from the dissociation fragments coming from the object point of the mass spectrometer only.

In principle the transformation from laboratory to center of mass is quite simple and straightforward. Figure 10 shows the vector diagram for a

318 Chapter 6

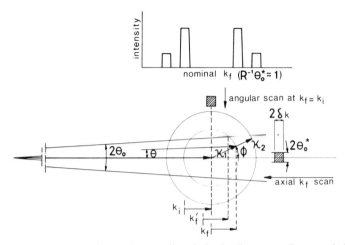

Fig. 10. Momentum vector diagram for a predissociation leading to two discrete κ (e_d) values. The influence of the angular aperture θ_0 upon the peak shape is shown in a qualitative way. The shaded squares and the momentum distribution in the upper part of the figure refer to the case in which $R^{-1}\theta_0^* \approx 1$ (see text).

dissociation leading to a fragment with a momentum vector κ, ϕ in the center of mass of the molecule. We note that

$$k_f = [(k_i + \kappa \cos \phi)^2 + \kappa^2 \sin^2 \phi]^{1/2} \tag{3.6}$$

$$\theta = \arcsin\left[\left(\frac{\kappa}{k_f}\right) \sin \phi\right] \tag{3.7}$$

where k_i is the momentum associated with the motion of the fragment before dissociation (but after the collisional excitation step).

The leading term in the Jacobian, D, which transfers the differential cross section in the center-of-mass frame to that in the laboratory system, is $(k_f/\kappa)^2$:

$$\sigma_{\text{lab}}(k_f, \theta) \simeq \sigma_{\text{c.m.}}(\kappa, \phi)\left(\frac{k_f}{\kappa}\right)^2 \tag{3.8}$$

The application of this transformation equation gives rise to two major difficulties. The first one is inherent to the experimental method: only the momentum of one fragment is measured, so that the direction and magnitude of k_i, and thus κ and ϕ [cf. Eqs. (3.6) and (3.7)], are not unambiguously established. In the lab → c.m. transformation one usually assumes that the center of mass is not deflected, i.e., that k_i lies along the primary beam axis. This is only strictly true for spontaneous dissociation, where the primary beam is defined after the excitation process, and in photodissociation, where the momentum transfer is completely negligible. The assumption is reasonable for dissociation via electronic excitation or through polarization forces (the Russek mechanism), because the momentum transfer in these processes is small. For nuclear vibrational dissociation we have already remarked that this condition is severely violated, and this is the reason why we speak in terms of an apparent center-of-mass distribution (Section 2.3).

Even if the angular deflection of the center of mass is negligible, dissociation measurements involving heavy-particle collisions are not completely unambiguous because of the slowdown of the incident molecule in the excitation step. We have already mentioned that this effect is less important at high collision energies. Nevertheless it is desirable to correct for the change in speed of the center of mass, and this can be done by estimating the inelastic energy loss from the released kinetic energy and an assumption about the electronic state by which the dissociation proceeds. For the Russek mechanism this correction is of minor importance.

The second major difficulty in the transformation is of a purely experimental nature. Taking into account the restrictions mentioned in the preceding paragraph, one may write as a first approximation

$$\sigma_{c.m.}(\kappa, \phi) \propto I_{lab}(k_f, \theta)\left(\frac{k_f}{\kappa}\right)^{-2} \qquad (3.9)$$

where $I_{lab}(k_f, \theta)$ is the measured intensity of fragments having a nominal momentum k_f and scattering angle θ in the detection plane (the median plane of the spectrometer). I_{lab} is assumed to be corrected for the change in detector response as a function of k_f. It is seen that Eq. (3.9) cannot be applied when $\kappa \to 0$, since then $\sigma_{c.m.} \to \infty$ for any finite I_{lab}. The proper use of the transformation Jacobian requires one to account for the finite momentum resolution and acceptance angle of the spectrometer. If $\Delta\kappa$ is the range of c.m. momenta falling within the corresponding acceptance interval, this means that Eq. (3.9) can certainly not be used when $\kappa \lesssim \Delta\kappa$.

The effects of finite resolution can be clarified by means of Fig. 10, which represents a case in which fragments are formed with discrete center-of-mass momenta κ_1 and κ_2 corresponding to discrete energy releases $(e_d)_1$ and $(e_d)_2$. It is seen that even in the limit of perfect momentum resolution, the collection of the fragments within a finite acceptance angle, θ_0, will give rise to peaks which extend, for example, from k_f to k'_f on the momentum scale. This also means that when fragments are detected at a nominal momentum k_f, the corresponding value of κ will be uncertain by an amount $\delta\kappa$. This uncertainty is smallest for fragments scattered forward or backward with respect to the center of mass, in which case it may be approximated[33] by

$$\frac{\delta\kappa}{\kappa} \approx \frac{1}{2}\frac{k_f^2}{\kappa^2}\theta_0^2 \qquad (3.10)$$

This approximation holds as long as $\theta_0 \approx \phi_0$, where ϕ_0 is the center-of-mass angle corresponding to θ_0. It is seen that the angular broadening is most serious at high beam energies and low separation energies in the center of mass, and that it rapidly deteriorates with increasing θ_0. If we now consider that the momentum resolution too is finite and that the beam has a finite energy width, resulting in an effective momentum resolution $R = \delta k/k$, the smallest κ which we may hope to resolve is $\kappa \approx \delta \kappa \approx Rk_f$. Substituting in Eq. (3.10) we find that the product $R^{-1}\theta_0$ should be on the order of unity or less in order to ensure that our ability to distinguish between small values of κ is not limited by the finite angular resolution of the apparatus. The acceptance interval corresponding to $R^{-1}\theta_0^* \approx 1$ is reported by the shaded square of Fig. 10. It will be recognized that when $R^{-1}\theta_0^* \approx 1$, a k_f scan along the beam axis can be replaced by one in which the transmitted momentum value is fixed at k_i while deflecting plates are used to bring fragments ejected perpendicular to the beam within the acceptance aperture of the analyzer.

The effects discussed in the preceding paragraphs have been recognized by a number of workers, and led Fournier[33] and Anderson[64] to the derivation of a "spectrometer efficiency function." This function is obtained by integration of the transformation Jacobian over the laboratory acceptance interval, $\Delta k_f \, \Delta\Omega$:

$$A = \int_{\Delta k_f} \int_{\Delta \Omega} D \, d\Omega \, dk_f \tag{3.11}$$

and relative center-of-mass cross sections are then estimated from

$$\sigma_{c.m.}(\kappa, \phi) \propto \frac{I_{lab}(k_f, \theta)}{A} \tag{3.12}$$

Such a procedure is only acceptable if $\sigma_{c.m.}(\kappa, \phi)$ does not change much within the center-of-mass equivalents of Δk_f and $\Delta\Omega$. When there are discrete peaks in the center-of-mass distribution, however, or, in general, when $\sigma_{c.m.}(\kappa, \phi)$ is a strong function of κ, this approximation is not allowed. Maas et al.[55] therefore adopted a convolution method rather than a deconvolution procedure. In order to obtain the quasi-discrete κ distribution characteristic of the rotational predissociation of H_2^+,[55] for example, these authors start with a distribution in which the location of the various peaks is obtained from theoretical arguments and they then evaluate the height of these peaks by iterative comparisons of the computed laboratory distributions with the observed one. A similar method has been developed by Terwilliger et al.[56] The integral which is evaluated in this simulation is

$$I_{fr} = \sum_n \int_{\Delta \kappa_n} \int_{\Delta \omega_n} N_n(\kappa) \, d\omega \, d\kappa' \tag{3.13}$$

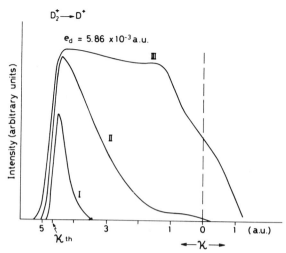

Fig. 11. Computer-simulated peak shapes for the predissociation of 10 keV D_2^+ with an e_d value of 8.58×10^{-3} a.u. The corresponding κ value is indicated by κ_{th}. The curves I, II, and III are peak shapes for dissociations taking place at distances of 100, 50, and 1 cm before the collimating aperture in front of the analyzing magnet.[55]

where I_{fr} is the number of particles arriving at the detector at a fixed setting of the momentum selector, and $N_n(\kappa)$ is the current per unit solid angle in the center-of-mass system of fragments originating from the predissociating level n with a momentum vector κ. The c.m. intervals $\Delta \kappa_n$ and $\Delta \omega_n$ depend on κ, on the instrument settings, and on the corresponding laboratory intervals Δk_f and $\Delta \Omega$.

An example of this computer simulation is given in Fig. 11. The three peaks which are displayed are due to spontaneous predissociation with a discrete excess energy of $e_d = 0.22$ eV, taking place at distances of 100, 50, and ≈ 1 cm from the aperture in front of the magnetic analyzer. Only fragments scattered backward in the center of mass are shown. Apart from the smearing out of the peak with increasing acceptance angle—decreasing distance from the aperture—its position is also slightly shifted.

This convolution should also be employed in the case of collision-induced dissociation when the cross section varies rapidly with κ and when κ tends to zero. Moreover it should be recognized that deconvolution is not allowed if several processes with different Q produce fragments with the same k_f, because dividing through D (or A) is not additive.

4. Direct Dissociation in Heavy-Particle Collisions

As already discussed in Section 2.4, two limiting cases can be distinguished in collision-induced dissociation: a dissociation caused by excitation to a purely dissociative state and one in which a predissociative intermediate state is excited. This section deals with the first case, the direct dissociation. The decisive difference in the experimental spectra which makes it possible to determine which of the two limiting cases is involved is that an intermediate predissociative state gives rise to discrete peaks in the momentum distribution, the peaks reflecting the energy levels of the quasi-bound state. A direct dissociation in general will give rise to a smooth momentum spectrum, for the transition moment as well as the Franck–Condon factors will vary rather smoothly as a function of the excitation energy. As a consequence, no abrupt changes in the distribution as a function of the excess energy e_d should generally be expected.

In converting the experimental momentum distributions into differential cross sections $\sigma(R, \phi)$ which are a function of the internuclear distance R of the molecule and the angle ϕ between the beam direction and the internuclear axis, the reflection approximation is used. First the laboratory distribution is transformed to the center-of-mass distribution in which it is necessary to correct for the (estimated) excitation energy, since this causes a displacement of the moving coordinate system. The internuclear separation of the molecule at the moment of impact is then taken to be equal to the turning point associated with the excess energy e_d (the reflection model; cf. Section 2.2). The orientation ϕ of the collisionally excited molecule with respect to the beam direction is assumed to be identical with that of the internuclear axis at the moment of impact. It has already been emphasized that this leads to completely erroneous conclusions if the center of mass of the molecule is deflected as in the case in violent vibrational dissociation (Section 2.3). The occurrence of such a process can be recognized from characteristic features in the fragmentation pattern, as discussed later. In some cases it is possible to estimate this contribution roughly and to subtract it from the fragment spectrum. It then becomes a reasonable approximation to neglect the deviation of the molecular center of mass in the remainder of the analysis.

As has been mentioned earlier, in actual experiments there will be almost no pure examples of a collision-induced dissociation which is due to only one clearly defined mechanism. There may be a single exception, namely the dissociation of $HeH^+ \rightarrow He + H^+$, for which there are strong indications that the dissociation is caused by polarization-induced forces only. In some other cases the different contributing mechanisms can be clearly recognized in the experimental fragment distributions, for they show distinct features which are predicted by one or another theoretical model.

4.1. The Dissociation of H_2^+

There is convincing experimental evidence that the major contribution to the collision-induced dissociation of H_2^+ molecules in the keV energy

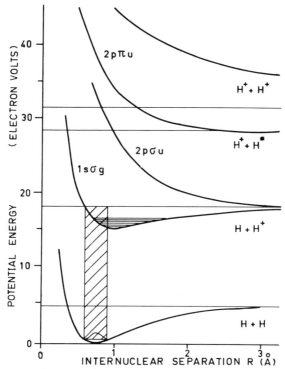

Fig. 12. Potential-energy curves of the three lowest states of H_2^+, together with the ground-state potential of H_2 and the potential of the doubly ionized $H^+ + H^+$ state.

range is due to electronic excitation of the $1s\sigma_g$ ground state to the first excited $2p\sigma_u$ state. This is particularly true for the lighter target atoms, such as He and Ne. Especially decisive in this respect are those measurements in which the angular as well as the velocity distributions have been measured, i.e., the complete momentum distribution of the proton fragments.

In Fig. 12 the potential energy curves of the three lowest states of H_2^+ are shown, together with the ground state potential of H_2 and the completely ionized state $H^+ + H^+$. Ionization of H_2 in a mass spectrometer source with electron energies on the order of 100 eV, yields a Franck–Condon distribution over all 19 vibrational levels of H_2^+ which is peaked around $v = 3$. This implies that the internuclear distances of the molecules range from the minimum value $R = 1.5a_0$ (0.79 Å) to very large values $R > 15a_0$ (7.94 Å). Vertical transitions from the ground state to the $2p\sigma_u$ state give rise to dissociations in which the energy release e_d ranges from ~20 eV to zero. The excitation energy and as a consequence the minimum momentum transfer K_{min} varies drastically as a function of the internuclear distance. At $R = 2a_0$ (1.06 Å), K_{min} at 10 keV incident energy is then of the order of 0.8 a.u. At

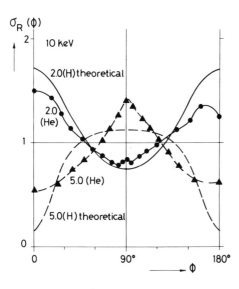

Fig. 13. Angular dependence of the experimental center-of-mass dissociation cross section, $\sigma_R(\phi)$, for 10-keV H_2^+ incident on He,[35] compared with theoretical Born calculations for an H-atom target.[57] The reflection model was used under the assumption that the dissociation is due to the $1s\sigma_g \to 2p\sigma_u$ transition. The parameter R is indicated in units of a_0.

$R = 5a_0$, however, this value has been reduced to 0.16 a.u. We therefore expect that for e_d values corresponding to $R \approx 2a_0$ the distribution is peaked in the forward–backward direction, whereas at $R = 5a_0$ the distribution should show a maximum at right angles to the primary beam (see Section 2.2). In Fig. 13 the experimental results of Gibson et al.[35] are shown, together with theoretical predictions of Peek and Green (cited by McClure[57]). It is obvious that there is a good qualitative agreement between the theoretical predictions based on a Born calculation for the $1s\sigma_g \to 2p\sigma_u$ transitions and the experimental results. Only the $1s\sigma_g \to 2p\sigma_u$ transition shows this large relative variation of K_{min} with internuclear distance. Excitation to $2p\pi_u$ or $2s\sigma_g$ requires an additional 10 eV excitation energy, which means that K_{min} for these cases is large. Although the value of K_{min} for these excitations still varies as a function of internuclear distance, the variation in the dissociation pattern is negligible as $K_{min} \geq 1$ a.u. These considerations are confirmed by the theoretical Born calculations of Green and Peek.[12] In the first place they calculated that the transitions to the two higher excited states only amount to about 10% of the $1s\sigma_g \to 2p\sigma_u$ transition. Moreover, the angular distributions, although not isotropic, are almost independent of the internuclear distance. Experimental verification of the relatively low cross sections for excitation to the higher states is given by Sauers et al.[58] They measured the laboratory angular distributions of the metastable H(2s) fragments, together with the H(1s) and H^+ fragments for collision-induced dissociation of H_2^+ on various target gases (He, Ar, H_2, and N_2) in the energy range 4–12 keV. Indeed, the laboratory cross section for H(2s) formation is a factor of 10 smaller than that for H(1s) or H^+ formation. The experiments of Sauers confirm the results of Jaecks et al.[53] who measured the laboratory angular distributions of H(2p) and H(2s) separately. These are of equal magnitude. Moreover, the observations of Sauers are in very good agreement with the first laboratory angular measurements of McClure[57] for H_2^+ dissociating by collision with H_2.

The influence of K_{min} on the center-of-mass distribution of the proton fragments is also very apparent from the experiments of Gibson et al.[35] performed at 3 keV on a He target gas (see Fig. 14). Since the minimal required momentum transfer is inversely proportional to the initial velocity of the primary beam, K_{min} has increased by a factor of 1.8 compared with an incident energy of 10 keV. At 10 keV, for example, dissociative excitation at $R = 3a_0$ yields an almost isotropic fragment distribution, whereas at 3 keV the distribution definitely is peaked in the forward–backward direction.

The experimental results of the Amsterdam group are fully confirmed by the measurements of Vogler and Seibt.[37] They measured with a parabola mass spectrograph the laboratory momentum distribution, angle and velocity, for collision-induced dissociation of H_2^+ at an incident energy of 20.4 keV. The target gases used were the noble gases, H_2 and D_2. Moreover a comparison was made between D_2^+ at 10.2 keV and H_2^+ at 20.4 keV on argon. Vogler and Seibt also concluded from their experiments that the major contribution to the dissociation came from the $1s\sigma_g \rightarrow 2p\sigma_u$ excitation. They inferred, however, that ionization of H_2^+ also was important. It has

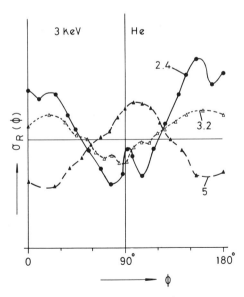

Fig. 14. Differential cross sections $\sigma_R(\phi)$ for 3-keV H_2^+ incident on He.[35] The internuclear distance R in atomic units is indicated for each curve.

been established experimentally by Guidini[59] that 10% of the total cross section for dissociation is due to ionization. This is of the same order of magnitude as contributions from excitation to higher repulsive states of H_2^+. As remarked by Caudano and Delfosse[60] the tails of the velocity distribution, measured in the forward–backward direction, can be explained by ionization. These tails correspond to release energies which are higher than can be reconciled with excitation to any excited state.

Vogler and Seibt[37] were the first to draw attention to the fact that nuclear vibrational dissociation might explain a very marked anomaly in the differential cross sections. It is found that the differential cross section shows a narrow peak at angles perpendicular to the beam. The magnitude of this peak increases markedly with increasing mass number of the target atom. This is very apparent in Fig. 15, where the experimental results of Gibson et al.[35] are shown for 10-keV H_2^+ incident on He, Ne, Ar, and Kr at $R = 2a_0$, together with the theoretical curve of Peek et al.[57] for a $1s\sigma_g \to 2p\sigma_u$ transition. The peak at large scattering angles is not expected from the Born calculations. Moreover, the intensity of this peak increases with decreasing initial velocity as is deduced from the 3-keV experiments of Gibson et al.[35] The same trend can be observed in the experimental results of Vogler and Seibt. Figure 16 shows the measurement of Vogler and Seibt[37] of 20.4 keV H_2^+ on Ar. They plotted their results as a function of $v_\perp$ and $v_\parallel$, the velocity components of the fragment proton perpendicular and parallel to the primary beam, respectively. The parallel-velocity component is defined

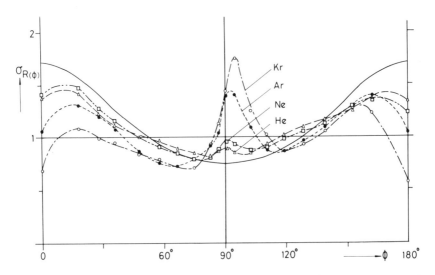

Fig. 15. Differential cross sections $\sigma_R(\phi)$ at $R = 2a_0$ for 10-keV H_2^+ incident on He, Ne, Ar, and Kr.[35] The full curve is calculated for the electronic transition $1s\sigma_g \to 2p\sigma_u$.[57]

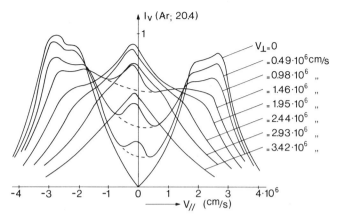

Fig. 16. Center-of-mass proton velocity distribution for the collision-induced dissociation of 20.4-keV H_2^+ incident on Ar.[15,37] For the lower three nonzero values of the velocity $v_\perp$ perpendicular to the direction of the primary beam the contribution from vibrational dissociation is indicated by the dashed lines.

relative to the initial center-of-mass velocity of the molecule. These plots clearly exhibit an additional contribution to the fragment distribution for $v_\parallel \approx 0$, which for lower values of $v_\perp$ can be estimated by graphical subtraction (dashed line). For higher values of $v_\perp$ vibrational dissociation dominates the cross section.

As explained in Section 2.3, a vibrational dissociation due to impulsive momentum change of one or both protons might lead to a distribution of fragments perpendicular to the beam with almost no velocity change parallel to the beam. In fact, all the experimentally observed features can be explained assuming a contribution from this type of vibrational dissociation. In Fig. 17 the theoretical results of Green[14] are shown for 3-keV H_2^+ incident on He, which can be compared with the experimental results of Fig. 14. The differential cross sections are a convolution of the true differential cross sections due to excitation to the $2p\sigma_u$ state, and the apparent differential cross sections due to vibrational dissociation. The results of Green, however, are only of a qualitative nature, as the fit of the theoretical calculations to the experimental data required a scaling of the vibrational contribution with a factor 0.1. As has been pointed out, however, the choice of the proper screening constants is very important. The results of the classical calculation by Meierjohann[15] are given in Fig. 3; these beautifully illustrate that vibrational dissociation through impulsive momentum change leads to velocity distributions which are sharply peaked at $v_\parallel \approx 0$. These calculations even confirm that the sharp peaks are slightly displaced to lower velocities, a shift which increases with larger $v_\perp$ values. These shifts are also observed in the experiments of Gibson *et al.* and of Vogler and Seibt. It is clear that these

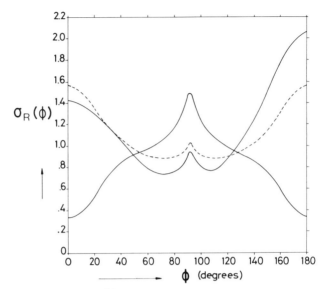

Fig. 17. Theoretical calculations[14] of the center-of-mass (apparent) differential cross sections for the collision-induced dissociation of 3-keV H_2^+ incident on He through a combination of electronic excitation to the $2p\sigma_u$ state and vibrational dissociation.

shifts are not due to an electronic excitation of the target as originally proposed, but merely to the momentum change of the molecule. Meierjohann compared his calculations with the experimental results obtained by graphical subtraction. Introducing, as a consequence of the Heisenberg uncertainty relations, a broadening of the sharp peaked distributions, he found excellent agreement for the three cases indicated in Fig. 16.

The collision-induced dissociation of H_2^+ on noble-gas targets, leading to velocities or apparent velocities in the H_2^+ center-of-mass frame which correspond to e_d values larger than 1 eV, can thus be satisfactorily explained. Electronic excitation and dissociation by impulsive momentum transfer are the mechanisms responsible for the dissociation. However, there are a number of experiments in which the velocity distribution of the proton fragments has been measured in forward–backward direction, which show that the central peak region, i.e., the region of very small e_d values, cannot be explained by a $1s\sigma_g \rightarrow 2p\sigma_u$ transition only. Valckx and Verveer[61] and Caudano and Delfosse[60] observed that the height of the central peak increased more rapidly than the side wings at larger e_d values, with increasing mass number of the target atom. Gibson and Los[62] and Vogler and Seibt[37] also noticed the anomalous intensity of the central peak region. Several authors suggested that vibrational dissociation by impulsive momentum change also would be responsible for this contribution to the dissociation fragment pattern.

This central peak region has been investigated very thoroughly by Durup and co-workers.[31] By applying the appearance-potential method on one hand, and by measuring the energy position of the peak on the other, these authors came to very interesting conclusions.

In Fig. 18 are shown the center-of-mass distributions in the forward–backward direction as a function of e_d, for 5 keV H_2^+ incident on a krypton target at various energies of the ionizing electrons. The surprising result of these curves is that a secondary maximum appears at e_d values of ~0.1 eV, besides the main maximum at $e_d = 3$ eV which is due to electronic excitation. These secondary peaks do not appear in He and Ne; for Ar they depend on the primary beam energy and for Kr and Xe and some molecular target gases they appear in the whole energy range which has been explored from 3 to 5 keV. Anderson and Swan[63,64] recently investigated these secondary maxima to lower primary beam energies. They concluded that these contributions indeed are due to vibrational dissociation, but of the Russek type, that is, caused by polarization-induced forces.

An even more surprising observation has been made by Durup et al.[31] by measuring very carefully the appearance potential at different e_d values. They found that for the heavier noble gases and for molecular gases, at the lowest measurable ionizing electron energy, there were indeed fragments with $e_d \simeq 0$. Now at these low electron energies only the lowest vibrational levels of H_2^+ can be populated. Vertical electron transition to the $2p\sigma_u$ state would result in dissociations where ~7 eV is released as kinetic energy of the

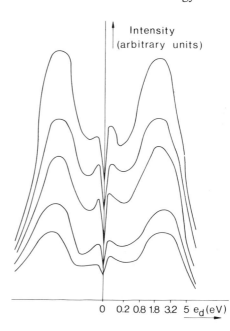

Fig. 18. Center-of-mass distribution of the kinetic energy e_d of forward–backward scattered proton fragments for 5-keV H_2^+ incident on Kr; ionizing electron energies, from bottom to top: 18.4, 18.8, 20.4, 21.4, and 22.8 eV.[31]

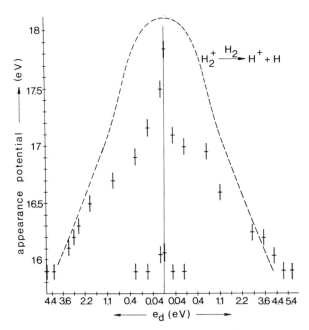

Fig. 19. Appearance potentials as a function of e_d for H^+ from the collision-induced dissociation of 3-keV H_2^+ incident on H_2.[31] The dashed curve is the theoretical appearance potential in the case of a vertical Franck–Condon transition.

separating fragments. Figure 19 gives the results of these appearance potential measurements as a function of e_d, together with a theoretical value expected from vertical Franck–Condon excitation, which can be calculated from the potential energy diagram. The figure clearly shows that part of the central peak appears at the expected electron energy of 17–18 eV, but that another part is already appearing at electron energies as low as 16 eV. For higher e_d values the appearance potential is in agreement with the expected value for electronic excitation. For the lighter noble-gas atoms only a single appearance potential has been observed, and this agrees very well with the theoretical values.

Careful measurement of the peak position revealed that at electron energies of 30 eV the inelastic energy loss for H_2^+ ions leading to fragments with $e_d \simeq 0$ was very small. For electron energies of 17 eV, however, an inelastic energy loss was observed from the shift in peak position, which for Kr and Xe amounted to ~10 eV, and for N_2 to ~6 eV.

Fournier[65] suggested the correct explanation for these observations. He inferred that charge transfer to high Rydberg states would yield highly excited H_2 molecules which then could dissociate to $H^+ + H^-$ by a curve-crossing mechanism. He also measured the production of H^- with very low

e_d values in collisions of H_2^+ with various target gases. Definite proof of this process has been given by Brouillard et al.,[66] who actually measured the simultaneous production of H^+ and H^- in collisions of 6-keV H_2^+ incident on Kr and Xe, by means of coincidence techniques.

4.2. The Dissociation of HeH$^+$

The second molecule to be discussed in this section is HeH$^+$, the simplest heterogeneous molecule. The collision-induced dissociation of HeH$^+$ shows several interesting features, some of which are still partly unexplained, whereas others have recently been explained by the mechanism of vibrational dissociation by polarization-induced forces. The HeH$^+$ molecule has a ground state, dissociating into He and a proton, with a rather narrow and deep potential well of about 2 eV and an equilibrium internuclear distance of $1.44a_0$ (0.76 Å). Potential energy curves of the first 14 states of HeH$^+$ have been calculated by Michels[67] and more recently by Green et al.[68] The first two excited states, $A\ ^1\Sigma$ and $a\ ^3\Sigma$. separating into He$^+$ and H both have very shallow potential wells with equilibrium distances of $5.65a_0$ and $4.47a_0$. These potential wells sustain a few vibrational levels. At internuclear distances smaller than $\sim 2a_0$ both potential curves rise very steeply and at the equilibrium internuclear distance of the ground state they are strongly repulsive. The excitation energy for a vertical transition from the bottom of the ground-state potential well to the two lowest excited states is on the order of 10 eV. The next higher states lie about 10 eV higher and are all strongly repulsive at internuclear distances on the order of the equilibrium ground-state separation.

Because of these features one might expect that dissociative electronic excitation would yield very broad distributions of the fragment ions, whether H^+ or He^+, in the center-of-mass frame of the molecule. Surprisingly, however, the center-of-mass momentum distribution of proton fragments in collisional dissociation is extremely narrow; e_d ranges from zero to ~ 1 eV. The forward–backward center-of-mass momentum distribution[69] of proton fragments for 10-keV HeH$^+$ incident on He is shown in Fig. 20. Electronic excitation definitely cannot account for this dissociative process, and neither can a vibrational dissociation by impulsive momentum change of one of the nuclei of HeH$^+$. The latter mechanism would yield strong contributions perpendicular to the beam with relatively high apparent e_d values, which were not observed. The actual mechanism responsible for this dissociation process has been suggested by Russek,[16] and has been discussed in Section 2.3. The Russek mechanism predicts a proportionality of the dissociation cross section to α^2/a^6 for a dipolar molecular ion and to α^2/a^8 for a homonuclear molecule, where α is the polarizability of the target atom, and a a screening constant for the polarization-induced potentials, which

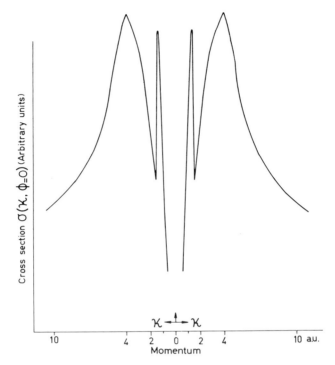

Fig. 20. Center-of-mass momentum distribution of proton fragments for 10-keV HeH$^+$ incident on He.[69]

depends on the target atom. Schopman and Los[69] determined relative total cross sections for different noble-gas targets. Since the polarizabilities are well known, these measurements yield relative values for the screening parameters a which compare favorably with the values of Schoenebeck[70] obtained from elastic scattering of Li$^+$ on the noble gases. This is not surprising, because the value of the cutoff parameter will be determined mainly by the noble-gas collision partner.

Concerning the shape of the distribution in Fig. 20 we observe a minimum at $\kappa = 0$, a very sharp peak at $\kappa = 1.43$ a.u., and a maximum at $\kappa = 4$ a.u. The sharp peak is caused by rotational predissociation from a quasi-bound state, which is mainly populated by collisions in which purely rotational excitation with $\Delta J = 1$ takes place. This is discussed in Section 5.2 together with other sharp peaks which appear in the momentum distribution and which have been smoothed out in Fig. 20. The Russek mechanism would predict a center-of-mass momentum distribution for the proton fragments which is inversely proportional to κ^2. The experimental results show first an increase in intensity with increasing κ and then a maximum at $\kappa = 4$ a.u. This is caused by the fact that molecules with larger rotational quantum numbers, upon excitation just above the dissociation threshold do not dissociate immediately because of the centrifugal barrier. Such slow rotational predissociation is not detected efficiently, as discussed in Section 5.2.

The He^+ fragment distribution from collision-induced dissociation of HeH^+ behaves as expected from the repulsive character of the excited $A\ ^1\Sigma$ and $a\ ^3\Sigma$ states. Schopman et al.[71] measured a broad distribution of He^+ fragments in the forward–backward direction, an observation which was confirmed by Houver et al.[72] In both experiments a sharp peak at small center-of-mass velocities was observed and attributed to vibrational dissociation from the $a\ ^3\Sigma$ state, the presence of which in the primary beam was demonstrated by Schopman and Los.[73] This is due to the Russek mechanism as discussed earlier. Both groups[71,72] made the observation that, contrary to the dissociation of H_2^+, there was a large asymmetry in the intensity distribution of forward–backward scattered fragments, the intensity in the forward direction being much larger. The same results have been obtained with collision-induced dissociation of the asymmetric molecules CO^+ and NO^+ [74] in contrast to dissociations of symmetric molecules such as N_2^+ and O_2^+ which again are symmetric the the forward–backward direction. It was even observed by Pham Dông and Durup[75] that the dissociation of HD^+ shows a slightly more pronounced forward scattering of the D^+ fragment.

In Fig. 21 the center-of-mass angular distributions, measured by Schopman et al.,[71] are shown for two different kinetic energies of the He^+

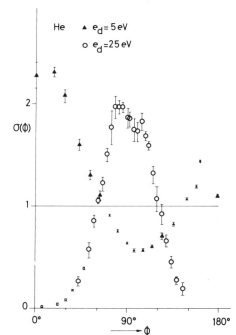

Fig. 21. Doubly differential center-of-mass cross sections at two different e_d values for the collision-induced dissociation into He^+ and H of 10-keV HeH^+ incident on He.[71]

fragment, a low e_d value of 5 eV, and a high one of 25 eV. At $e_d = 5$ eV, in accordance with Dunn's angular selection rules, we observe maxima in the forward–backward direction and a minimum at right angles to the primary beam. However, an asymmetry is superimposed with respect to the 90° direction, the intensity in the forward direction being considerably larger than that in the backward direction. The explanation for this asymmetry in the fragment distributions of asymmetric molecules has not yet been given. Certainly the first Born approximation is unable to explain such asymmetric shapes. The curve for $e_d = 25$ eV consists only of a rather broad peak, the maximum of which lies slightly below 90°. A vertical electronic excitation from the ground state does not lead to dissociations in which the kinetic energy of the fragments would amount to 25 eV. It is clear that in this case vibrational dissociation due to impulsive momentum change of one of the nuclei of HeH$^+$ is observed, as has been discussed in the preceding section on H$_2^+$. The magnitude of this contribution leading to high apparent e_d values is rather small, as can be concluded from Fig. 22, where the cross section integrated over angle is plotted as a function of the (apparent) e_d values. Only a few percent of the total cross section is due to these more violent collisions.

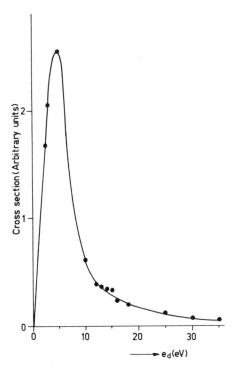

Fig. 22. The dissociation cross section integrated over angle, of 10-keV HeH$^+$ incident on He and dissociating into He$^+$+H, as a function of the (apparent) kinetic energy of the fragments e_d. The tail at large energy reflects the contribution from vibrational dissociation in violent collisions.[71]

4.3. Discussion

In the two preceding sections, we have discussed the experimental results obtained for the two lightest molecular ions, H_2^+ and HeH^+. The results obtained with these two molecules are most informative with respect to experimental evidence concerning the different mechanisms which are responsible for collision-induced dissociation. However, more refined experiments are needed in order to obtain data which can be attributed exclusively to one single mechanism. It is therefore necessary that both the initial-state preparation and the final-state analysis be improved. As already discussed in Section 3, work is in progress to accomplish these refinements. For the initial-state preparation the laser offers the most attractive possibilities, either by means of photoionization or by selective depopulation of well-defined molecular levels by photodissociation. For the final-state analysis it is a necessity that besides the momentum distribution of the fragment ions a second observable quantity should be measured in coincidence with the fragment ion.

In this section we have only touched slightly upon the possibilities which are offered by the appearance-potential method, and neither have we discussed the possibilities which experiments on collision-induced dissociation offer in determining molecular structure, e.g., identification of dissociative states and identification of bound excited states in the primary beam. These aspects are treated in more detail in the following discussion on translational spectroscopy. The reader is also referred to discussions in the introductory chapter and Chapter 7.

5. Translational Spectroscopy

The expression "translational spectroscopy" generally designates the measurement of kinetic energy (or momentum) distributions of fragments resulting from the (pre)dissociation of a parent molecule or molecular ion. In the following paragraphs we consider a few examples in which the fragmentation of a diatomic ion was studied by means of high-resolution mass spectrometry. The designation "translational spectroscopy" fits these studies particularly well, because the measurements are of sufficiently high resolution to localize vibrational and, in the most favorable cases, rotational energy levels of the parent molecular ion.

In the first two examples considered herein, fine structure is observed in the momentum distribution of dissociation fragments due to a predissociation mechanism. The two possible cases, predissociation by curve crossing and rotational predissociation, have been discussed briefly in Section 2.4. They will be illustrated by the translational spectroscopy of dissociation processes in N_2^+ and HeH^+, respectively. Because of the close connection with the analogous collision-induced process, we include a brief discussion of the unimolecular dissociation of HeH$^+$ ions. By "unimolecular" we mean that the parent ion was brought to an excited state within the mass spectrometer ion source and that fragmentation of the accelerated ion occurred without any subsequent collision.

The third example deals with the combination of translational spectroscopy with another promising technique: the photodissociation of diatomic ions by monochromatic light.

5.1. Collisional Dissociation of 4–10 keV N_2^+ Ions

To our knowledge, the collisional dissociation of fast N_2^+ ions was first reported by Smyth[76,77] in 1923. He observed a peak at an apparent mass of 7 in the mass spectrum of N_2. It appeared at low ionizing electron energies and its intensity was very sensitive to the pressure in the instrument.[76] Smyth concluded "that there are N_2^+ ions, probably those formed at the first ionization potential, which dissociate on collision after acquiring large kinetic energies."[77] This conclusion was confirmed some 20 years later by Hagstrum and Tate[78] and by Hipple et al.,[79] and numerous publications have appeared since then (see, e.g., Refs. 32, 40, 80–82).

The information provided by these experiments has been summarized by Fournier et al.[32] Figure 23 is reproduced from this last reference and shows the laboratory momentum distribution of forward- and backward-scattered N^+ arising from collisions of 4.4 keV N_2^+ on He, the target gas corresponding to the largest cross section for N^+ production in this energy region. The apparatus used for these measurements comprised an electron impact source of the Nier type, a collision chamber, and a 60° magnetic sector analyzer with a momentum resolution of 5×10^{-4}. The acceptance angle subtended at the collision chamber was 2.8×10^{-3} rad.

The sharp peak at apparent mass 7 results from N^{2+} ions produced in the ion source, and appears at electron energies above $\approx$60 eV. The broad

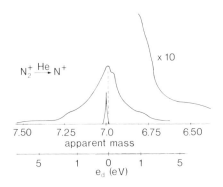

Fig. 23. Laboratory distribution of forward–backward scattered N^+ fragments from collision-induced dissociation of 4.4-keV N_2^+ incident on He.[32] The sharp peak at $m/e = 7$ is due to N^{2+} which is used to calibrate the mass scale.

N^+ Aston band is nearly symmetric about a main peak whose displacement with respect to the N^{2+} marker corresponds, according to the two-step model of Section 2.1, to an energy loss of the parent N_2^+ of $\Delta E = 10 \pm 2$ eV for $e_d = 0$. Most of the central region of the Aston band is associated with ΔE values of this magnitude. Appearance potential measurements show that most of the parent N_2^+ ions are in the $X\,^2\Sigma_g^+$ ground state (ionization potential 15.6 eV). An energy loss $\Delta E = 10 \pm 2$ eV then implies that these ions are excited to a state lying at $15.6 + (10 \pm 2) = 25.6 \pm 2$ eV above the ground state of N_2 and from which N_2^+ dissociates with small excess energy. Higher-lying metastable N_2^+ states with appearance potentials around 22 eV become apparent at lower collision energies.[32]

The secondary maximum at apparent mass ~6.95 in Fig. 23 is due to N_2^{2+} from the process

$$N_2^+ \xrightarrow{He} (N_2^+)^* \to N_2^{2+} + e$$

the energy loss in the excitation step being $\Delta E = 32 \pm 2$ eV. This process was identified through isotopic substitution: with $^{14}N\,^{15}N^{2+}$ the peak shifts to an apparent mass of 7.25, whereas N^+ fragments appear at an apparent mass of 6.76 or 7.76. The symmetrical wings of the Aston band are probably due to

$$N_2^+ \xrightarrow{He} N_2^{2+} \to N^+ + N^+$$

($e_d \approx 6.5$ eV).[32,83]

Figure 24 shows a more detailed scan of the central region of the N_2^+ Aston band as obtained at 10 keV with the Amsterdam apparatus.[82] The primary beam was collimated within a half-angle of 1.3×10^{-3} rad and the half-angle for acceptance of N^+ formed in the collision chamber was 1.6×10^{-3} rad. Such a high degree of collimation and careful alignment were essential in observing the structure shown in the figure. Peaks and shoulders labeled with the same letter are seen to be very nearly centered about a maximum labeled j whose position with respect to the N^{2+} marker corresponds to an energy loss $\Delta E = 8 \pm 2$ eV. The structural features correspond to (sub)maxima in the production of N^+ fragments at specific values of e_d; forward-scattered N^+ appears on the right-hand side of the spectrum, and backward-scattered N^+ on the left. The sharp peak marked i near the center is N_2^{2+} from

$$N_2^+ \xrightarrow{He} N_2^{2+}$$

The primary N^{2+} peak does not appear in Fig. 24 because this spectrum shows ions formed within the collision chamber only; this separation was

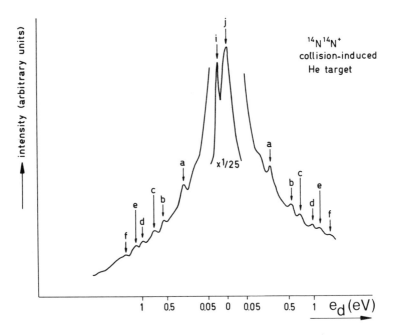

Fig. 24. High-resolution scan of the Aston band resulting from 10-keV N_2^+ incident on He.[82]

achieved by applying a biasing potential to the collision chamber, as outlined in Section 3.1.

In Table I we list the e_d values corresponding to the labeled peaks and shoulders in Fig. 24. Also shown in this table are the excess energies of the vibrational levels $v' = 4$–9 of the $C\,^2\Sigma_u^+$ state of N_2^+, calculated relative to the first dissociation limit using spectroscopic data compiled by Albritton et al.[84] The agreement between the two sets of data is a first indication that the

Table 1. Comparison between the Energy Release (e_d) Values Corresponding to the Labeled Structure in Fig. 24 and Those Calculated for the Vibrational Levels $v' = 4$–9 of the $C^2\Sigma_u^+$ State of N_2^+ Using Spectroscopic Data [a]

Observed (eV)		Calculated (eV)	
Label a	0.27 ± 0.02	$v' = 4$	0.289 ± 0.005
b	0.55 ± 0.03	5	0.530 ± 0.005
c	0.73 ± 0.04	6	0.767 ± 0.005
d	0.98 ± 0.03	7	1.000 ± 0.005
e	1.19 ± 0.06	8	1.228 ± 0.005
f	1.44 ± 0.07	9	1.452 ± 0.005

[a] From Ref. 84.

labeled structure in Fig. 24 can be identified with the collisional excitation and subsequent predissociation of the levels $v' = 4$–9 of the C state.[82]

This assignment is supported by the ΔE measurement and by lifetime considerations. Although the use of a unoplasmatron ion souce precludes systematic measurements of appearance potentials, earlier measurements[32] and the $\approx 10^{-6}$ sec flight time to the collision chamber suggest that N_2^+ ions whose excitation gives rise to the spectrum of Fig. 24 are mostly in the $X\ ^2\Sigma_g^+$ ground state, with a small fraction in the lower levels of the long-lived $A\ ^2\Pi_u$ state.[82] The experimental energy-loss value $\Delta E = 8 \pm 2$ eV is in agreement with the first dissociation threshold of both states. It is therefore probable that this structure is produced by ground-state fragments, as required by the proposed C-state assignment: all the vibrational levels in question lie below the second dissociation limit of N_2^+. Collisional excitation from the $X\ ^2\Sigma_2^+$ and the $A\ ^2\Pi_u$ state to the repulsive part of the $D\ ^2\Pi_g$ state of N_2^+ has been suggested[82] to account for the continuous spectrum underlying the structure of Fig. 24.

The lifetime argument proposed by Fournier et al.[82] is based on the observation that $\geq 90\%$ of these "C-state" fragments are formed within the collision chamber. This fraction is given by

$$f_L = \left[1 - \frac{v^*\tau}{L}(1 - e^{-L/v^*\tau})\right] \tag{5.1}$$

where L is the length of the collision chamber, and v^* and τ are the speed and lifetime of the excited parent ion, respectively. In the present case, $L/v^* \approx 10^{-7}$ sec, so the foregoing observation indicates that $\tau \lesssim 10^{-8}$ sec. Optical measurements yielded $(9 \pm 3) \times 10^{-8}$ sec as the lifetime of the levels $v' \leq 2$ of the C state[82]; these levels cannot predissociate and so-called second negative radiation due to the transition $C\ ^2\Sigma_u^+ \to X\ ^2\Sigma_g^+$ apparently constitutes their only decay mode.[85] Given Albritton et al.'s estimate of about 20 for the ratio between the rates of $(C, v' = 4)$ predissociation and radiative decay,[86] it was concluded that the vibrational levels $v' \geq 4$ have lifetimes on the order of 5×10^{-9} sec, in agreement with the upper limit of 10^{-8} sec just given.

In this manner, Fournier et al.[82] were able to interpret their data as "direct and unambiguous evidence for the predissociation of the C state of N_2^+." Their results also implied that this predissociation, proposed many years ago[87,88] but never fully confirmed,[89] extends over a rather wide range of vibrational levels. More recent beam fluorescence experiments fully support these conclusions.[85,90] In these investigations, the C state was populated by controlled electron or ion impact on N_2, and the resulting $C\ ^2\Sigma_u^+ \to X\ ^2\Sigma_g^+$ "second negative" emission spectrum compared with that expected in the absence of predissociation. In this way, a quantitative comparison could be made between these two modes of C-state decay; the lifetimes for the levels of the $^{14}N_2^+$ isotope considered so far were found to vary from about 9×10^{-9} sec for $v' = 3$ to an estimated 4×10^{-10} sec for $v' = 9$, a result which supports the earlier lifetime argument.[82]

The predissociation of the N_2^+ C state is of interest because the reaction sequence

$$He^+ + N_2 \rightarrow He + N_2^+(C\ ^2\Sigma_u^+, v' = 3\text{-}4)$$
$$N_2^+(C\ ^2\Sigma_u^+, v' = 3\text{-}4) \rightarrow N^+ + N \quad (5.2)$$

constitutes an important source of N^+ in the high ionosphere.[91,92] The strong variation of the predissociation probability with vibrational quantum number and the strong isotope effect[85,90] are a sensitive probe of the potential interaction responsible for the predissociation. Lorquet and Lorquet were the first to consider this situation in detail, and they proposed an accidental predissociation mechanism to explain the observations.[18] More recent calculations by Tellinghuisen and Albritton[93] and Roche and Lefebvre–Brion[94] suggest that a direct predissociation resulting from the interaction of the C state with the continuum of the $B\ ^2\Sigma_u^+$ state constitutes a more likely explanation.

5.2. Collisional and Unimolecular Dissociation of 10-keV HeH$^+$ Ions

When there is introduced into an ion source such as a unoplasmatron a mixture of, say, 90% He and 10% H_2, HeH^+ ions are formed, the most likely reactions being

$$He + H_2^+ \rightarrow HeH^+ + H \quad (5.3)$$

and

$$He^* + H_2 \rightarrow HeH^+ + H + e \quad (5.4)$$

The excited He* in Eq. (5.4) is probably He($2\ ^1P$). Both reactions can produce HeH$^+$ ions in vibrationally and rotationally excited states close to the dissociation limit of the electronic ground state.[69] The dissociation of such ions after acceleration to several keV has been studied quite extensively in the last few years both at the F.O.M. Institute in Amsterdam[69,71,73] and at the University of Paris Orsay Campus.[72] Several recent results[95] were obtained through close collaboration between the two groups and involved all four stable isotopes: ^{4}HeH$^+$, ^{3}HeH$^+$, ^{4}HeD$^+$, and ^{3}H3D$^+$. In the following paragraphs we shall be concerned mainly with the fine structure which is observed in the momentum distribution of the H$^+$ (or D$^+$) fragments. Other features of interest have been discussed in Section 4.2. Both the collision-induced and the unimolecular fragmentations are considered, as these are closely interrelated.

The upper trace in Fig. 25 shows the laboratory intensity distribution of H^+ fragments scattered forward and backward along the primary beam axis when 10-keV $^4HeH^+$ ions collide with a (10^{-3} torr $\times$ 3.5 cm) He target[95]; the abscissa gives the magnitude of their momenta κ with respect to the molecular center of mass. The lower trace shows the result obtained when no gas is introduced into the collision chamber, i.e., at a residual pressure of $\approx 10^{-6}$ torr. Experiments performed at intermediate pressures indicate that the lower (background) spectrum is produced by unimolecular dissociation whereas the upper spectrum results from a superposition of the background spectrum and a contribution by fragments produced by single collisions. In both cases the fragments are formed with momenta κ which correspond to a rather narrow range of excess energies e_d: after transforming the upper

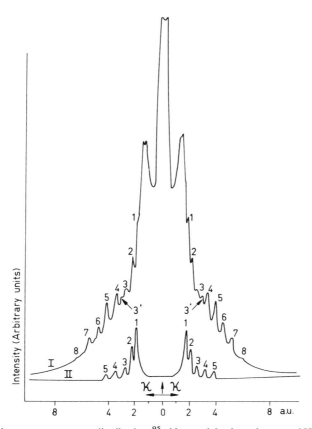

Fig. 25. Laboratory momentum distributions[95] of forward–backward scattered H^+ fragments from 10-keV $^4HeH^+$. Lower trace: distribution observed at residual gas pressure (unimolecular dissociation). Upper trace: distribution observed when 10^{-3} torr of He is introduced into the 3.5-cm-long collision chamber (unimolecular + collision-induced dissociation).

trace of Fig. 25 to a fragment distribution in the center of mass, one finds a curve with a half-width corresponding to $\kappa \approx 8$ a.u., or $e_d \approx 2.2 \times 10^{-2}$ Hartree (0.59 eV). Because the fragment-energy distribution is narrow, reliable angular measurements are not available. As discussed in Section 4.2, the adiabatic distortion mechanism proposed by Russek[16] can account for the narrow distribution (Section 2.3).

Although we may accept that unimolecular dissociation and collisions according to the Russek mechanism are responsible for the production of H^+ from fast HeH^+ ions, another obvious question remains. What is the origin of the intriguing fine structure in the H^+ momentum distribution, first reported by Schopman et al.[96] in 1969? A most convincing answer has recently been given by Peek[97] who identified this structure as due to the rotational predissociation (cf. Section 2.4 and Fig. 5) of quasi-bound HeH^+ ions in the electronic ground state, with well-defined vibrational and rotational quantum numbers. The difference between the collision-induced and the unimolecular contributions is due to the difference of the lifetimes of these rotationally highly excited ions.

Lifetime arguments indeed furnish an important clue in the interpretation of the experimental results. The spectra under consideration result from fragments produced in the collision chamber only. They were singled out by applying a potential to the collision chamber, as explained in Section 3.1. The "background" fragments result from the unimolecular decay of HeH^+ ions formed in the ion source. For them to contribute to the fragmentation peak centered about the appropriate mass position ($\frac{1}{5}$ for H^+ from $^4HeH^+$), the dissociation has to occur *after* the parent ion has reached the full acceleration potential. This of course puts a lower limit to the average time that the parent ion must exist before decaying to a state which cannot dissociate unimolecularly. On the other hand, this average lifetime, τ, has to be short enough for a measurable fraction of the molecules to dissociate during the time, L/v^*, which the excited parent ions need to traverse the collision chamber.[30] If this time is short compared with the interval, t_f, between the formation of the parent ion and its arrival in the collision chamber (and such is the case here), the value τ_{opt}, for which unimolecular decay in the collision chamber is most readily observed, approximately equals t_f. If one neglects the residence time in the ion source, the optimum value appropriate to the 10-keV $^4HeH^+$ data obtained with the Amsterdam apparatus is found to be $\tau_{opt} \approx 6.3 \times 10^{-7}$ sec (2.6×10^{10} a.u.). When the lifetime decreases to $\tau \approx t_f/10$ the fraction of the parent ions which dissociate within the 3.5-cm-long collision chamber is only 1% of that at $\tau = t_f$; the decrease in fragment yield with increasing τ is slower, the 1% level being reached at $\tau \approx 270 t_f$. By looking only at H^+ from fragmentation $^4HeH^+$ *inside the collision chamber* one thus defines the following "lifetime window" for observation of unimolecular dissociation:

10-keV ^{4}HeH$^+$ unimolecular:

$$2.6 \times 10^9 \lesssim \tau \lesssim 7.0 \times 10^{12} \text{ a.u.}$$

$$\tau_{opt} \approx 2.6 \times 10^{10} \text{ a.u.} \quad (6.3 \times 10^{-7} \text{ sec})$$

or $\quad 1.4 \times 10^{-13} \lesssim \Gamma \lesssim 3.8 \times 10^{-10}$ Hartree (5.5)

$$\Gamma_{opt} \approx 3.8 \times 10^{-11} \text{ Hartree} \quad (1.0 \times 10^{-9} \text{ eV})$$

where $\Gamma = \hbar/\tau$ is the energy width associated with the decay lifetime τ.

The situation is quite different when one considers fragments formed by collisions inside the collision chamber: the shorter the lifetime of the parent ion, the larger is the fraction of fragments observed. This fraction decreases to 1% when the lifetime for dissociation becomes about 50 times larger than L/v^*, the time needed by the parent ion to traverse the collision chamber. For 10-keV ^{4}HeH$^+$ ions and $L = 3.5$ cm, $L/v^* \approx 5.5 \times 10^{-8}$ sec, so that the 1% level is reached when $\tau \approx 2.8 \times 10^{-6}$ sec $(1.1 \times 10^{11}$ a.u.). A lower limit to τ can be estimated from the half-widths (FWHM), Δk_f, of the peaks in the momentum distribution of the collisional fragments. These widths lie between 0.2 and 0.3 a.u., and their relative values, $\Delta k_f/k_f$, are not measurably different from those of the parent ions in the mass spectrum. In other words, there is no measurable broadening due to the shortness of the parent ion's lifetime. Since the fragments under consideration all have center-of-mass momenta κ smaller than 10 a.u., this broadening, expressed as an uncertainty Γ in the value of e_d, is certainly less than $\kappa \, \Delta k_f/\mu = 10 \times 0.3/1470 \approx 2.0 \times 10^{-3}$ Hartree $(5.4 \times 10^{-2}$ eV), resulting in a lower limit of $\tau \gtrsim 500$ a.u. $(1.2 \times 10^{-14}$ sec). One thus finds

10-keV HeH$^+$ collision-induced:

$$500 \lesssim \tau \lesssim 1.1 \times 10^{11} \text{ a.u.}$$

$$9.0 \times 10^{-12} \lesssim \Gamma \lesssim 2.0 \times 10^{-3} \text{ Hartree}$$

(5.6)

These "lifetime windows" will of course be slightly different for the other HeH$^+$ isotopes. Although they are rather wide, they do put a stringent limit on the number of quasi-bound levels that could contribute to the present spectra. This reflects: (a) the strong dependence of a quasi-bound level's rotational predissociation probability on the local width of the barrier in the effective potential and (b) the relatively large rotational energy spacing due to the small reduced mass. These two characteristics result in a strong variation in rotational predissociation probability from one quasi-bound level to another. This has been shown quantitatively by Peek by means of a discrete-state method.[97] He found that quasi-bound levels having the same vibrational quantum number v, but different rotational numbers J, have widths (lifetimes) for predissociation which differ by several factors of 10. In the few cases in which more than one quasi-bound level with the same J exist, the different v values also give rise to very different widths. If rotational predissociation is the only decay mode of the quasi-bound levels that needs to be considered, one can compare the calculated widths with the lifetime criteria discussed in the preceding paragraph. One thus finds that from a total of 19 quasi-bound states of ^{4}HeH$^+$ only six fall within the unimolecular "lifetime window" (5.5). The corresponding data are listed in Table II together with energies and widths calculated for two different HeH$^+$ potentials $V_H(R)$ and $V_G(R)$. The first of these was obtained by Helbig et al.[98] by fitting an analytical expression to Wolniewicz's data[99] and extrapolating to large R. This potential differs somewhat in the long-range region from the Green–Michels potential, $V_G(R)$, which was scaled to connect smoothly to Wolniewicz's result and to the correct long-range R^{-4} polarization interaction (cf. Ref. 97). Notice that the widths (lifetimes) of the predissociating levels are much more sensitive to the difference between $V_G(R)$ and $V_H(R)$ than the corresponding energy values, and that this sensitivity is larger at high v and low J. A discussion on the last point, with specific reference to BeH and HgH, has recently been given by Gottdiener and Murrell.[100]

Also shown in Table II are the experimental dissociation energies e_d associated with the sharp peaks observed in the unimolecular spectrum of Fig. 25. The comparison between these experimental values and the five higher values calculated for the V_G potential is quite satisfactory and leaves little doubt that the unimolecular spectrum of Fig. 25 results from the rotational predissociation of HeH$^+$ ions in the electronic ground state, the

Table II. Unimolecular Dissociation of ^{4}HeH$^+$: Comparison of the Experimental Energy Release e_d with Theoretical Values[a]

Assignment		Energy release (10^{-3} Hartree)				Width (Hartree)	
		Calculated					
v	J	V_H	V_G	Fit	Experimental	V_H	V_G
8	8	—	0.0240	—	—	—	4.9 (−13)
4	17	1.9969	1.2511	1.03	1.05	1.4 (−7)	9.2 (−11)
3	19	2.6584	1.7603	1.61	1.52 ± 0.04	1.2 (−8)	8.8 (−12)
2	21	3.6236	2.5747	2.40	2.31 ± 0.05	2.1 (−9)	3.3 (−12)
1	23	4.9893	3.7813	3.45	3.40 ± 0.09	6.7 (−10)	2.5 (−12)
0	25	6.8444	5.4417	4.84	4.96 ± 0.1	2.2 (−10)	1.8 (−12)

[a] Experimental data from Schopman et al.[95] The columns labeled V_H and V_G refer to the calculations of Peek[97] based on the potential of Helbig et al.[98] and Green and Michels (cf. Ref. 97), respectively. The column labeled "Fit" gives e_d values obtained by Bernstein by means of a "spectroscopic" fitting procedure.[101] Numbers in parentheses indicate the power of 10 by which the preceding number should be multiplied.

vibrational and rotational quantum numbers being those shown in Table II. Peek pointed out that the $V_G(R)$ potential should be less binding than the true potential, so that the calculated e_d values and, in particular, the widths will be too high. A slight lowering of the calculated e_d values would indeed improve the accord with experiments, and the (much more drastic) lowering of Γ would also explain why the ($v = 8, J = 8$) level is not observed experimentally. The calculations based on the $V_H(R)$ potential, on the other hand, do not fit the experimental results nearly as well. This suggests that the experimental data provide a useful probe of the HeH$^+$ potential (cf. Refs. 19 and 23), and in particular of the long-range part since the potential well is rather accurately known in the present case.[99,101] The energy widths, especially, vary strongly with the shape of the potential, so that it would be of interest to narrow the lifetime windows of Eq. (5.5), for example, by measuring the variation in the relative peak heights of Fig. 25 as a function of the flight time to the collision (observation) chamber. Such experiments would also provide information on the initial vibrational and rotational population of the quasi-bound HeH$^+$, and thus on the mechanisms by which they are formed.

Experimental data[95] and theoretical calculations[97] analogous to those just discussed for ^{4}HeH$^+ \rightarrow$ H$^+$ are also available for ^{3}HeH$^+$, ^{4}HeD$^+$, and ^{3}HeD$^+$, both for unimolecular and collision-induced dissociation. These data substantiate the interpretation given by Peek[97] and outlined earlier. The self-consistency of the observations and assignments has been verified by Bernstein.[101] Taking account of the assignments proposed by Peek, Bernstein adopted a "spectroscopic" approach and fitted the observed e_d values by means of a Dunham[102] double power series. The lowest-order Dunham coefficient was set equal to zero and three of the others, together with the binding energy, were fixed at values obtained on the basis of Wolniewicz's calculations of the potential well.[99] Six higher-order Dunham coefficients were calculated from a least-squares fit to the experimental data[95] for both unimolecular and collision-induced H$^+$ production from ^{4}HeH$^+$ and ^{3}HeH$^+$. The column entitled Fit in Table II illustrates the results Bernstein obtained by combining the ^{4}HeH$^+$ and ^{3}HeH$^+$ data using mass-reduced quantum numbers as proposed by Stwalley.[103] The agreement between these e_d values and the experimental ones substantiates the (v, J) assignments and, together with similar results for the collision-induced spectra, shows that the experimental data for the two isotopes are self-consistent and also consistent with the characteristics of the potential well. The Dunham coefficients obtained from the analysis[101] of the ^{4}HeH$^+$ and ^{3}HeH$^+$ results satisfactorily predict the energies of the quasi-bound levels observed with ^{4}HeD$^+$ and ^{3}HeD$^+$.[95]

As noted earlier, the difference between the lower "background" spectrum of Fig. 25 and the upper spectrum obtained when gas is introduced in the collision chamber, lies in the difference between the two lifetime windows of Eqs. (5.5) and (5.6), respectively. In the collision-induced experiment one is able to observe not only fragments from relatively long-lived HeH$^+$ ions but also fragments from HeH$^+$ ions which are much shorter-lived than those which fall within the unimolecular lifetime window, since the lower limits on τ are ≈ 500 a.u. (1.2×10^{-14} sec) in the former case, and $\approx 2.6 \times 10^9$ a.u. (6.3×10^{-8} sec) in the latter. If one selects the quasi-bound ^{4}HeH$^+$ states calculated by Peek[97] according to the criteria of Eq. (5.6), one finds 11 additional resonances, whose energies agree quite well with the additional features observed experimentally in the collision-induced spectrum.[95,97] These features can therefore be attributed to the collisional excitation and subsequent rotational predissociation of short-lived quasi-bound ^{4}HeH$^+$ ions in their electronic ground state. The data on the other isotopes agree with this interpretation, as discussed in Refs. 95 and 97; these references also give the assignment of the vibrational and rotational quantum numbers involved. Since the unimolecular decay spectra establish the presence in the beam of HeH$^+$ ions with high vibrational and rotational quantum numbers, v and J, and since the Russek mechanism[16] (Section 2.3) predicts the largest excitation cross sections for collisions characterized by a small momentum transfer, one expects that the collision-induced part of the dissociation spectrum is dominated by transitions which involve small changes in v and J. Bernstein[101] has concluded that the short-lived daughter ions arise mainly from transitions of the type

$$\text{HeH}^+(v, J) + M \rightarrow \text{HeH}^+(v, J+1) + M$$

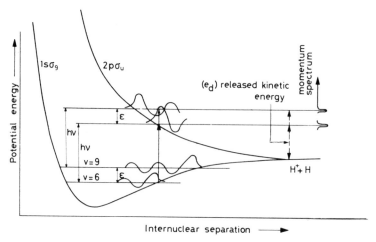

Fig. 26. Illustration of the photodissociation of H_2^+ ions by monochromatic excitation from the $1s\sigma_g$ to the $2p\sigma_u$ state ($h\nu = 2.410$ eV) and the momentum spectrum of the fragments. The spacing ε between any two vibrational levels of the ion is preserved in the momentum spectrum.

"Although simultaneous v, J (or even pure Δv transitions) cannot be ruled out," Bernstein suggested, "there is a strong propensity for pure rotational excitation, at least for the forward scattered HeH^+ arising from inelastic collisions of the 10-keV HeH^+ beam with a He target gas."[101] Thus the foregoing experiments on the unimolecular and collision-induced dissociation of HeH^+ and their theoretical analysis has provided information on the shape of the lowest HeH^+ potential and on the collision-induced dissociation of HeH^+ ions with high rotational and vibrational quantum numbers.

5.3. Photodissociation of H_2^+

In the following paragraphs we shall discuss some recent applications of translational spectroscopy to the photodissociation of H_2^+ and D_2^+ ions, as carried out by Ozenne et al. at Orsay[104] and van Asselt et al. at the F.O.M. in Amsterdam.[36,105]* In these experiments a well-collimated beam of fast (2–10 keV) H_2^+ (or D_2^+) ions is crossed at right angles by a laser beam; the momentum distribution of the resulting H^+ or D^+ fragments is then analyzed under high resolution as described in the preceding sections. Ozenne et al. used a pulsed ruby laser ($h\nu = 1.786$ eV) and van Asselt et al. a continuous Ar-ion laser ($h\nu = 2.410$–2.602 eV). These and related studies on photodissociation of ions have been reviewed by Durup.[106]

In the Orsay experiments, the H_2^+ ions are formed by fast electron impact on H_2, so that the vibrational levels of the $1s\sigma_g$ state of H_2^+ are expected to be populated according to the corresponding Franck–Condon factors. In the Amsterdam experiments the ions are extracted from a unoplasmatron source, in which ion–molecule reactions may also take place, so that it is difficult to predict the vibrational population of the parent ions. In the absence of subsequent collisions, the initial vibrational population will persist up to a point where the ion beam crosses the laser beam, since the vibrational relaxation of H_2^+ is negligibly slow. If the photon energy is high enough, H^+ fragments will now be produced by the process

$$H_2^+(1s\sigma_g, v) + h\nu \to H_2^+(2p\sigma_u) \to H^+ + H(1s) \quad (5.7)$$

as shown schematically in Fig. 26. The photon energy of 2.41 eV indicated in this figure corresponds to the 5145-Å line of the Ar-ion laser used by the Amsterdam group. For a given initial vibrational level v, the excess energy e_d carried off by the dissociation fragments will be the difference between the photon energy and the binding energy of the vibrational level under consideration. The dissociation by monochromatic light of a beam of

---

Note Added in Proof: Several groups have made impressive progress since this manuscript was completed (May, 1976). Two review papers on this subject will be published shortly in the literature: J. Durup, in *Etats Atomiques et Moléculaires Couplés à un Continuum. Atomes et Molécules Hautements Excités*, Colloques International du C.N.R.S. (Aussois, 1977), and J. Moseley, in *X ICPEAC, Invited Talks and Progress Reports*, North Holland, Amsterdam (Paris, 1977).

molecular ions with a certain distribution over vibrational levels will therefore produce a series of peaks in the momentum distribution of fragments which have the same energy spacing as the vibrational levels of the parent ion. The intensity of these peaks depends on the initial vibrational population and on the dissociative excitation cross section of each level. These excitation cross sections have been calculated by Dunn[107,108] as a function of the wavelength of the incident light for all the vibrational levels of ground state H_2^+ and D_2^+. Their strong variation with wavelength and vibrational quantum number can be understood by considering Franck's classical principle. This principle states that during an electronic transition in a molecule both the positions and the momenta (and thus the kinetic energies) of the nuclei *strongly* tend to remain constant.[109] This is often briefly and not quite correctly expressed as "only vertical transitions are allowed." The second aspect to be considered is that classically, the probability for the two nuclei to be at an internuclear distance R is inversely proportional to the local velocity of the vibrational motion. This strongly favors transitions between classical turning points. In the case illustrated in Fig. 26, these two propensity rules, combined with the fact that the excitation energy is fixed by the photon energy $h\nu = 2.41$ eV, result in a strong preference for photodissociation of the vibrational level $v = 6$, because its outer turning point lies closest to R^*, the distance at which the photon energy exactly matches the vertical energy difference between the upper and lower potential curves. Franck's principle does not exclude slightly nonvertical transitions, however, so that even $v = 5$ is accessible using 2.41 eV photons. Also the probability for higher levels to be photodissociated is considerably less, but certainly nonzero, because there is a finite probability for the nuclei to be at an internuclear distance close to R^*. This qualitative picture, as is well known, is refined and confirmed by the quantum treatment, the probability for a certain internuclear distance being determined by the amplitude of the wave function, which has a maximum at the turning points. Levels above $v = 6$ overlap best with continuum wave functions there where the wavelengths are equal, i.e., at equal kinetic energy (see Fig. 26). Since the vibrational wave functions do not vanish abruptly outside the classical turning points, lower levels can also be excited. For an up-to-date discussion of Franck's principle and its quantum mechanical equivalent, we refer to papers by Mulliken[109,110] and Murrel and Taylor,[111] who discuss the related problem of predissociation. The translational spectra of the photodissociation of H_2^+ for five Ar-ion laser lines as measured by van Asselt *et al.*[36] are shown in Fig. 27. The spectrum of the 5145 Å line indeed shows the features discussed. The peak corresponding to $v = 6$ has the strongest intensity. Intensities for the lower $v = 5$, as well as for the higher $v = 7, 8$ levels are one order of magnitude lower.

Since the value of the critical R^* will vary with the photon energy, the vibrational dependence of the photodissociation cross section will

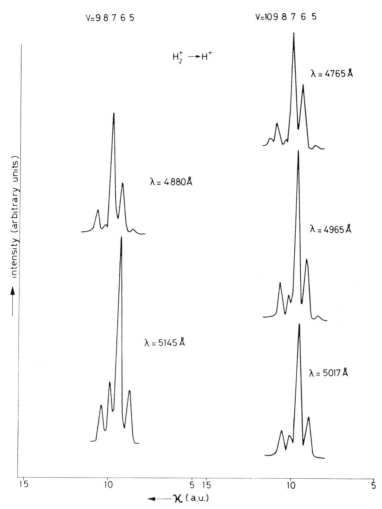

Fig. 27. Laboratory momentum distributions of backward-scattered H^+ fragments from the photodissociation of H_2^+ by selected lines from an Ar-ion laser (wavelength λ). The vibrational levels v of the H_2^+ $1s\sigma_g$ state which give rise to the peaks in the momentum distribution are identified in the upper part of the figure.[105]

differ, depending on the wavelength of the incident light. This is demonstrated in Fig. 27 which shows laboratory momentum distributions of backward-scattered H^+ ions from the photodissociation of H_2^+ using the five most intense lines produced by the Ar-ion laser. The energy releases corresponding to the various peaks in these spectra can be used to identify the vibrational levels of the parent ions as indicated in the figure. When the experimental e_d values are corrected for the ≈ 0.02-eV shift caused by the finite angular resolution of the experiment (cf. Ref. 55 and Section 3.4), they agree within experimental error (≈ 0.02 eV) with the values calculated from the photon energy and the binding energies of Ref. 112. Similar agreement is obtained for D_3^+.[36,113] The results illustrate the possibility of determining energy levels in stable or metastable gaseous ions by the combination of translational spectroscopy and monochromatic photon excitation. The agreement between the experimental and theoretical e_d values also indicates that most of the molecular ions have relatively low rotational quantum numbers J, since the theoretical e_d values were calculated for $J = 0$.

It is also possible to obtain information about the vibrational population in the parent ion beam from measurements such as those shown in Fig. 27. The measured relative intensities should then first be converted to relative intensities in the center-of-mass system, as discussed in Section 3.4. In the present case the e_d values are sufficiently large that this transformation may be approximated[36] by a multiplication by $(u/v)^2$. From these relative center-of-mass intensities the relative vibrational populations of the primary H_2^+ and D_2^+ ions can be calculated by means of Dunn's photodissociation cross sections.[107,108] Van Asselt et al.[36,105] found good agreement between the calculated populations and the vibrational populations deduced from the data obtained with the five different photon energies used in their experiment. This indicates that both the experimental results and Dunn's cross sections are at least self-consistent. The weighted average of these populations agrees closely with that obtained by von Busch and Dunn[114] and by Ozenne et al.[104] for ions produced by electron impact on H_2 at low pressures. It thus appears that the majority of H_2^+ and D_2^+ ions extracted from the unoplasmatron source are produced by electron-impact ionization and that the corresponding vibrational distribution remains nearly unaltered during the extraction of the ions from the source.

A third application of the present technique is based on the fact that the laser light is polarized. The spectra shown in Fig. 27 correspond to fragments which are ejected along the axis of the primary beam, the laser light being polarized in the same direction. Since the $1s\sigma_g \to 2p\sigma_u$ transition has its dipole moment parallel to the internuclear axis, the laser light will be mostly absorbed by ions which are oriented parallel to the beam axis. Since the rotational energy of the majority of the ions is small compared with e_d, the fragments will be ejected in the same direction (the axial recoil approximation; cf. Section 2.1). Ozenne et al. verified this by also polarizing the laser light perpendicular to the beam axis. The fragment spectrum measured along the beam axis indeed decreased in intensity[104,106] by a factor of ≈ 50. Van Asselt et al. kept the polarization of the laser light parallel to the direction of the primary beam and measured the angular distribution of the photofragments. From a comparison with a computer simulation of the experiment, these authors concluded that the decrease in fragment intensity with increasing angle between the molecular axis and the direction of the incident beam is slower than expected for an isotropic distribution of initial molecular orientations. They suggested that the incident molecules are preferentially oriented perpendicular to the beam axis.[36] Clearly, angular measurements on laser-induced photofragments offer interesting possibilities in regard to the study of polarization effects in molecular collisions.

6. Concluding Remarks

In the foregoing discussion we have tried to illustrate how mass-spectrometric studies of collision-induced dissociation in diatomic molecular ions can provide information on the dynamics of these collisions and on the (pre)dissociative states involved.

A very interesting feature of studies on collision-induced dissociation at high energies is the possibility which is offered for experimental determination of cross sections in very fine detail. As has been discussed, the fragment distribution gives the excitation probability as a function of the orientation of the molecular axis and as a function of the internuclear distance. For dissociation by electronic excitation the dynamics of the process are very analogous to the dynamics of photodissociation. In a very comprehensive review by Zare[115] on what is called "photoejection dynamics" it is shown that the angular distribution of photofragments can be used to learn about the photodissociation process. Many of the characteristic features of the photoejection process also apply to collision-induced dissociation by electronic excitation. Moreover, as is stated by Zare, the results obtained for photodissociation of diatomic molecules can be easily extended to polyatomic molecules. The advantage of photodissociation lies in the fact that the amount of inelastic energy involved in the transition is also specified. It might be envisaged that the disadvantage of collision-induced dissociation studies with respect to photodissociation—the fact that several dissociation mechanisms may contribute to the fragment distribution—will be circumvented by applying coincidence techniques, while the initial-state preparation might be improved substantially by the application of lasers. Although in this study we only discussed the $1s\sigma_g \rightarrow 2p\sigma_u$ transition in H_2^+ in terms of the dipole moment, higher-order multipole transitions, whether or not accompanied by simultaneous target excitation, are of great interest. Theoretically, higher-order multipole transitions for the excitation of H_2^+ have already been treated by Green and Peek.[12] Experimentally, however, the influence of these transitions has not yet been distinguished. Again the application of coincidence techniques might be invaluable, for instance by measuring the fragment distribution in coincidence with the excitation of the target atom.

Among the most directly applicable data when it comes to using mass-spectrometric data on collision-induced dissociation as a spectroscopic tool, are the excess energies associated with discrete structure in the fragmentation spectrum, appearance-potential data, measurements on inelastic energy loss, and information on the angular distribution of the fragments. Again we should mention the close similarity between some aspects of these "Aston-banding" techniques with the studies on photofragment spectroscopy as, for example, carried out by Hancock and Wilson,[116] and also with the time-of-flight translational spectroscopy by Freund and co-workers[117] on the dissociation of neutral molecules in very high Rydberg states.

The experiments on the rotational predissociation of HeH^+ illustrate that a close connection exists between the techniques used in, and the information provided by, mass-spectrometric studies on collision-induced dissociation on the one hand and unimolecular dissociation on the other. Other molecules where fine structure has been observed for both collision-induced and/or unimolecular fragments are N_2^+,[44,82] NO^+,[43,118] and CH^+.[119] In the unimolecular fragmentation spectrum of N_2^+, extensive fine structure was observed, which was attributed to the decay of unidentified metastable N_2^+ states. The dissociation limit involved could not be established, but very recent measurements of Locht et al.[120] indicate that these levels dissociate into ground-state fragments. In this case the contribution of translational spectroscopy and appearance-potential measurements has made it possible to localize molecular energy levels in ions with near-spectroscopic precision.

ACKNOWLEDGMENTS

Much of the work discussed in this chapter has been the result of cordial and rewarding collaboration between the research group at the F.O.M. Institute in Amsterdam, Professor Durup's group at Orsay, and Dr. J. M. Peek and Dr. T. A. Green of the Sandia Laboratories, Albuquerque, New Mexico. To all members of these groups we express our sincere appreciation. We are especially grateful to Dr. J. Schopman and Dr. P. G. Fournier for their thorough and imaginative experimental work, and to Dr. J. M. Peek and Professor A. Russek for providing much of the theoretical framework which is necessary to understand and exploit the experimental data. To Dr. J. G. Maas we owe many thanks for careful work on the apparatus function.

One of us (T. R. G.) wrote part of his contribution to this chapter while a postdoctoral teaching fellow in the Department of Chemistry, University of Waterloo, Canada. He would like to thank the members of this department and also Professor J. Kistemaker and Professor J. Los and their co-workers at the F.O.M. Institute, Amsterdam, for their kind hospitality.

This work is part of the research program of the Stichting Fundamenteel Onderzoek der Materie (F.O.M., Foundation for Fundamental Research on Matter). It was made possible by financial support from the

Nederlandse Organisatie voor Zuiver-Wetenschappelijk Onderzoek (Z.W.O., Netherlands Organization for the Advancement of Pure Research) and by an exchange fellowship which was granted to T. R. G. by Z.W.O. and the Centre National de la Recherche Scientifique, France.

References

1. G. W. McClure and J. M. Peek, *Dissociation in Heavy Particle Collisions*, Wiley-Interscience, New York, 1952.
2. J. Durup, in *Recent Developments in Mass Spectrometry* (K. Ogata and T. Hayakwa, eds.), University of Tokyo Press, Tokyo, 1970, p. 921.
3. J. Los, *Ber. Bunsen-Gesell. Phys. Chem.* **77**, 640 (1973).
4. *Proceedings of the Workshop on Dissociative Excitation of Simple Molecules* (L. J. Kieffer, ed.), JILA Information Center Report 12, University of Colorado, Boulder, Colorado, 1972.
5. J. Schöttler and J. P. Toennies, *Chem. Phys. Lett.* **12**, 1615 (1972).
6. P. F. Dittner and S. Datz, *J. Chem. Phys.* **49**, 1969 (1968).
7. P. F. Dittner and S. Datz, *J. Chem. Phys.* **54**, 4228 (1971).
8. H. van Dop, A. J. H. Boerboom, and J. Los, *Physica* **54**, 223 (1971).
9. J. Schöttler and J. P. Toennies, *Z. Phys.* **214**, 472 (1968).
10. R. N. Zare, *J. Chem. Phys.* **47**, 204 (1967).
11. G. H. Dunn, *Phys. Rev. Lett.* **8**, 62 (1962).
12. T. A. Green and J. M. Peek, *Phys. Rev.* **183**, 166 (1969).
13. R. K. Preston and J. C. Tully, *J. Chem. Phys.* **54**, 4297 (1971).
14. T. A. Green, *Phys. Rev. A* **1**, 1416 (1970).
15. B. Meierjohann, *Physica* **65**, 41 (1973).
16. A. Russek, *Physica* **48**, 165 (1970).
17. G. Herzberg, *Molecular Spectra and Molecular Structure, I. Spectra of Diatomic Molecules*, 2nd ed., Van Nostrand–Reinhold, Princeton, New Jersey, 1950.
18. A. J. Lorquet and J. C. Lorquet, *Chem. Phys. Lett.* **26**, 132 (1974).
19. R. B. Bernstein, *Phys. Rev. Lett.* **16**, 385 (1966).
20. M. Cavallini, G. Gallinaro, L. Meneghetti, G. Scoles, and U. Valbusa, *Chem. Phys. Lett.* **1**, 303 (1970).
21. V. Henri, *Compt. Rend.* **177**, 1037 (1923).
22. A. G. Gaydon, *Dissociation Energies and Spectra of Diatomic Molecules*, 3rd ed., Chapman & Hall, London, 1968.
23. R. J. Le Roy, in *Specialist Periodical Reports: Molecular Spectroscopy*, Vol. 1 (R. F. Barrow, D. A. Long, and D. J. Millen, eds.), The Chemical Society, London, 1973, p. 113.
24. K. R. Way and W. C. Stwalley, *J. Chem. Phys.* **59**, 5298 (1973).
25. M. S. Child, in *Specialist Periodical Reports: Molecular Spectroscopy*, Vol. 2 (R. F. Barrow, D. A. Long, and D. J. Millen, eds.), The Chemical Society, London, 1974, Chapter 7.
26. F. W. Aston, *Proc. Cambridge Phil. Soc.* **19**, 317 (1920).
27. E. Friedländer, H. Kallmann, W. Lasareff, and B. Rosen, *Z. Phys.* **76**, 60, 70 (1932).
28. W. McGowan and L. Kerwin, *Can. J. Phys.* **41**, 316 (1963).
29. R. G. Cooks, J. H. Beynon, R. M. Caprioli, and G. R. Lester, *Metastable Ions*, Elsevier, Amsterdam, 1973.
30. W. A. Chupka, *J. Chem. Phys.* **30**, 191 (1959).
31. J. Durup, P. Fournier, and Pham Dông, *Int. J. Mass Spectrom. Ion Phys.* **2**, 311 (1969).
32. P. Fournier, A. Pernot, and J. Durup, *J. Phys. (Paris)* **32**, 533 (1971).

33. P. G. A. Fournier, in *Méthodes de Spectroscopie sans Largeur Doppler de Niveaux Excités de Systèmes Moléculaires Simples*, Colloques Internationaux du C.N.R.S., No. 217, Centre National de la Recherche Scientifique, Paris, 1973, p. 12.
34. J. H. Beynon, R. G. Cooks, J. W. Amy, W. E. Baitinger, and T. E. Ridley, *Anal. Chem.* **45**, 1023A (1973).
35. D. K. Gibson, J. Los, and J. Schopman, *Physica* **40**, 385 (1968).
36. N. P. F. B. van Asselt, J. G. Maas, and J. Los, *Chem. Phys.* **5**, 429 (1974).
37. M. Vogler and W. Seibt, *Z. Phys.* **210**, 337 (1968).
38. A. S. Newton and A. G. Sciamanna, *J. Chem. Phys.* **50**, 4868 (1970).
39. T. F. Moran, F. C. Petty, and A. F. Hedrick, *J. Chem. Phys.* **51**, 2112 (1969).
40. H. Wankenne and J. Momigny, *Int. J. Mass Spectrom. Ion Phys.* **7**, 227 (1971).
41. T. O. Tiernan and R. E. Marcotte, *J. Chem. Phys.* **53**, 2107 (1970).
42. R. F. Mathis, B. R. Turner, and J. A. Rutherford, *J. Chem. Phys.* **49**, 2051 (1968).
43. T. R. Govers and J. Schopman, *Chem. Phys. Lett.* **12**, 414 (1971).
44. P. G. Fournier, T. R. Govers, C. A. van de Runstraat, J. Schopman, and J. Los, *J. Phys. (Paris)*, **33**, 755 (1972).
45. J. W. McGowan and L. Kerwin, *Can. J. Phys.* **42**, 972 (1964).
46. J. Berkowitz and W. A. Chupka, *J. Chem. Phys.* **51**, 2341 (1969).
47. V. Cermák and Z. Herman, *Nucleonics* **19**, 106 (1961).
48. F. H. Field, *Accounts Chem. Res.* **1**, 42 (1968).
49. E. Lindholm, in *Ion–Molecule Reactions in the Gas Phase* (P. J. Ausloos, ed.), Adv. in Chem. Series No. 58, Am. Chem. Soc. Publ., Washington, 1966, p. 1.
50. C. B. Richardson, K. B. Jefferts, and H. G. Dehmelt, *Phys. Rev.* **165**, 80 (1968).
51. J. B. Ozenne, private communication.
52. M. Vogler and B. Meierjohann, *Abstracts IXth ICPEAC* (J. S. Risley and R. Geballe, eds.), University of Washington Press, Seattle, 1975, p. 711.
53. D. H. Jaecks, W. de Rijk, and P. J. Martin, *Abstracts VIIth ICPEAC* (L. Branscomb et al., eds.), North-Holland Publ., Amsterdam, 1971, p. 424.
54. P. Erman, *Physica Scripta* **14**, 51 (1976), and references therein.
55. J. G. Maas, N. P. F. B. van Asselt, and J. Los, *Chem. Phys.* **8**, 37 (1975).
56. D. T. Terwilliger, J. H. Beynon, and R. G. Cooks, *Proc. R. Soc.* **341**, 135 (1974).
57. G. W. McClure, *Phys. Rev.* **140**, A769 (1965).
58. I. Sauers, R. L. Fitzwilson, J. C. Ford, and E. W. Thomas, *Phys. Rev. A* **6**, 1418 (1972).
59. J. Guidini, *C. R. Acad. Sci. (Paris)* **253**, 829 (1961).
60. R. Caudano and J. M. Delfosse, *J. Phys. B* **1**, 813 (1968).
61. F. P. G. Valckx and P. Verveer, *J. Phys. (Paris)* **27**, 480 (1966).
62. D. K. Gibson and J. Los, *Physica* **35**, 258 (1967).
63. S. J. Anderson and J. B. Swan, *Phys. Lett.* **48A**, 435 (1974).
64. S. J. Anderson, *J. Chem. Phys.* **60**, 3278 (1974).
65. P. G. Fournier, *Bull. Am. Phys. Soc. Ser. II* **19**, 447 (1974).
66. F. Brouillard, W. Claeys, J. Delfosse, A. Oliver, and G. Poulaert, *Abstracts IXth ICPEAC* (J. S. Risley and R. Geballe, eds.), University of Washington Press, Seattle, 1975, p. 713.
67. H. H. Michels, *J. Chem. Phys.* **44**, 3834 (1966).
68. T. A. Green, H. H. Michels, J. C. Browne, and M. M. Madsen, *J. Chem. Phys.* **61**, 5186, 5198 (1974).
69. J. Schopman and J. Los, *Physica* **48**, 190 (1970).
70. H. K. Schoenebeck, *Z. Phys.* **177**, 111 (1964).
71. J. Schopman, J. Los, and J. Maas, *Physica* **51**, 113 (1971).
72. J. C. Houver, J. Baudon, M. Abignoli, M. Barat, P. Fournier, and J. Durup, *Int. J. Mass Spectrom. Ion Phys.* **4**, 137 (1970).
73. J. Schopman and J. Los, *Physica* **51**, 132 (1971).

74. W. Seibt, *Abstracts VIth ICPEAC* (I. Amdur, ed), MIT Press, Cambridge, Massachusetts, 1969, p. 803.
75. Pham Dông and J. Durup, *Chem. Phys. Lett.* **5**, 340 (1970).
76. H. D. Smyth, *Proc. R. Soc. A* **104**, 121 (1923).
77. H. D. Smyth, *Rev. Mod. Phys.* **3**, 347 (1931).
78. H. D. Hagstrum and J. T. Tate, *Phys. Rev.* **59**, 354 (1941).
79. J. A. Hippel, R. E. Fox, and E. U. Condon, *Phys. Rev.* **69**, 347 (1946).
80. T. F. Moran, F. C. Petty, and A. F. Hedrick, *J. Chem. Phys.* **51**, 2112 (1969).
81. S. E. Kuprianov, *Sov. Phys. Tech. Phys.* **9**, 659 (1964).
82. P. G. Fournier, C. A. van de Runstraat, T. R. Govers, J. Schopman, F. J. de Heer, and J. Los, *Chem. Phys. Lett.* **9**, 426 (1971).
83. W. Schultz, B. Meierjohann, W. Seibt, and H. Ewald, in *Recent Developments in Mass Spectrometry* (K. Ogata and T. Hayakwa, eds.), Univ. of Toyko Press, Tokyo, 1970, p. 939.
84. D. L. Albritton, A. L. Schmeltekopf, and R. N. Zare, *Diatomic Intensity Factors*, John Wiley, in preparation.
85. C. A. van de Runstraat, F. J. de Heer, and T. R. Govers, *Chem. Phys.* **3**, 431 (1974).
86. D. L. Albritton, A. L. Schmeltekopf, and E. E. Ferguson, *Abstracts VIth ICPEAC* (I. Amdur, ed.), MIT Press, Cambridge, Massachusetts, 1969, p. 331.
87. A. E. Douglas, *Can. J. Phys.* **30**, 302 (1952).
88. P. K. Carroll, *Can. J. Phys.* **37**, 880 (1959).
89. R. F. Holland and W. B. Maier, II, *J. Chem. Phys.* **55**, 1299 (1971).
90. T. R. Govers, C. A. van de Runstraat, and F. J. de Heer, *Chem. Phys.* **9**, 285 (1975).
91. T. R. Govers, F. C. Fehsenfeld, D. L. Albritton, P. G. Fournier, and J. Fournier, *Chem. Phys. Lett.* **26**, 134 (1974).
92. E. E. Ferguson, *Rev. Geophys. Space Phys.* **12**, 703 (1974).
93. J. Tellinghuisen and D. L. Albritton, *Chem. Phys. Lett.* **31**, 91 (1974).
94. A. L. Roche and H. Lefebvre-Brion, *Chem. Phys. Lett.* **32**, 155 (1975).
95. J. Schopman, P. G. Fournier, and J. Los, *Physica* **63**, 518 (1973).
96. J. Schopman, A. K. Barua, and J. Los, *Phys. Lett.* **29A**, 112 (1969).
97. J. M. Peek, *Physica* **64**, 93 (1973).
98. H. F. Helbig, D. B. Millis, and L. W. Todd, *Phys. Rev.* **2**, 771 (1970).
99. L. Wolniewicz, *J. Chem. Phys.* **43**, 1087 (1965).
100. L. Gottdiener and J. N. Murrell, *Mol. Phys.* **25**, 1041 (1973).
101. R. B. Bernstein, *Chem. Phys. Lett.* **25**, 1 (1974).
102. J. L. Dunham, *Phys. Rev.* **41**, 721 (1932).
103. W. C. Stwalley, in *Energy, Structure and Reactivity* (D. W. Smith and W. B. McRae, eds.), Wiley, New York, 1973, p. 259.
104. J. B. Ozenne, Pham Dông, and J. Durup, *Chem. Phys. Lett.* **17**, 422 (1972).
105. N. P. F. B. van Asselt, J. G. Maas, and J. Los, *Chem. Phys. Lett.* **24**, 555 (1974).
106. J. Durup, 21st Annual Meeting on Mass Spectrometry and Allied Topics, San Francisco, May 1973, p. 109.
107. G. H. Dunn, *Phys. Rev. A* **5**, 1726 (1972).
108. G. H. Dunn, JILA Report No. 92, JILA, Boulder, Colorado, 1968.
109. R. S. Mulliken, *J. Chem. Phys.* **55**, 309 (1971).
110. R. S. Mulliken, *J. Chem. Phys.* **33**, 247 (1960).
111. J. N. Murrel and J. M. Taylor, *Mol. Phys.* **16**, 609 (1969).
112. S. Cohen, J. R. Hiskes, and R. J. Riddell, Jr., *Phys. Rev.* **119**, 1025 (1960).
113. G. H. Dunn, *J. Chem. Phys.* **44**, 2592 (1966).
114. F. von Busch and G. H. Dunn, *Phys. Rev. A* **5**, 1726 (1972).
115. R. N. Zare, *Mol. Photochem.* **4**, 1 (1972).

116. G. Hancock and K. R. Wilson, in *Fundamental and Applied Laser Physics*, Proc. Esfahan Symposium, Esfahan, Iran, 1971 (M. Feld, W. Kurnit, and A. Javan, eds.), Wiley, New York, 1972, and references cited therein.
117. K. C. Smyth, J. A. Schiavone, and R. S. Freund, *J. Chem. Phys.* **60**, 1358 (1974), and references cited therein.
118. Pham Dông and M. Bizot, *Int. J. Mass Spectrom. Ion Phys.* **10**, 227 (1972/73).
119. M. Roussel and A. Julienne, *Int. J. Mass Spectrom. Ion Phys.* **9**, 463 (1972).
120. R. Locht, J. Schopman, H. Wankenne, and J. Momigny, *Chem. Phys.* **7**, 393 (1975).

7

Collision-Induced Dissociation of Polyatomic Ions

R. G. Cooks

1. Introduction

1.1. Scope of Chapter

This chapter deals with reactions between polyatomic ions and collision targets which lead to fragmentation of the ion. Most but not all of the work involves gaseous targets. Impact energies fall in the range from ~100 eV to ~100 keV. Virtually all studies have employed detection of the fast product ion at near zero scattering angle. Many have been made in some type of mass spectrometer. The reactions of interest involve inelastic collisions, the internal energy of the ion increasing as a result of the collision. The reactions of fast ions with which we are concerned are mechanistically dissimilar from collision-induced dissociations (CID) at low energy (electron volt range) which proceed via long-lived complexes. The latter reactions are well known[1] but fall outside the scope of this work.

The prototype reaction is that in which a singly charged polyatomic ion m_1^+ interacts with a gas atom or molecule N to yield the fragments m_2^+ and m_3

R. G. Cooks • Department of Chemistry, Purdue University, West Lafayette, Indiana 47907.

as shown:

$$m_1^+ + N \rightarrow m_2^+ + m_3 + N \qquad (1.1)$$

There are a number of related CID processes which will be dealt with later in the chapter, involving multiply or negatively charged ions. There are also related reactions which involve charge transfer as well as dissociation.

Collision-induced dissociation is related more to traditional mass spectrometry and probably has more application to chemical analysis than have those high-energy ion–molecule reactions which involve only charge-changing collisions (see Chapter 5). Since fragmentation of polyatomic ions usually occurs by a number of competitive pathways, CID yields relative ion-abundance data which, as in other forms of mass spectrometry, reflect the structures and energies of the species involved. In contrast to most mass-spectrometric techniques, CID is complicated by the fact that the fragmenting ion is prepared in two steps, i.e., by ionization and then collision, rather than directly. However, this is also the source of some of its applications in that it allows the chemistry of the nonfragmenting ions prior to collision to be followed. It also means that the internal energy of the energized ion can, in principle, be determined, a significant advantage over many methods of ionization and one which leads to applications in kinetics and thermochemistry (discussed later).

The features just noted, especially the potential usefulness of CID as a means of deducing molecular and ionic structures, make it appear that CID will soon become widely applied in organic mass spectrometry and elsewhere. Because most efforts in this direction have only been made recently, the available data are limited in scope and are more qualitative than might be wished. It is hoped that this chapter will help catalyze the wider study of the CID phenomenon and especially its applications to chemical problems. To this end we have set out both to collect the available data and to analyze them. Criticisms are made and new interpretations, experiments, and potential applications suggested.

1.2. Comparison with Metastable Ions

Any discussion of collision-induced dissociation of polyatomic ions invites comparison and contrast with the much more thoroughly studied phenomenon of metastable ions.[2-4] The appropriateness of this juxtaposition follows from the fact that both concern the fragmentations of high-velocity ions and that similar methods can be used to generate the reactant ion and to analyze the products. Moreover, both share the valuable characteristic that the reaction under study can be specified by mass and ion kinetic energy measurements. (This is in contrast to most mass spectrometric studies in which reaction occurs in the ion source and only the product is specified.) Although reactant and product ions are identified in both techniques, intermediate steps are not, and this is more particularly a problem with CID where the higher-average energies make more extensive reaction sequences accessible than is the case for metastable ions.

Contrasts between the two techniques are to be found in the methods used to excite the ions and in the internal energies so acquired. Since internal energy is the only factor determining the subsequent behavior of an isolated species, the chemistry involved in collision-induced dissociations can be quite different from that obtaining for metastable ions. The latter ions are characterized by low internal energies (typically a few tenths of an electron volt) covering relatively narrow ranges whereas collision-induced dissociation involves excited ions which may have much larger internal energies. Moreover, for CID the instrumental limitation on the range of ion lifetimes—and thus the internal energy distribution—leading to observable products is not narrowly circumscribed as it is for metastable ions.

For these reasons the term collision-induced metastable ions is unsatisfactory. Moreover, the ion produced on collision is itself not necessarily metastable in the sense of having a lifetime which is comparable to the instrument transit time. It may have a lifetime which is many orders of magnitude shorter. The term metastable does not seem adequately to convey this situation in which there are actually two quite distinct forms of the reactant ion—the long-lived low energy ion prior to collision and the variable energy/variable lifetime postcollision species which undergoes fragmentation. We shall therefore employ the traditional term, by which the entire process is known as collision-induced dissociation (CID). If the ion–molecule reaction is specifically intended, the term collisional excitation or collisional activation (CA) is appropriate.

1.3. Development of CID Studies

In view of the present standing of collision-induced dissociation vis-à-vis metastable ions as applied to polyatomic species, the latter being more widely known and used, it is of some note that the discovery of collision-induced dissociation considerably anteceded that of metastable ions. It was quite natural, therefore, that when Hipple and Condon[5] discovered metastable ions in 1945, considerable attention was given to proving that the observed peaks were indeed due to unimolecular reactions and not to collision-induced dissociation.[6] It is one of the ironies of the development of mass spectrometry that as a result of the subsequent widespread use of metastable ions the error which these workers avoided has been repeated by many. Recent data which show the value of collision-induced dissociation in the study of polyatomic and particularly organic ions should prevent such neglect in the future.

Signals due to the products of collision-induced dissociation were observed by the first mass spectroscopist, J. J. Thomson.[7] He noted that the parabola due to hydrogen ions (H^+) observed in his crossed electric and magnetic field instrument contained two components with velocities in the ratio $1:\sqrt{2}$. Thomson surmised that the slower ions were due to fragmentation of an $H_2^{+\cdot}$ ion after its acceleration through the hollow cathode but before deflection by the electric or magnetic analyzing fields. Thomson did not specify that the fragmentation was due to collision although he had previously suggested collisions as being responsible for the analogous charge-transfer processes.

Aston[8] followed up this work with a clear explanation of the phenomenon in $H_2^{+\cdot}$, including a demonstration of the pressure dependence. Using his newly constructed mass spectrograph he was able to measure the position of the hydrogen CID peak and also that at apparent mass $5\frac{1}{7}$ in CO.[9] In recognition of Aston's work, these peaks were and often still are widely referred to as Aston bands.

After these early observations made on photoplate instruments the subject was taken up by several groups using electrical detection. Improvements in vacuum technology greatly advanced this endeavor and in 1925 Smyth[10] showed a mass spectrum of hydrogen which included peaks at $m/e\ \frac{1}{2}$ and $\frac{1}{3}$. His explanation of the phenomenon, although correct, was arrived at reluctantly, since, in his words, "it may seem absurd to suppose that ions can be disrupted by collision... and yet retain their speed and direction." Similar observations were made by Hogness and Lunn on NO^+ which was found to fragment on collision at an impact energy of 650 eV.[11]

The collision-induced dissociation of hydrogen was again studied by Dorsch and Kallmann[12] and shown to occur only for high velocity $H_2^{+\cdot}$ ions. A number of other examples were also collected about this time[13] although the mechanism was not discussed.

The possibility of translational energy being lost by the ion on collision was recognized and led to experiments in which the exact positions of CID peaks were measured. The first such precise measurement was made[14] for the reaction $N_2^{+\cdot} \rightarrow N^+ + N\cdot$, but within experimental error no energy loss could be detected.

Later a detailed study of a large number of collision-induced dissociations was undertaken by Mattauch and Lichtblau[15] using a double-focusing mass spectrometer capable of excellent sensitivity. These authors made a unified presentation of the subject of secondary reactions—those which occur in the analyzer tube of the mass spectrometer—and gave the following general formula for the apparent mass (m^*) of the product of any such reaction:

$$m^* = \frac{m_2^2}{m_1} \frac{x}{y^2} \qquad (1.2)$$

where m_1 and m_2 are the masses of the reactant m_1^{x+} and the analyzed product ion m_2^{y+}, respectively. CID peaks were recorded both electrically and photographically and their widths and exact positions were measured. The peak positions were generally found to be somewhat displaced to lower apparent mass and this was interpreted as evidence for the loss of a small amount of kinetic energy in the collision.

The reason for the widths of the Aston bands was not clarified, however, although the phenomenon of energy loss without band formation was observed in several types of charge-transfer reactions at high energy. Later, after Hipple, Fox, and Condon's discovery of metastable ions[5,6] and their suggestion that the characteristic widths of these peaks were, at least in part, due to kinetic energy release upon dissociation, all the facts needed to understand collision-induced dissociations had been assembled.

With the revival of interest in ion–molecule reactions which occurred during the 1950s, high-energy ion–molecule reactions were studied in several laboratories, notably those of Henglein, Melton, White, and Kerwin. Henglein[16,17] employed a parabola mass spectrograph capable of accelerating voltages of 20 kV. This enabled him to achieve good sensitivity combined with a unique feature of the parabola instrument, namely that all high-energy ion–molecule reaction products are recorded simultaneously on a single photographic plate. Thus he observed not only collision-induced dissociation but also more complex reactions where dissociation was accompanied by charge inversion, as in the process

$$CH_3^+ + N \rightarrow C^- + (H_3) + N^{2+} \qquad (1.3)$$

where N is the neutral target. Dissociation accompanied by charge exchange of doubly charged ions was also observed.

White and co-workers, working at impact energies in the 40 keV region, observed triple peaks in the translational energy distributions for products of collision-induced dissociation.[18] Figure 1 shows a typical energy spectrum recorded on a triple-sector mass spectrometer. The observed peak shapes were explained as the result of two processes: one accompanied by a small energy release in the center-of-mass system, and the other by a much larger release. (Instrumental discrimination causes the broad peak for this latter process to appear as two separate peaks.) These authors also derived a relationship between the peak width and the kinetic energy released.

The high accelerating voltages used in White's experiments minimized the fractional shift in the CID peak position due to kinetic energy loss

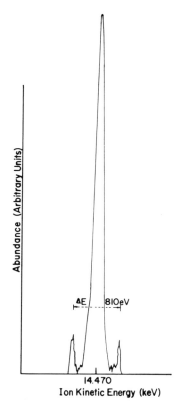

Fig. 1. Kinetic energy spectrum due to CH_3^+ ions formed by collision-induced dissociation of $C_3H_7^+$ ions of 41.5 keV energy. This experiment was done using a mass-analyzed beam of $C_3H_7^+$ ions and a short collision chamber.

accompanying excitation. It thus remained for McGowan and Kerwin[19] to measure, apparently for the first time, both the kinetic energy loss and kinetic energy release in collision-induced dissociations. This completed the developments which laid the foundation for our present detailed knowledge of collision-induced dissociation of diatomic ions, which is described in Chapter 6 of this book.

Meanwhile, detailed study of CID of organic ions had commenced. Kolotyrkin et al.[20] estimated the cross section for collision-induced dissociation of the methane molecular ion as 10^{-16} cm^2 and Melton and Rosenstock obtained a similar result.[21] These latter authors compared the relative abundances of the products of the various CID channels in this ion and noted their similarity to the abundances of the corresponding fragment ions in the electron-impact mass spectrum. They concluded that the energy transferred upon collision was similar to that deposited upon electron impact. In the years that followed Melton and co-workers greatly extended these investigations.[22] A number of papers by Kupriyanov and co-workers on the CID of polyatomics also appeared in which many of the main features of the collision-induced dissociation of polyatomic ions were explored. These authors measured the cross sections for the collision-induced dissociations of various organic ions over a range of kinetic energies,[23,24] they explored the effects of the nature of the target and the electron energy used in the formation of the ion,[25] and they emphasized the similarities between fragmentation spectra of ions and the mass spectra of the corresponding molecules.[26,27] Kupriyanov suggested that two types of mechanisms might be operative, electronic excitation at higher kinetic energies and direct vibrational excitation at lower energies.[28] Later work has confirmed this view. This group also initiated the use of fragmentation spectra to characterize ions prepared from different sources.[29]

The recent upsurge of interest in collision-induced dissociation of organic ions was catalyzed by a paper of Jennings in 1968.[30] Using a commercial mass spectrometer, this author showed that it was possible to study readily the collision-induced dissociations of various ions in the benzene mass spectrum. Thereafter McLafferty and his co-workers[31,32] began an extended investigation of collision-induced dissociation of organic ions. In this work emphasis was on the use of collision-induced dissociation to study ions with energies intermediate between those of metastable ions and ions which fragment in the ion source and to characterize ion structure.

More recently, modifications of commercial instruments which allow collision gas introduction have facilitated these studies. The development of the reversed-sector mass spectrometer[33,34] has considerably simplified experiments in which the various modes of reaction of an ion of interest are followed. This instrument achieves what had previously been done using double mass spectrometers since it allows reactant ion selection to be performed independently of product-ion analysis. Using the methods of ion kinetic energy analysis developed for the study of metastable ions,[4] routine determinations of the energy spectra associated with individual CID processes in polyatomic ions are now being made. When taken at high energy resolution such measurements—mass-analyzed ion kinetic energy (MIKE) spectra—provide detailed thermochemical data on individual CID reactions whereas low-resolution data taken with the same instrument give relative cross section data for the various competitive reaction channels.[35] The term collisional activation (CA) spectra has been widely applied to low-resolution MIKE spectra corrected for contributions from unimolecular fragmentations.

2. The Reaction

2.1. The Basic Phenomenon and Its Experimental Characterization

Of concern here is the process whereby a gaseous ion interacts at high relative velocity with a target and yields lower-mass fragment ions. Also of interest are the types of experimental measurements possible and the information they yield.

Characterization of collision-induced dissociation requires as a minimum that the *masses* of the reactant and product ions be specified. Both ion masses can be measured directly in an instrument which employs two successive mass analyzers. Alternatively, the nature of the reactant ion may merely be inferred from the nature of the neutral sample and the method of ionization. Methanol, for example, if ionized by 11-eV electrons, yields exclusively $CH_4O^{+\cdot}$ ions. The most frequently used experimental methods, however, are those based on single- or dual-analyzer mass spectometers. By giving the reacting ions a well-characterized translational energy, the measurement of the momentum or kinetic energy of the product serves to provide the ratio of the masses of the reactant and product ions. By combining such a measurement with that of the mass of either of the ions the reaction is unambiguously defined in terms of the masses of the species involved. If only the momentum analysis is done, as is characteristic of single-focusing mass spectrometers, ambiguities in mass assignments exist and they increase as the complexity of the ion increases; the method is not recommended. The various methods of operating dual-analyzer mass spectrometers are discussed in Section 3.

Questions regarding the rate of reaction are considered in detail in Section 2.3. The minimum requirement for any type of study of CID is that the product-ion abundance be measured. Since a high-velocity species is formed standard methods of particle detection, specifically the use of electron multipliers, can be employed. The signal due to the product ion (the

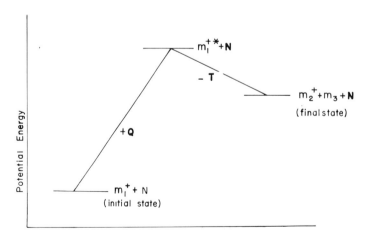

Fig. 2. Energetics of collision-induced dissociation. The energy of excitation Q represents the endothermicity of the ion–molecule reaction and T represents the kinetic energy release accompanying fragmentation.

ion abundance) can be expressed relative to the reactant ion signal or, as is common for polyatomic ions, as an abundance relative to the signals for other products of collision-induced dissociation.

The simple methods just outlined allow one to characterize the reaction under study and to measure its rate. Elementary precautions, usually involving measurements over a range of target pressures, must also be taken to ensure that the process is indeed bimolecular.

Collision-induced dissociation involves two separate processes, bimolecular excitation followed by unimolecular fragmentation (Section 2.2). Although the overall reaction may be endothermic, exothermic, or thermoneutral, collisional activation is necessarily endothermic and fragmentation is necessarily exothermic. Thus, translational energy must be converted to internal energy in the bimolecular collision process whereas internal energy must be released as relative translational energy of separation of the fragments in the unimolecular fragmentation process (Fig. 2). Both phenomena have characteristic effects upon the translational energy of the product ions.

For the case of near-zero scattering angles, which is of interest here, the total kinetic energy lost by the reactant ion is only slightly greater than the endothermicity Q of the collisional activation step. The small difference appears as translational energy of the target, and the ratio of the changes in the translational energies of the ion and the target has been referred to as the disproportionation factor.[4] It depends on the masses of the ion and the neutral m_1 and N, the initial translational energy Ve, and the endothermicity Q. In the limit of negligible scattering of the fast ion it is given (cf. Appendix 1, Chapter 4) as

$$D = \left(\frac{4N}{m_1}\right)\left(\frac{Ve}{Q}\right) \qquad (2.1)$$

The experimentally accessible quantity is the difference between the translational energy of the reactant ion and that of the product ion. Correcting the latter for the mass associated with the neutral fragment(s) gives the difference between reactant and product kinetic energies, that is, the kinetic energy loss. In virtually all the experiments with which this chapter is concerned the measured energy loss is, within experimental error, indistinguishable from the endothermicity Q.

The average translational energy of the product ion is thus determined by the reaction endothermicity. In general, for polyatomic ions, the reactant ion beam will include ions in several vibrational and perhaps even electronic states. Thus the activation step will be characterized by a distribution of kinetic energy losses, the nature of which will depend on the reaction in hand. In this regard CID reactions are analogous to charge-transfer reactions. Moreover, in both situations the kinetic energy of the ion is measured relative to its expected value in the absence of energy loss or gain. For the fragmentation $m_1^+ \to m_2^+ + m_3$ this reference energy is m_2/m_1 of the energy of the ion m_1^+. Measurements of energy loss, to be useful, require an accuracy which is better than ±1 eV. Hence the energy resolution of an instrument to be used for energy-loss measurements must be on the order of 1 eV in several kilovolts, depending on the accelerating voltage employed. This desideratum is considerably more severe than that required merely to identify the reaction where an energy (or momentum) resolution of one part in m_1, or typically about 10^2, suffices.

In addition to the energy loss, the product-ion translational energy distribution will reflect the translational energy T released in the fragmentation. This energy release occurs when an isolated ion, moving at high velocity, fragments. Fragmentation is isotropic in the center of mass, with the product energy in the laboratory system as likely to be reduced as it is to be increased. Although the maximum range of translational energies of m_2^+ as a result of the release of a kinetic energy T can be readily calculated, the probability distribution within this range is a complex function of instrumentation parameters which is only accessible through numerical integration procedures.[36,37] One fact is of primary significance here: the energy release in the center-of-mass (c.m.) system is amplified on converting coordinates so that small c.m. energy releases can be accurately measured. This velocity amplification causes the m_2^+ energy distribution to be very broad relative to that for m_1^+. The effect is of course identical to that which operates when metastable ions fragment and the same equations apply.[4] Thus the amplification factor, namely, the range of kinetic energies of m_2^+ in the laboratory versus the c.m. system, is given by

$$A = 4\frac{(m_2 m_3)^{1/2}}{m_1}\left(\frac{Ve}{T}\right)^{1/2} \quad (2.2)$$

where Ve is the translational energy of m_1^+ which fragments to give m_2^+ and m_3 with release of energy T. An extra complication of CID is that two distributions, one due to energy release and the other associated with energy loss, are convoluted together in the experimental peak shape. The kinetic energy release for collision-induced dissociation can readily be calculated, via Eq. (2.2) if the peak broadening is assumed to be entirely due to the kinetic energy release.

That this will often be a good approximation is seen by considering a typical reaction studied at 10 keV, namely, $78^+ \to 52^+ + 26$ in benzene. The amplification factor is 770 so that the width of the peak due to release of 0.06 eV, the average value for metastable ions, is 46 eV in the laboratory system. Excitation energies are typically an order of magnitude smaller. Added to this is the fact that the range of energy losses associated with polyatomic ions is expected to be relatively small since breakdown curves indicate that the energy range of stability for ions which possess more than a few modes of further fragmentation seldom exceeds a few electron volts. One can therefore expect that the measurement of the kinetic-energy-release distribution accompanying collision-induced dissociation will not be severely compromised by the distribution of kinetic energy losses. On the other hand, the distribution of energy losses is not accessible although the most probable energy loss will correspond to the shift in the peak position. The broadness of the peak due to energy release will of course make the energy loss more difficult to determine than is the case for simple charge-changing reactions. It should also be noted that the determination of the kinetic energy release T presupposes a single-step fragmentation; the problem of multistep processes has not yet been addressed.

Figure 3 summarizes the three types of measurement just discussed and the information obtainable from each.

2.2. Mechanism

2.2.1. Basic Type

The resurgence of interest in collision-induced dissociation in the past several years has emphasized the study of polyatomic ions. Much of this work has been qualitative or application-oriented but some mechanistic insights are available and analogies with the much more thoroughly studied diatomic ions (Chapter 6) are also proving helpful.

It is important at this point to distinguish between two quite different ion–molecule interactions which can result in fragmentation of an ion and which are both, therefore, examples of collision-induced dissociation. The processes which occur at higher kinetic energies involve minimal momentum transfer, small angles of scattering, and, usually, electronic interactions.

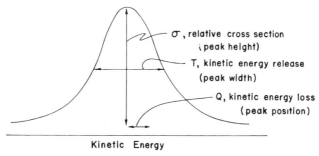

Fig. 3. Illustration of the types of measurements possible for CID peaks. In the absence of a kinetic energy loss the average translational energy of m_2^+ is (m_2/m_1) that of m_1^+.

At lower energies nuclear interactions occur, all or much of the kinetic energy of the system being made available as internal energy through formation of a long-lived complex.[1,38] Thus, these latter reactions are adiabatic in character and can occur at relative kinetic energies of just a few electron volts. Their cross sections are typically smaller than those for the higher energy reactions which belong to the general class of stripping reactions. In the latter processes, one part of the species interacts with the target independently of the rest of the molecule. The term, as introduced into chemistry, has generally referred to reactions in which atoms or groups were removed. Here it refers to an interaction between the electrons of the ion and the target which occurs without affecting the relative motion of the ion and target except as the conservation laws dictate. The processes of primary interest in this chapter are of the stripping type, namely, nonadiabatic high-energy collision-induced dissociations.

An intermediate possibility between complex formation and the spectator stripping interaction will also be encountered to some extent in what follows. In this mechanism, only part of the ion is a spectator, while some atom or group is intimately involved in a direct momentum-sharing collision with the target. This collision is, for an instant, just an elastic binary collision between two species, a part of the ion and the target. The change in energy of the relevant atom or group in the ion rapidly appears as vibrational or rotational energy of the entire ion. Since a hard-sphere collision is involved, these vibrational excitation processes can also be expected to have smaller maximum cross sections than the electronic excitation reaction. Electronic excitation typically requires high relative velocities of ion and target, the cross section falling rapidly at low velocities. By contrast the vibrational excitation has its maximum cross section at lower ion energies.[39] In Section 2.2.6 this and other less well-known CID mechanisms are discussed further.

The collision-induced dissociation of a polyatomic ion can be considered as involving two quite distinct steps, collisional excitation [Eq. (2.3)] and unimolecular dissociation [Eq. (2.4)] of the excited ion:

$$m_1^+ + N \rightarrow m_1^{+*} + N \quad (2.3)$$

$$m_1^{+*} \rightarrow m_2^+ + m_3 \quad (2.4)$$

This separation is justified if the relative velocity of the ion and target is such that they are well separated before fragmentation occurs. If this were not the case, the dissociation would be influenced by the target. If one assumes a large interaction radius of 20 Å for perturbation by the target, then the relative velocity must be such that the separation is at least 20 Å before fragmentation occurs. Since the fastest fragmentations cannot occur faster than the fastest-bond vibrational frequencies, the minimum time allowed is of the order of 10^{-14} sec. Hence a center-of-mass velocity of 2×10^7 cm/sec suffices, even in extreme cases, to cause the excited ion to behave as an isolated system. Provided the target mass is not large compared to that of the projectile (in the experiments discussed here it is usually smaller), the corresponding velocity in the laboratory system will be on the order of 2×10^7 cm/sec or less. An 8-keV ion of mass 100 has a velocity of just over 10^7 cm/sec so the separation of the components (2.3) and (2.4) of the CID reaction is justified for kilovolt beams and common organic ions except for the fastest fragmentations (frequency factors approaching or exceeding 10^{14} sec^{-1}) where the possible influence of the target on the process must be considered. Polyatomic targets are more likely to show these effects than the rare gases usually employed, particularly if they are much more massive than the reactant ion or if they are highly polar.

We have so far considered the fragmentation step and its timing in relation to relative motion of the two particles. The excitation process must be complete within an even shorter time, an estimated interaction radius of 10 Å (see data on cross sections, Section 2.3) being appropriate to this reaction. All but the fastest nuclear motions will therefore be almost completely frozen out in the kilovolt energy range. This leads one to expect quasi-Franck–Condon transitions, although it must be pointed out that confirmatory data are as yet available only for diatomic ions and particularly $H_2^{+\cdot}$.

The time required for excitation and that allowed during which subsequent fragmentation of the isolated excited ion can lead to detectable products are typically orders of magnitude different. Fragmentation may occur, in typical instruments, 10^{-6} or 10^{-5} sec after excitation so that the range of ion lifetimes sampled is 10^{-13} to 10^{-5} sec, which is comparable to that sampled in mass spectrometer ion sources. This is one of several reasons for the similarity between the reactions induced by electron impact and those which follow collisional excitation. The other major reason is the similarity in the internal energies deposited in the excited species by the two methods (cf. Section 2.2.5).

Collisional activation represents an unusual reaction in that the reactants can both initially be electronically relaxed. *Translational* relaxation accompanies electronic excitation. Commonly the mechanism involves direct electronic excitation so that the target can be thought of as a collection of electrons, available for inelastic collisions with the electrons of the ion. The relative translational energy of the projectiles is both the source of the energy of reaction and a significant determinant of its cross section (Section 2.3). However, the relative translational velocity of ion and target is not the immediate source of the relative kinetic energies of the electrons involved in the electron–electron collision. This follows from the fact that the kinetic energy of an electron associated with a 10-keV ion of mass 100 is only 0.05 eV. Thus, it is the orbital energy of the electrons of ion and target which allows electron–electron collisions to occur with high (>10 eV) relative kinetic energies.

The mechanism of collisional activation can be thought to proceed via a transient complex in which electronic reorganization occurs. This electronic reorganization can lead, as the species separate, to regeneration of reactants or on to products (Fig. 4). In the course of these changes energy conservation demands that appropriate adjustments in translational energy occur. This energy change is expressed almost entirely in the kinetic energy of the fast species (Section 2.1).

2.2.2. Ions Subject to Collision-Induced Dissociation

Since collision-induced dissociation is a consequence of two distinct excitation steps, a primary step, often electron impact, and a secondary atomic- or molecular-impact event, the nature of the ion population which is subject to atomic or molecular collision is important. Considering only a mass-analyzed ion beam and ignoring the presence of small quantities of fast neutrals which may have been formed by charge exchange, the ions present in the collision chamber are of two types: (a) stable ions which, in the absence of collision gas, will reach the detector and be recorded, and (b) metastable ions, the internal energies and rates of fragmentation of which will, in the absence of collision gas, cause them to fragment in the field-free region or in later parts of the instrument. For typical instruments in which ions traverse the field-free region in times on the order of microseconds, commencing some few microseconds after formation, the total metastable ion abundance due to reaction of a given ion is typically 10^{-2} to 10^{-4} of the signal due to stable ions. Even allowing for the possibility that excitation of a metastable ion (as broadly defined) is more likely to result in fragmentation in the collision chamber than is excitation of a stable ion (excitation by the same amount may leave the ion energy below the critical threshold for reaction in the latter case), it is probable that essentially all collision-induced dissociation is a result of excitation of ions which are stable to unimolecular

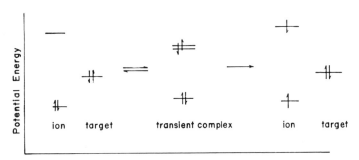

Fig. 4. Schematic illustration of electronic changes accompanying collisional activation.

fragmentation. Exceptions could be found if the primary ionization were affected by transfer of a discrete energy, as in charge exchange, but they are not expected with broadband energy-transfer techniques such as electron and photon ionization. Since stable ions have energies below the critical (activation) energy for the lowest-energy reaction, the uncertainty in the internal energy of the ions subject to collisional excitation is approximately half the critical energy for the unimolecular reaction of lowest critical energy. A typical value of this uncertainty might be ±0.7 eV although in some ions, particularly stable aromatic ions, it may be greater. Naturally, this uncertainty is further reduced when independent information is available regarding the energy states of the stable beam, say by photoelectron spectroscopy, photoelectron–photoion coincidence studies, charge stripping, or otherwise.

These arguments do not imply that all stable ions will be uniformly sampled in CID; on the contrary, a case can be made for the view that the lowest-energy ions will have the smallest dissociation cross sections. This view and its consequences for ion structure determination are discussed in Section 4.1.

2.2.3. Behavior of the Excited Ion

We have so far not considered in detail the events which follow collisional activation. Aside from instrumental factors, such as time spent in the collision chamber, the internal energy acquired in the collision will govern subsequent events. This parameter is rather inaccessible (Section 2.3.5) but it is known that energies in the range 1–10 eV are represented. This energy range is similar to that produced by 70-eV electron impact and, as noted previously, the time scale over which an integrated view of events is obtained is also similar. It follows that most fragmentation will occur soon after collisional activation, namely at the low end of the 10^{-13}–10^{-5}-sec time scale referred to earlier.

If the statistical theory is assumed and use is made of the body of fragmentation data built up through mass-spectrometric studies, it can be predicted that a network of competing and consecutive unimolecular reactions will follow collisional activation. There are likely to be more primary reaction paths for highly excited ions than for those of lower energy since competition between different channels is controlled largely by the frequency factor rather than the critical energy as the internal energy increases. Moreover, rate constant curves show less and less energy dependence in the high-energy range. It therefore seems that simple cleavage reactions will be more important in collision-induced dissociations than in metastable ion reactions, a fact which has been applied in several studies utilizing CID (Section 4). It is also evident that many of the product ions will retain sufficient internal energy to fragment further and, just as in conventional mass spectrometry, the sequence of steps leading to the observed fragment ion may be lost. Certain types of reactions can be ascribed to single or multiple fragmentations on the basis of the neutral eliminated, e.g., CH_3 is lost as an entity, $CH_5O^.$ as $CH_3^.$, and H_2O separately. Experiments in which a time lag is allowed between excitation and fragmentation would entirely resolve such questions although at the expense of loss in signal.

2.2.4. Effect on Target

The dynamics of the collision will govern the translational energy acquired by the target; this is typically very small under the conditions of interest here.

Electronic excitation, including ionization, occurs quite generally in the course of ion–molecule collisions and processes in which the target is excited are inherently as probable as is excitation of the reactant ion. The widespread use of a rare gas, particularly helium, as collision target has the fortunate consequence that target excitation reactions are expected to be minimized due to their high energy requirements.

For other targets, particularly those with low-lying excited states, simultaneous excitation or ionization of the target may well occur during CID although this will increase the energy loss associated with the reaction. Under the conditions of interest in this chapter any such increase in the endothermicity of the reaction will decrease its cross section (Section 2.3). The situation regarding target excitation accompanying collisional activation can be contrasted with that pertaining to charge-changing collisions where target excitation *may* reduce the exothermicity of the reaction and so increase the cross section. The empirical observation (Section 2.4.1) that quite different targets often function with similar effectiveness in CID supports the view that the target has only secondary effects on collision-induced dissociation spectra. Nevertheless, target excitation is known[40] for collision-induced dissociation of H_2^+ and the extent of its occurrence in the reactions of polyatomic ions has yet to be ascertained.

2.2.5. Kinetics of Fragmentation

The argument of Section 2.2.1 for the separation of the excitation and fragmentation steps in collision-induced dissociation suggest that the dissociation should be describable in terms of standard unimolecular reaction theory. This normally implies that complete internal energy randomization occurs in the energized species, the method of energization retaining significance only insofar as it produces the characteristic internal energy distribution of the ions. Methods providing quantitative evidence for this description of CID reactions are still being developed, the main difficulty being that the internal energy distribution after collision is difficult to characterize. Nevertheless, some progress in this area has been made, and it is discussed in this section.

The hypothesis that most CID reactions are describable in terms of statistical unimolecular reaction theory finds qualitative support in comparisons of mass spectra associated with electron-impact and collision-induced dissociations. In one of the earliest of such studies propane was examined and its mass-spectra reactions and collision-induced dissociations were shown to be similar.[41] In a succeeding study[29] the cross sections for the CID reactions of the propane molecular ion and many of its fragment ions were measured. For example, that for H˙ loss from $C_3H_8^{+\cdot}$ was found to be 3.5×10^{-16} cm^2. The fragmentations undergone by particular ions upon collision were found to be predictable in terms of standard rules for interpreting mass spectra. Thus, odd-electron ions were found to dissociate more than even-electron ions and to do so preferentially by loss of radicals, e.g., H, whereas even-electron ions tended to lose neutral molecules, particularly H_2. Furthermore, the dominant fragmentations were those of lowest activation energy. Thus, qualitative agreement with quasi-equilibrium theory (QET) was observed, but the lack of information on internal energies limited quantitative comparisons. The observed behavior did establish, however, that the ion energies after collision were considerably greater than those of metastable ions and comparable to those formed by 70-eV electron impact.

Qualitatively the other available results also suggest that the energized ion behaves as it would if it had been similarly energized by some other agent. The large number and variety of fragment ions which typically result from the dissociation of complex polyatomic ions with energies above the threshold for fragmentation make this type of comparison particularly valuable. Jennings[30] compared the fragmentation pattern of the benzene molecular ion as generated from the molecule by 70 eV electron impact with that resulting when long-lived molecular ions were energized by collision. Both in this case and in those of other aromatic molecular ions examined, the spectra were qualitatively very similar.

Using a rather different approach for comparing energy deposition by the two techniques, Futrell and co-workers[42] compared the isotope effects for H˙ versus D loss from partially deuterated methanes and concluded that the energy of ions excited by collision was similar to that of ions fragmenting directly after electron impact. These authors also noted a marked propensity for multistep fragmentation sequences on CID.

Comparisons of the dissociation spectrum for a selected ion with the mass spectrum of the corresponding neutral molecule have also been made by others. On the basis of a number of such studies McLafferty and co-workers have suggested[43] that the average energy of the ion after collision is somewhat less than that deposited by electron impact (50 eV),

374   Chapter 7

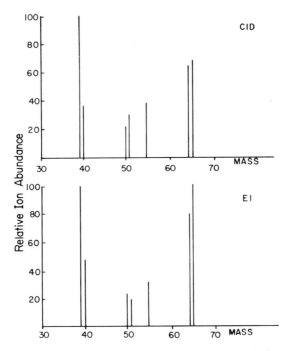

Fig. 5. Comparison of major fragment ions in the electron impact (EI) mass spectrum of phenol and the collision-induced dissociation (CID) spectrum of the phenol molecular ion. (Data taken from Ref. 43.)

the lower activation energy reactions showing relatively higher abundances in CID than in electron impact. The spectrum of phenol, taken in these two ways, is shown in Fig. 5 to illustrate the similarity between electron impact and ion dissociation spectra.

Given that there may be little resemblance in the exact form of the internal energy distributions associated with each method of energization, the conclusion from the foregoing qualitative studies is that ions of similar average internal energy are involved. There is also no indication that fragmentation following collisional excitation is describable in terms other than that of statistical unimolecular reaction theory.

An alternative approach to characterizing the kinetics of fragmentation by CID involves measuring the kinetic energy loss so as to determine the internal energy of the fragmenting ion.[44] These data have been used in two ways. First, for ions with known breakdown curves (variation in product ion distribution with internal energy), it has been possible to show that the energy losses measured for particular ions correlate with their internal energies given in the breakdown curves (Fig. 6). This confirms that ions of the same internal energy formed by molecular impact and by electron

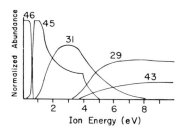
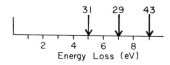

Fig. 6. Correlation observed between the breakdown curve of n-propanol and the kinetic energy losses associated with CID of the molecular ion leading to fragment ions 31^+, 29^+, and 43^+.

impact undergo the same reactions, and it is therefore probable that collisional activation results in fragmentation from the same states as are involved in ion-source reactions. The second test of the kinetics of CID reactions compares the kinetic energy release due to collisionally activated ions with that due to metastable ions undergoing the same reaction. The difference between these energies represents the kinetic energy partitioned from the nonfixed energy $\varepsilon^{\neq}$ of the activated complex, any contributions from the reverse activation energy (ε'_0) canceling. The experimental result is then directly comparable to statistical calculations regarding partitioning of the nonfixed energy of the activated complex. Figure 7 illustrates the composite origin of the kinetic energy release, breaking it down into a contribution (T^e) from the reverse activation energy and one $(T^{\neq})$ from the nonfixed energy, $\varepsilon^{\neq}$.

Consider, for example, the metastable ion reaction (2.5) in methanol, which is accompanied by an average kinetic energy release of ~ 10 meV:

$$CH_3OH^{+\cdot} \rightarrow CH_2OH^+ + H^{\cdot} \qquad (2.5)$$

The corresponding collision-induced dissociation was found to release 400 meV and to be accompanied by an energy loss of 1.7 eV.[44] From the known thermochemistry of this process a nonfixed energy of 1.2 eV was calculated for the collisionally activated ion. The kinetic energy release of almost 0.4 eV associated with this nonfixed energy is consistent with the predictions of statistical partitioning. Thus, measurements on kinetic energy loss and energy release allow the study of energy partitioning in highly excited ions.

It is expected that under appropriate circumstances instances of non-statistical fragmentation following collisional activation might be encountered. This follows both from the large energy transfers which may occur on collision (up to 10 eV) and from the fact that such processes have been

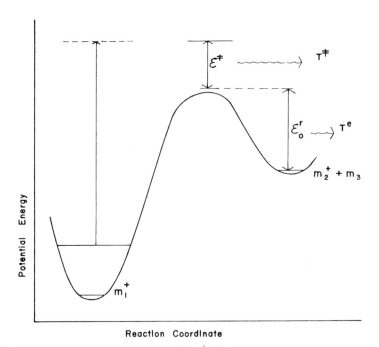

Fig. 7. Origin of the kinetic energy release ($T = T^+ + T^e$) accompanying dissociation after a collision. The ion m_1^+ is excited from the vibrational state shown and acquires an internal energy in excess of the dissociation threshold by an amount ε^+. The excitation energy acquired by the ion plus that acquired by the target (if any) is approximately equal to the kinetic energy loss.

observed for other methods of excitation. Two possibilities may be distinguished: First, fragmentation may proceed more rapidly than theory would predict for an ion of a given energy. Data on neutral molecules, particularly chemical activation experiments, have shown that incomplete energy randomization prior to fragmentation can occur in very highly excited species. The general expectation seems to be that rate constants of $\geq 10^{11}$ sec^{-1} are required and these may be accessible on collisional activation. The second possibility is that fragmentation may occur from electronic states which are not in equilibrium with the ground state of the ion. In these cases the rate of fragmentation is nevertheless describable in terms of the statistical theory, provided the energy of the activated ion is referenced to the isolated state not the ground state.

Although the intervention of isolated electronic states has been invoked not infrequently as a possible means of reconciling otherwise contradictory data, cases in which there is good evidence for such states are rare.[46,47] A primary indication of the intervention of isolated states is slow fragmentation from a highly energized ion. Such an observation requires (a) that the internal energy of the ion can be characterized and (b) that the rate of fragmentation required by quasi-equilibrium theory can be predicted with sufficient accuracy. The fulfillment of the latter condition has been greatly facilitated by advances in RRKM methods[48] and by the experience which has been gained in assigning reasonable activated complex structures for polyatomic ions. Experimental tests of these calculations and comparisons with results on energized neutral species have increased the reliability of these calculations. Angular momentum effects remain a source of difficulty, a point of note since the rotational energy of an ion activated by collision cannot be assumed to be negligible as may be done for electron impact. Methods of sufficient accuracy for characterizing the internal energy of ions which may react from isolated electronic states are available— notably photoelectron–photoion coincidence and charge-exchange ionization. They are currently providing the best evidence for isolated states in the reactions of polyatomic ions.

The three experimental measurements which are possible for CID reactions (see Fig. 3): (a) an ion-abundance or cross-section measurement, (b) the kinetic energy loss associated with the excitation step, and (c) the kinetic energy release associated with the fragmentation step, suggest several methods of recognizing reactions occurring from isolated electronic states:

1. If the breakdown curve for the reaction in question has been measured or calculated, then the fractional ion abundance carried by a particular product is known as a function of internal energy. The observation of a product ion at an internal energy for which the breakdown curve predicts zero abundance would indicate the intervention of isolated states or prior emission of radiation.
2. The relative cross sections for two or more reactions can be measured and compared with the breakdown-curve data. This supposes that multiple reactions with similar energy losses can be observed.
3. If the kinetic energy release measured from the CID peak width is smaller than the value calculated assuming statistical theory for an ion energy given by the measured energy loss, then an isolated electronic state is indicated. It should be noted for this last case, that the effects of the critical energy for the reverse reaction and of the distribution of energy losses will be to broaden the peak, so that the method will not be compromised by ignoring these factors.

The same approaches can be suggested to probe fragmentations associated with incomplete energy randomization. Thus, unexpected modes of fragmentation or inappropriate relative ion abundances might occur or the kinetic energy release associated with a particular reaction might be observed to be much greater than theory would allow, allowance being made for the factors noted in item 3.

2.2.6. Vibrational Excitation and Other Mechanisms

We have so far considered CID in terms of an electronic excitation mechanism, a process in which momentum transfer to the target is minimal, the fast species suffering virtually no change in momentum perpendicular to its direction of flight. There is evidence, from the dissociation of diatomic species, for the operation of a second type of mechanism in which one atom or group of atoms in the ion suffers an elastic collision with the target with the result that the overall reaction is inelastic, the ion being internally excited as well as appreciably deflected. For a polyatomic ion the overall result, vibrational excitation, is the same as that achieved in excitation by the electronic interaction mechanism where rapid radiationless transitions convert electronic into vibrational energy. Differences would be expected in the energy deposition function and in the degree of rotational excitation of the ion.

As yet there has been only tenuous evidence for the operation of this direct vibrational excitation mechanism in polyatomic ions because most work has employed low-energy resolution and scattering angles have not been measured. Moreover, the instruments capable of high-energy resolution which might detect the energy losses associated with momentum transfer to the target in these processes are typically operated at about 10 keV where this adiabatic reaction would have a low cross section. Durup[39] has suggested that the process might be important in the 100–1000 eV energy range with a maximum cross section at an energy a few times the dissociation energy, and that mixed reactions which have the characteristics of both electronic and vibrational excitation may also occur. These processes can be described as nonvertical electronic excitations, the non-Franck–Condon transition being the result of momentum transfer to one of the nuclei. However, for most reactions the available data, including energy-loss measurements, are still satisfactorily accounted for in terms of the simple electronic excitation mechanism. Possible exceptions have been noted where an alternative to electronic excitation is inferred, for example, from a decrease in cross section with kinetic energy.[27b] One of these cases is the reaction

$$CH_4^{+\cdot} \xrightarrow{air} CH_3^+ + H^{\cdot} \qquad (2.6)$$

It is significant that in this case the lowest energy (metastable ion) fragmentation has been shown to involve rotational predissociation.[49–51] It is therefore noteworthy that the vibrational excitation mechanism is expected to be much more efficient than the electronic mechanism in increasing the rotational energy of the ion.

Excitation on collision leading to an increase in the rotational energy of the ion and so producing fragmentation by a predissociation mechanism is to be contrasted with collision-induced predissociation where the role of the target is to effect a perturbation and hence to promote curve crossing to a dissociative state. Predissociation of this type, accomplished essentially without excitation, has been suggested[52] to be one of the mechanisms responsible for the collision-induced dissociation of acetylene:

$$C_2H_2^{+\cdot} \rightarrow C_2H^+ + H^\cdot \qquad (2.7)$$

The absence of any measurable energy loss suggests that the reaction should be relatively more sensitive to collision than are analogous endothermic reactions. Thus a large impact parameter can be expected and the collision-induced reaction should be observable at very low pressures. A mass-spectrometric study of the metastable ion reaction corresponding to (2.7) gave results in agreement with these expectations (Section 4.3.1).

Another mechanism of collision-induced dissociation which should be noted since it has been reported in diatomic ions is a polarization-induced vibrational/rotational excitation.[53] An important characteristic of this mechanism is that the kinetic energy release is not isotropic in the center of mass.

In the remainder of this section some suggestions are made regarding the possible occurrence of an as yet uncharacterized mechanism of collisional activation. This is a multicollision mechanism whereby the ion increases its internal energy in a sequence of small increments. Double collisions have been reported[54,55] as leading to the dissociation of $H_2^{+\cdot}$ but it is in polyatomic species with their large state densities that such processes seem most likely. Since this mechanism would achieve in a number of steps the energy increment normally associated with a single collision, it is apparent that the individual collisions would be weaker than normal. Thus, a number of long-range collisions having large cross sections might be involved. A target pressure corresponding to single-collision conditions for the overall reaction or for the single-collision reaction might, however, correspond to a large number of collisions of larger cross section (Section 2.4.1). Another consequence of such a multicollision or staircase mechanism is that the fragmentation reaction(s) of lowest critical energy might be relatively favored over the single-collision case, the sequence of excitations ceasing as soon as a species with sufficient energy to fragment is formed. The small energy transfer associated with each step would, of course, increase the probability that the lowest-energy reaction would predominate. The observation that small kinetic energy releases are associated with the double-collision process in $H_2^{+\cdot}$ is consistent with this. Because of the large impact parameters associated with the individual collisions in the proposed mechanism it is also expected that the reaction would have a tendency to be adiabatic rather than involve vertical Franck–Condon transitions.

The evidence for multicollision reactions as a method of collision-induced dissociation of polyatomic ions is at present very limited, although some of the effects of pressure on CID cross sections (Section 2.4.1) are suggestive of this possibility.

2.3. Cross Section

This section covers some general points regarding the cross sections for CID reactions; the next section deals at greater length with the effects of various experimental parameters on cross sections. The wide-ranging utility of the CID technique is dependent on the fact that cross sections are typically large. Values of 10^{-15} cm^2 have been reported, and it has been noted that the cross section can be an order of magnitude greater than that for charge stripping,[29] which is one of the more favorable of the high-energy ion–molecule reactions.

An immediate point to be made is that the cross section for a collision-induced dissociation is not that for a single elementary reaction, rather it is the resultant of the rates of an ion–molecule reaction and a unimolecular dissociation. This considerably complicates the interpretation of cross-section measurements. For certain purposes, such as ion-structure characterization, the measured overall cross sections serve well but in other applications they may even be misleading if their composite origin is not borne in mind. Thus,

$$\sigma_{CID} = \sigma_{exc} \cdot \sigma_{frag} \tag{2.8}$$

but σ_{frag} is dependent on the internal energy of the ion after excitation (controlled by the state of the ion prior to collision and the nature of the collision) and on the rates of any competitive fragmentations. This is illustrated in Fig. 8 for the reaction

$$m_1^+ \rightarrow m_2^+ + m_3$$

where this is the lowest-energy fragmentation of m_1^+ ions. Competitive reactions are assumed to commence at some energy above the onset for m_2^+ formation and to dominate the fragmentation pattern soon after their onset. The cross section for the collisional activation step is assumed to decrease with increasing energy defect which should be a good approximation under the conditions of interest (see the discussion of the resonance criterion later in this section). The consequences of the different and quite independent effects of the energy loss upon the rates of excitation and fragmentation lead to the overall behavior shown. Thus, the largest cross section for excitation will not necessarily result in any product being formed while the same may be true of the excitation reaction of smallest cross section. The relative cross sections for excitation and fragmentation for cases in which the product m_2^+

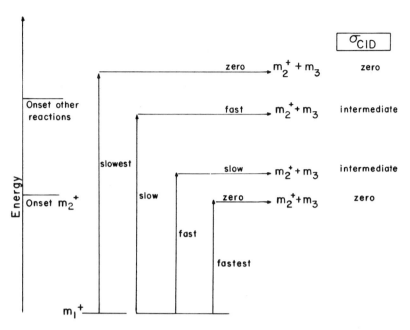

Fig. 8. Illustration of the dependence of the cross section σ_{CID} for collision-induced dissociation on (i) the probability of excitation of the ion by an appropriate amount and (ii) the rate of unimolecular fragmentation of the excited ion.

is formed will, *even when no other fragmentation is occurring competitively*, tend to oppose each other. Thus the greater the cross section for excitation, the lower is the energy of the collisionally excited ion and the smaller its rate of fragmentation.

Clearly, if cross sections for excitation are of interest additional data besides σ_{CID} will be required. One possible approach would be to characterize the internal energy of the excited ion since σ_{frag} could then be determined by standard mass-spectrometric techniques. This could involve the measurement of the energy loss accompanying excitation although sufficiently accurate measurements are not yet being made. In the usual situation where the reactant ion has a wide and unknown distribution of internal energies, the measurement of meaningful cross sections becomes even more difficult.

Cross sections which are averaged over all the states of the reactant ion can be measured by standard methods. Thus if it is assumed that the fragment ions are collected with the same efficiency as the reactant ions, the cross section for a particular collision-induced dissociation (leading to a given product ion) can readily be estimated under single-collision conditions. The cross section (σ, in square centimeters per target atom or molecule) is given by

$$I = I_0 e^{-\sigma l N} \quad (2.9)$$

where I and I_0 are the final and initial beam intensities, l is the path length in centimeters and N is the target density in atoms or molecules per cubic centimeter. Under conditions of low total conversion of I_0, Eq. (2.9) can be written

$$\frac{I}{I_0} \approx 1 - \sigma l N \quad \text{or} \quad \frac{I_0 - I}{I_0} \approx \sigma l N \quad (2.10)$$

Several authors have thus estimated the CID cross sections for typical reactions to lie in the range 10^{-16}–10^{-15} cm^2.[20,21,29a]

Consider, as an example, the reaction

$$C_6H_6^{+\cdot} + He \rightarrow C_4H_3^+ + He + (C_2H_3^\cdot) \quad (2.11)$$

studied in the first field-free region of the RMH-2 mass spectrometer (see Section 3 for typical experimental conditions and a brief description of the instrument). At an estimated pressure of 3.5×10^{-5} torr the CID peak has an intensity of 43 units while the main beam of $C_6H_6^{+\cdot}$ ions is 3000 units. This single data point yields a cross section of 1.95×10^{-16} cm^2 using Eq. (2.10) and the value 2.8×10^{-17} to convert from gas density in atoms or molecules per cubic centimeter to pressure in torrs.

Measurements made in this way may yield low values. Because of the two-step nature of the reaction the direction of motion of the product will be controlled by both the scattering angle upon collision and the angular properties of the fragmentation step. It is not apparent that the collection efficiency for the fragment ion will necessarily resemble that for the reactant ion. Indeed, a low collection efficiency would be predicted from two further considerations unique to collision-induced dissociation. First, although the collision can be confined to the collision chamber, the subsequent fragmentation can occur at any subsequent point. In most experiments collection efficiency from such points is low even when these are in the same field-free region as the collision chamber since the latter is typically located at a focal point of the subsequent mass or energy analyzer. Second, kinetic energy is released upon fragmentation and the discrimination which results for ions in the kiloelectron volt range is well known through its effects on the shapes of metastable peaks. Most equipment has not been designed with these factors in mind.

In spite of the caveats raised in this section, relative cross sections for the overall collision-induced dissociation sequence show dependences upon reaction channel and experimental variables which add considerably to our understanding of the reaction and allow its application in the many ways detailed in later sections.

Before terminating this general discussion on cross sections, some comments on the effects of the internal energy defect are important. The energy defect for a particular transition is only one factor controlling the cross section. Franck–Condon overlaps, symmetry, and spin selection rules are all important. However, the high density of states in the species and energy ranges of interest here mean that these latter effects may tend to average out. The difficulties in describing excitation cross sections for even one-electron systems and the paucity of experimental data on polyatomics mean that we have to limit ourselves to making some tentative generalizations.

Collision-induced dissociation of polyatomic ions typically involves a minimum energy loss which ranges from several tenths of an electron volt to several electron volts, depending on the internal energies of the reactant ions and on the critical energy of the particular process. The experimental results of Kupriyanov and of McLafferty and their co-workers for collision in the low kilovolt-energy region show that the probability that the ion will be excited by 5 eV or more is lower than that for smaller excitations. The available measurements of the energy loss for CID of polyatomic ions[44] confirm this conclusion. Two principles can be considered in accounting for these results. First, the behavior of the cross section as a function of the internal energy defect Q may be described in terms of the criterion of nearest approach to resonance which applies to charge transfer reactions under adiabatic conditions.[56,57] According to this model the cross sections fall off as the absolute magnitude of the energy defect Q increases, as given by

$$\sigma = ce^{-|Q|/k} \qquad (2.12)$$

where c and k are constants.

The second general principle which can be expected qualitatively to relate cross sections to ion energy or velocity is the adiabatic criterion of Massey.[58,59] This criterion considers three ranges of ion velocity: (a) At low velocity the electrons can adjust adiabatically to the perturbation resulting from the interaction between ion and target, making a transition unlikely; (b) at high velocity the electronic transition time is long compared to the collision time, which again makes a transition unlikely; (c) at intermediate velocities the collision and electronic transition times are comparable, which maximizes the transition probability. The maximum cross section is assumed to result when these times are equal, i.e.,

$$\frac{h}{|Q|} = \frac{a}{v} \qquad (2.13)$$

or

$$v = \frac{a|Q|}{h} \qquad (2.14)$$

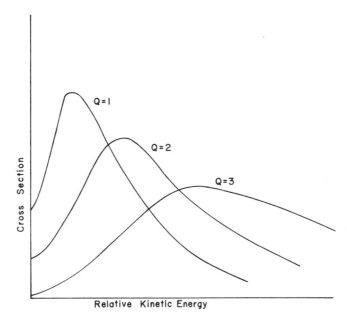

Fig. 9. Illustration of the effects of the resonance and adiabatic criteria on the cross sections for ion–molecule reactions having different endothermicities. σ is a function of the velocity of approach of ion and target.

where a, the adiabatic parameter, is the distance over which reaction occurs and is of the order of 7 Å for a large number of reactions involving transfer of a single charge. Employing this value, the cross section maximum occurs for 8-kV ions at

$$Q_{max} = 7.3 m^{-1/2} \qquad (2.15)$$

where m is in atomic mass units.

For energies of 10 keV and below and for ion masses of 100 amu, the Massey criterion leads to the prediction that cross sections should be at a maximum for energy losses of ~1 eV or less. Thus, both criteria lead to the conclusion that small energy transfers will be favored. It should also be emphasized that the criteria are not mutually exclusive. At an ion energy which maximizes the cross section for a reaction with a particular energy loss, other processes which require smaller energy losses may have still higher cross sections, even though they have already passed through their maxima (Fig. 9).

Perhaps it needs to be emphasized that these descriptions apply to reactions which occur following a direct electronic excitation. It has already been noted that other mechanisms contribute to CID; those which are most likely to contradict the generalizations made here are those associated with curve crossing. In particular, perturbation by the target or vibrational excitation may lead to pseudocrossing to a product state reached with only a small excitation energy. Hence reactions with low Q would have higher cross sections than expected. Such pseudo-curve-crossing processes are thought to be more common for excitation than for charge transfer reactions. Offsetting this tendency in favor of smaller energy transfers in some reactions is the fact that the density of excited states increases with increasing energy transfer. Qualitatively, this would tend to shift the most probable energy loss, under particular conditions, to greater values. The magnitude of this purely statistical effect is not known. A most probable energy loss of approximately 1 eV for 10-keV polyatomic ions is reasonable in terms of the foregoing considerations and experimental results. At lower-accelerating voltages somewhat smaller energy losses are expected.

The foregoing qualitative description of the relationship between internal-energy defect and ion energy suggests, in the cases of interest in this chapter, that a strong bias exists in favor of collision-induced dissociation of ions which already possess internal energy. Another important consequence of the inverse relationship between cross section and energy defect seen in Eq. (2.12) is that the stable ions sampled in a reaction of low critical energy will contain a higher proportion of those ions which initially had higher internal energies than will be the case for a high-critical-energy reaction. Hence, the sampling of the stable ion population is dependent on Q. Thus, in ion-structure studies where the structures of ground-state ions are required, free if possible from low-energy isomerizations, there is reason to prefer the use of a reaction, whether CID or other endothermic ion–molecule reaction, in which Q is large. For charge-transfer reactions it should be possible by proper choice of target to examine selectively different regions in the internal energy distribution of the reactant ion. The use of reactions which are slightly endothermic would be most suitable for sampling ground state ions. These points have been made use of in the applications of ion–molecule reactions to ion structure (Section 4).

2.4. Effects of Experimental Variables

In this section we consider the effects on cross section, reaction channel, and reaction mechanism of the chief experimental variables in the collision-induced dissociation experiment.

2.4.1. Pressure

If single-collision conditions are maintained, there should be a linear dependence of the intensity of a CID peak upon target pressure. However, the occurrence of spontaneous or surface-induced dissociations (Section

5.5) which gives rise to products with the same mass, requires that data be taken over a range of pressures in order to detect these processes. Relative abundances of ions due to processes with different endothermicities can be expected to vary with pressure, the less endoergic reactions being relatively favored at lower pressures by analogy with effects seen in other ion–molecule reactions.[60] This effect has been observed in the work of McLafferty and colleagues[43] working within the pressure range believed to correspond to single collisions. At higher pressures, where multiple collisions become likely, fragmentation sequences involving highly excited ions were observed to become more important, consistent with larger energy transfers.[61]

There are two further effects of target pressure which follow from its relationship to the relative probability for the occurrence of collisions involving different energy losses. First, transitions to repulsive excited states (where potential energy is strongly dependent on internuclear distance) occur with energy losses which vary with the target pressure. This effect can be quite marked, as attested to by results for the collision-induced dissociation

$$H_2O^{+\cdot} + N \rightarrow O^{+\cdot} + H_2 \quad (2.16)$$

the measured most probable energy loss increasing with pressure from 20 to ~50 eV.[62] Second, experiments done at higher target pressures, by favoring larger energy transfers, will tend to minimize the effects of different internal energy states in the reactant ion beam. By contrast, experiments at lower pressures will enhance the selectivity in favor of ions which have higher internal energies.

Another pressure-dependent phenomenon observed in polyatomic ions is the occurrence of upward breaks in the plots of CID ion abundances against pressure. This effect has been suggested to be due to preferential inelastic reactions and hence loss from the beam of higher-mass ions.[43] An alternative explanation of the upward breaks in the pressure plots is that they represent the onset of a new process leading to the fragment ion. A change in mechanism with change in pressure is not usually entertained but the excitation of polyatomic ions upon collision might be subject to such an effect. The new mechanism might be one in which excitation is effected by multiple collisions (Section 2.2.6), only the last of which raises the ion energy above the activation energy for fragmentation. In the simplest two-collision case, the process would be

$$m_1^+ + N \rightarrow m_1^{+*} + N \rightarrow m_1^{+**} + N$$
$$\qquad\quad \downarrow \qquad\qquad \downarrow$$
$$\qquad m_2^+ + m_3 \quad m_2^+ + m_3 \quad (2.17)$$

The excited ion formed in the first step, if formed by an electronic excitation, would be expected to undergo radiationless transitions to the ground state prior to the second collision which would, therefore, closely resemble the first except that the ion would possess more internal energy. The fact that the observed effects occur at high pressures and that the most probable degree of excitation is, under common mass spectrometric conditions, insufficient to effect fragmentation is consistent with the suggested multistep mechanism. Thus the majority of collisional excitations may lead to excited but stable ions as indicated in Eq. (2.17). The larger cross sections for each of the individual collisional excitations as compared to that for the corresponding one-step excitation could therefore, under appropriate pressure conditions, make the two-step sequence competitive.

In the ion–molecule reactions of polyatomic ions there are an almost infinite number of distinct interactions possible, leading, for example, to excited products whose energies differs only slightly. If the process of interest has a cross section which is much smaller than that for some other reaction, then multiple collision conditions with respect to the larger cross section process will mean that most ions which react by the path of interest will also undergo the process of larger cross section. For example, it is possible that some process such as symmetric charge transfer could have the highest cross section; if this were so, then above a certain pressure most fast species subject to collisional activation could be neutral molecules although their intermediacy might not even be suspected from the fragment ions examined. Another possibility is that the pressure may be such as to make some subsequent reaction of the fragment ion formed in CID mandatory (in the sense that its cross section is so large as to make the collision probability approach or exceed unity). Such a situation could, if a change in ion mass or charge were involved, lead to the nonobservation of products from the CID reaction even when this has a large cross section. Erroneous conclusions could certainly follow. Neutralization of the ion, which has a high cross section, could be a particular problem. Most commonly, the product might be excited or deflected, and although these processes are more subtle, they could invalidate a study on the detailed thermochemistry of the CID reaction.

These effects are particularly significant for collision-induced excitation reactions occurring in competition with each other or with charge transfer. This arises because it is not possible to discern from the product which is analyzed, whether or not is has been formed as a result of a single collision or whether extra collision–excitation steps have also occurred. *There is not a one-to-one relationship between the products and the sequences by which they are formed.* Reducing the pressure to the point where the interfering process of larger cross section is occurring under conditions of less than unit probability will solve the problem although it will reduce signal strength.

The foregoing considerations establish the importance of using low pressures in quantitative work. The large cross sections of some CID reactions means that particular care must be taken when studying other competing processes. It also means that the limiting pressures may not be particularly high. Pressure can conveniently be judged by the attenuation of the stable ion beam and in some laboratories, including this one, CID measurements have routinely been made under low-beam-attenuation conditions (e.g., at 95% of original strength). In others, higher pressures corresponding to attenuation of the reactant ion beam to as little as 10% of its original value have been the standard operating condition. Under high-beam-attenuation conditions the total CID yield comprises only a few percent of the decrease in ion current of the reactant beam.

If there are processes such as scattering or charge exchange with much larger cross sections than CID, then reactions run under approximately single-collision conditions for the dissociation are studied with these processes superimposed. If this combination does not, of necessity, lead to the nonobservation of the CID fragment, it will greatly influence the interpretation of the results. Until the high cross section reactions are characterized the magnitude of this difficulty cannot be assessed. However, it should be emphasized that these comments do not reflect upon the practical use of CID to characterize ion structure.

2.4.2. Ion Kinetic Energy

Some of the effects of ion kinetic energy upon cross sections for excitations involving different energy losses were discussed in Section 2.3. Experimentally it has usually been found that cross sections increase with increasing kinetic energy through the low-keV range to at least 20 keV and often beyond.[27b,52] Variations in the time during which fragmentation can occur and of fragmentation rate in response to the possible variations in energy deposited in the ion are also bound up in the kinetic energy effects. The results do show, however, that the cross section for electronic excitation increases with increasing energy, and the reason for this increase in cross section appears to lie in an increase in average energy transferred on collision.

This is illustrated by results such as those[63] on the ion $C_2H_n^+$ ($n = 2$–4), taken over a wide range of kinetic energies (4–40 keV), which show that increasing the ion kinetic energy continuously increases the average energy deposited. In particular, fragment ions which are formed by lower-activation-energy processes show a leveling off and then a decrease in yield at lower energies than do ions which require higher activation. Figure 10 shows some typical results. Ions which require the same energy loss will not necessarily be formed in similar yields, however, since ion yield is also

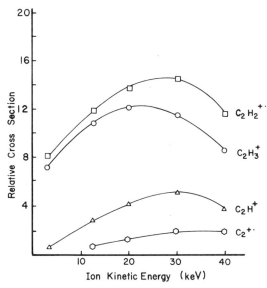

Fig. 10. Dependence of the cross section for collision-induced dissociation of $C_2H_4^{+\cdot}$ on the kinetic energy of the ion. Reactions leading to the different fragment ions indicated show different energy dependences, those with lower activation energies, such as formation of $C_2H_3^+$, passing through a maximum at lower kinetic energies.

controlled by the relative rates of competitive fragmentation from the excited ion and further fragmentation of the product ion. These data indicate that in the low-keV range many ions are probably excited insufficiently to cause them to fragment. However, the decrease in cross sections for some reactions at high energy (~30 keV) means that the most probable energy transfer is greater than the corresponding activation energy under these conditions. This is in agreement with the conclusions reached from the adiabatic criterion (Section 2.3).

The manner in which variation in the accelerating voltage affects competition between different fragmentation reactions of the energized ion, through its effect on the energy of the excited ion, can be further illustrated. Thus, for the 1-propanol molecular ion, it has been shown[43] that the high-activation-energy reaction leading to 31^+ increases in abundance relative to the lower-energy reactions leading to 42^+ and 59^+ as the accelerating voltage is increased. Similarly, the relative abundance of the product ion due to the loss of H$^{\cdot}$ from toluene has been found,[43] under particular conditions, to increase from ~0.1% at 1.2 keV, to 0.5% at 2.4 keV, to more than 1% at 3.6 keV.

Cases in which the cross section for collision-induced dissociation does not show the steady increase with kinetic energy in the low-to-middle-keV energy range illustrated in Fig. 10 deserve special study for the possible occurrence of mechanisms other than direct electronic excitation. Thus, changes in accelerating voltage can sometimes change the mechanism by which a particular fragment ion is formed as well as changing the relative rates of competitive fragmentation channels in response to changes in the internal energy of the energized ion. An instance in which this effect may operate is hydrogen atom loss from methane. At 7 keV, using nitrogen as target, the most probable kinetic energy release was observed to be ~0.3 eV.[64] An experiment[18] at 41.5 keV indicated both a central component in the energy distribution associated with a small kinetic energy release, and a new process for which the energy release was 3.0 eV. Apparently a new state becomes accessible with the larger energy transfers possible at high energy, and from this a reaction releasing a large kinetic energy occurs (cf. Section 5.2 where it is suggested that a state of the doubly charged ion may fragment). Cross-section data for a specified reaction as a function of ion kinetic energy can sometimes be expected to show complications due to the operation of different mechanisms in this way.

2.4.3. Nature of Target Species

Much of the work on collision-induced dissociation of polyatomic species has been done using as the target a rare gas (especially helium), nitrogen, or even air. In other experiments the compound used to generate the ions has been employed as the target. Where comparisons between targets have been made they have most frequently involved relative cross-section measurements and, less frequently, comparisons of peak profiles under conditions of high-energy resolution. Simple considerations suggest that the total cross section for all excitation reactions induced by xenon or a polyatomic target might be greater than that for helium merely because of the differences in electron cloud sizes. However, the energy transferred in helium collisions might be greater because the electrons involved are of higher energy.

In agreement with the latter conclusion, the effects of the mass of the target on the cross sections for collision-induced dissociation have been reported to show that lower-mass targets favor processes which require larger energy transfers.[39] For example, it has been shown that the dissociation ($-H^{\cdot}$) of the acetylene molecular ion from its ground state, which requires 5–6 eV, is accomplished much more efficiently by helium than by xenon.[52]

For a variety of polyatomic ions He and H_2 have been found to give CID products with about twice the relative abundance obtained with argon. These measurements[43] were made at pressures which maximized fragment-ion intensities, i.e., under conditions of high conversion of reactant ions to products (of all types). Data on the effects of target upon relative cross sections for processes requiring different degrees of excitation are apparently not yet available for polyatomic ions.

Another target effect is on the propensity for excitation by the electronic versus the vibrational mechanisms. For diatomic ions, it has been shown that large targets favor the vibrational excitation mechanism in the appropriate kinetic energy range. No data of this type are available for polyatomic ions.

In the discussion so far we have assumed that the target is not excited. For the rare-gas targets this should be a good assumption, the lowest excited state of the target requiring considerably more energy than the activation energy for the lower-energy fragmentation reactions. However, since excitation of the target in an ion–molecule reaction in the keV energy range is a well-known reaction (10/10*), it can be expected to occur in conjunction with excitation of the ion in some collisions. Direct evidence for this in polyatomic ions is lacking. The occurrence of this process, if its extent were unknown, would limit the use of energy-loss measurements in characterizing the excitation energy of the ion. At present, therefore, the only safe course of action in such determinations is to use a number of targets.[44] Differences in target mass may, of course, change the cross section for the reaction but this would not affect the energy-loss measurement. Targets such as helium, where the first excited state (20 eV) lies higher in energy than the energy loss in most CID reactions, can safely be used to measure energy losses. For other targets, the *minimum* energy loss will frequently correspond to reaction without excitation of the target. It should also be noted that target excitation, by increasing the endothermicity of the reaction, can be expected to lower the cross section. In simple systems, selection rules may obviate this conclusion since target excitation might allow or facilitate an alternative reaction mechanism such as an otherwise inaccessible predissociation. This occurrence might be inferred from the quite distinct characteristics of the product-ion energy distribution. Figure 11 shows the operation of target effects in dissociation of a diatomic ion $D_2^{+\cdot}$, where nitrogen and benzene have been used as collision gases. The peak shapes are strikingly different. At similar pressures the nitrogen signal is considerably (about two times) more intense. The nitrogen signal consists of a composite peak due to a vibrational predissociation (narrow peak) and excitation to a repulsive state (wide peak). With benzene the energy release in the latter process is greatly decreased due, possibly,[65] to formation of the target in an excited electronic state. (Compare Chapter 6, Section 4.1.)

In the few studies[44,52] in which thermochemical (energy loss) measurements have been made for polyatomic ions the results are accounted for without invoking signignificant target excitation.

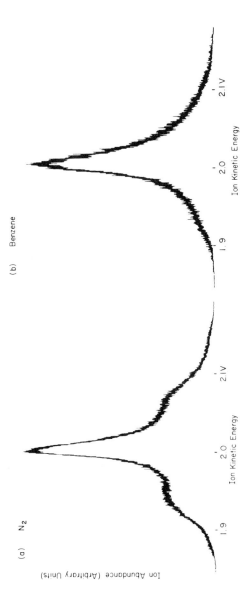

Fig. 11. Pronounced effect of the nature of the target on the peak shape for collision-induced dissociation of $D_2^{+\cdot}$. (a) The energy spectrum recorded by scanning the ion accelerating voltage V using nitrogen as target (Hitachi RMH-2 mass spectrometer, short collision region). (b) The same peak scanned at twice the sensitivity with benzene as target. The abscissa is calibrated in terms of the fractional ion-accelerating voltage relative to that required to transmit the stable ion beam.

2.4.4. Reactant Ion Internal Energy

The internal energy of the reactant ion will affect the minimum energy transfer required for fragmentation, the degree of excitation of the ion after collision, and the relative rates of fragmentation by the various channels open to the excited ion. In addition to the simple additive effect of the internal energy already possessed by the ions, Franck–Condon factors will be different for different vibrational states, and isomerization between different structures becomes increasingly likely in more complex systems. More subtle effects could also occur, since changes in the vibrational and rotational state of the reactant could alter the relative cross sections for different mechanisms of excitation and so influence the relative abundances of fragment ions.

A beginning has been made in the study of collision-induced dissociation of ions prepared with known internal energies by photoelectron–photoion coincidence techniques.[66] However, this technique is most readily applied to the study of low-energy reactions and all work with polyatomic ions in the kilovolt energy range has been done using methods, chiefly electron impact, which allow only course control and characterization of the excitation of the ion prior to collision. By increasing the electron energy used to form the molecular ion, nonfragmenting ions of increasing energy can be studied in CID experiments. As the electron energy is increased the minimum energy transfer needed to cause fragmentation will decrease and this approach can be used to obtain information on the energy-deposition function as the target, relative kinetic energy, pressure, and other experimental conditions are varied.

Experiments of this type support the conclusion of Section 2.3 that the cross section for CID is at a maximum for processes which involve near-zero energy defects (near-resonance and adiabatic maximum criteria). Careful work of this type can be used to reveal the presence of long-lived ions in isolated electronic states (e.g., above the dissociation limit) and cases of such behavior have been noted by several authors.[52,67,68] McGowan and Kerwin have used this approach to measure the cross sections for CID for the different states of the ion present in the beam. This method of characterizing long-lived isolated states is particularly attractive since the higher energy states are relatively easily distinguished since they have larger cross sections for excitation than the ground state.

McLafferty and co-workers[43] report changes in the abundance of the CID product ion relative to the reactant of as much as 15 times on changing the ionizing electron energy. These authors have also used other methods of changing the internal energy distribution of the ion prior to collision and shown the expected increase in the CID cross section with increasing internal energy of the reactant ion. In particular, fragment ions formed from homologous molecules have been subjected to CID. The internal energy of these ions has been shown by independent methods to be a function of the size of the molecular ion from which it is formed. If this has a large number of degrees of freedom, then the fragment ion will retain a relatively smaller fraction of the total energy. Since homologous compounds are employed the energy deposition function upon electron impact is not strongly dependent on the particular molecule. This procedure, therefore, allows the same fragment ion to be generated with a range of internal energies. It has been shown[68] that the logarithm of the yield of the CID fragment is approximately linearly related to the reciprocal of the number of degrees of freedom of the molecular ion. Figure 12 shows this behavior for the reaction

$$C_3H_6O^{+\cdot} \xrightarrow{Ar} C_2H_3O^+ + CH_3^{\cdot} \quad (2.18)$$

This is just the behavior expected if the cross section for collisional excitation is proportional to the internal energy of the ion prior to collision.

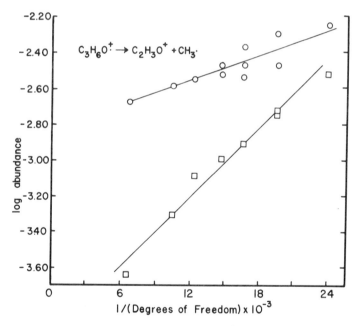

Fig. 12. Effect of the internal energy of the ion $C_3H_6O^{+\cdot}$ upon the abundance of $C_2H_3O^+$ formed in the field-free region, either spontaneously or upon collision. The $C_3H_6O^{+\cdot}$ internal energy increases with the reciprocal of the number of degrees of freedom in the ketone which yields $C_3H_6O^{+\cdot}$ as a fragment ion. The collision-induced dissociation (○) shows a much smaller energy dependence than does the unimolecular (metastable) reaction (□). The $C_2H_3O^+$ ion abundance is plotted on a logarithmic scale relative to the total $C_3H_6O^+$ ion current formed in the ion source.

It is suggested that it may be the tendency for cross sections for electronic excitations to increase as the energy defect decreases, rather than a contribution from the vibrational excitation mechanism, which accounts for the observed[43] increase in CID reaction yield when the electron energy approaches that corresponding to the appearance potential of the fragment ion (see also Section 2.4.1).

These examples emphasize that collisional excitation at high energy has a rate which is dependent on the internal energy of the reactants. The *relative* rates of different CID reactions (which in general require different energy transfers in the collision) can, however, be expected to be less dependent on internal energy differences in the reactant ion. Clearly the largest difference in CID yields will occur for products associated with the smallest activation energies. Evidence for this has been found. The dissociation spectra of particular ions as a function of internal energy show most pronounced differences in the abundances of the ions of lowest apperance potential. The otherwise lack of a strong dependence of CID dissociation spectra on internal energy is an advantage in ion structural studies (see Section 4.1). Nevertheless, the limitations of this relationship must be recognized, particularly the fact that if the ions being sampled contain a distribution of internal energies, those of higher energy which require the smaller energy transfers will be preferentially sampled. This preference will be greater for small energy loss reactions and smaller for reactions leading to ions of high activation energy. For the same reason, such difficulties will be less acute for charge-transfer reactions, such as charge stripping, which require energy losses on the order of 15 eV, than for collision-induced dissociations.

3. *Experimental Procedures*

There is a close similarity between the experimental techniques used to record the ions generated in collision-induced dissociations and in metastable ion fragmentations. As a result this section is brief and supplements and updates material given previously.[4]

3.1. *Instrumentation and Scanning Methods*

The study of collision-induced dissociation requires a facility for producing ions of a suitably high and well-defined translational energy, a collision chamber, and a detector. Useful data can be obtained using only a single mass or energy analyzer although relatively little work has been done this way because of the possibility of overlap of peaks due to different processes and of misassignment of the reaction. The study of CID in a single-focusing-sector mass spectrometer involves use of the field-free region, or some portion of it, as the collision chamber. Scanning the magnetic field then serves to identify the relative masses of reactant and product ions just as for the corresponding unimolecular (metastable ion) reactions. This identification is often unambiguous in simple systems although peak overlap with the much more intense signals due to stable ions may be a problem. A single scan records all reactions of all the ions introduced into the field-free region. In addition, the magnetic sector can be used for accurate momentum analysis and thus for kinetic-energy-loss and energy-release measurements.[19,52] Both the complete record of all reactions occurring and the energy loss information are also obtainable from an instrument employing a single electric sector for direct translational energy analysis. This has the advantage that the energy scale is more readily calibrated, hysteresis effects are minimized, the field strength is more readily controlled, and interference from signals due to stable ions is avoided. Figure 13 shows a kinetic energy spectrum taken on all the ions formed by bombardment of isobutyl ethyl ether with 70-eV electrons. The spectrum includes various charge-transfer reactions and metastable ion decompositions as well as collision-induced dissociations. Metastable peaks can be distinguished from those due to collision-induced dissociations by taking data at several pressures.

Although ion kinetic energy spectra taken as just described but in the absence of collision gas have found considerable use in characterizing isomeric molecules and in surveying the ion chemistry of various compounds,[4] little use has yet been made of the corresponding CID spectra. On the basis of the considerations developed in this chapter, they can be expected to be at least as useful in such applications.

The use of instruments employing two analyzers allows a single reaction to be isolated for study and to be uniquely characterized. Several arrangements have been employed. A dual magnetic sector instrument (double mass spectrometer) with an intermediate collision chamber has been used.[29a] This allows selection of the reactant ion by mass in the first stage of the instrument and mass analysis of the product ions in the second stage. This represents both the most direct and also one of the earliest methods of determining the fragmentation spectra of individual ions. In one recent version of a double mass spectrometer, the second stage of mass analysis is achieved using a quadrupole mass analyzer.[69]

If unit mass accuracy is all that is required, the use of a double mass spectrometer is to be recommended. On the other hand, for high-energy resolution data, including accurate determinations of energy losses, better results are obtainable if the second stage of the instrument is an energy rather than a mass analyzer. Several instruments employing energy analyzers following mass analyzers have been described.[34,35] Their major feature is that they allow all the reactions of a single ion to be followed in a single scan. The resulting ion kinetic energy spectrum is that associated with a single

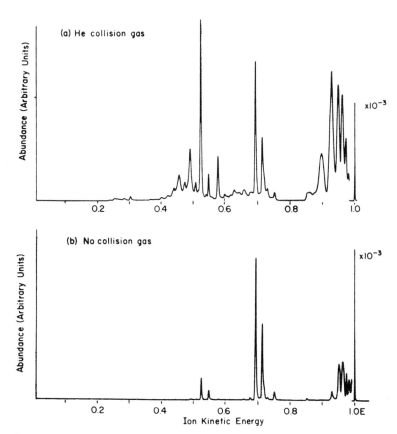

Fig. 13. Ion kinetic energy (IKE) spectrum of isobutyl ethyl ether (molecular weight 102) taken without mass analysis: (a) employing helium as collision gas at 2×10^{-5} torr and (b) without collision gas (2×10^{-7} torr). The main beam is attenuated by 35% on adding collision gas. With the exception of the peaks at $0.69E$ and $0.71E$ the observed peaks are chiefly due to collision-induced dissociation of ions represented in the mass spectrum of the ether. For example, $59^+ \to 31^+$ occurs at $0.525E$, $102^+ \to 73^+$ at $0.716E$, and $41^+ \to 39^+$ at $0.951E$, where E represents the kinetic energy of ions forming the main beam. It is a feature of IKE spectra that they record product ions due to all ion–molecule reactions which result in a change in the energy/charge ratio of the reactant ion.

mass and is commonly called a mass-analyzed ion kinetic energy (MIKE) spectrum. The chief difference between instruments of this type lies in their energy resolution and hence in the accuracy with which kinetic energy distributions can be characterized. If this is high, the useful thermochemical measurements of kinetic energy loss and kinetic energy release (Section 2.1) can be made from the CID peak position and width, respectively. The criteria necessary for such performance include a narrow energy distribution in the reactant ion beam and good angular resolution, as well as high-quality

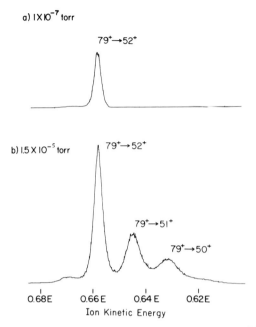

Fig. 14. A region of the spectrum of the molecular ion of pyridine taken (a) in the absence and (b) in the presence of collision gas on a reversed-sector mass spectrometer. The energy scale (abscissa) is given in terms of the energy E of the stable ion beam.

analyzers and adequate ion optics. An illustration of typical instrumental performance is given in Fig. 14 which shows a portion of the MIKE spectrum of the pyridine molecular ion taken with and without collision gas.

The ion source/mass analyzer/energy analyzer/detector geometry just described is the reverse of that usually associated with double-focusing mass spectrometers. The latter can, by the technique of scannning the ion-accelerating voltage,[70,71] provide equally good kinetic energy measurements. However, these are in a less convenient form, a single scan recording all transitions giving rise to a selected product ion. In addition, it is experimentally more convenient to scan and track a low voltage, as in electrostatic energy analysis, than to scan the accelerating voltage.

Instruments of conventional geometry have been used to simulate MIKE spectra. If the fragmentation spectrum of the molecular ion is desired, it is possible to obtain this from the peaks recorded in the conventional mass spectrum if the ionizing energy is kept low enough so as to ensure that fragment ions are not formed in the ion source. This method has been successfully employed.[30] It is also possible to follow the fragmentation spectrum of a selected ion on a conventional geometry double-focusing mass spectrometer by simultaneously scanning the accelerating voltage V and the electric sector voltage E such that E^2/V is a constant.[72,73] Under these conditions one transmits fragment ions formed in the preelectric sector region of the instrument. Simultaneous solution of the momentum-analyzing requirements of the mass analyzer is not achieved for fragments which carry excess kinetic energies. The result is that the procedure is limited to low-energy-resolution measurements and quantitative applications are affected by the discrimination against ions carrying excess kinetic energy. The technique should, nevertheless, be of use in many of the analytical applications of collision-induced dissociation.

Most studies on polyatomic ions have been done with instruments such as those already described which are appropriately characterized as mass spectrometers. They typically operate in the 2–10-keV range of ion energies. Tandem mass spectrometers, which include an additional stage of energy analysis, could also be used in studying high-energy ion–molecule reactions. Their chief application, however, has been to the study of ion–molecule reactions at low kinetic energy.[74] Even more specialized instruments operating at considerably highly accelerating voltage have also been used for a few studies on polyatomic ions.[37]

3.2. Energy Resolution and Kinetic Energy Measurement

The resolution achievable in an ion kinetic energy spectrum taken on either the conventional or the reversed geometry instrument is affected by (a) the finite energy distribution in the reactant ion beam, which is instrumentally determined and in good instruments on the order of 1 in 10^4; (b) the stability of the accelerating and analyzing fields which can be made better than 1 in 10^5; (c) the angular resolution of the apparatus; and (d) the inherent energy distribution of the product ions. This latter has two sources, of which one, the distribution of energy losses, is usually small compared to the other, the distribution of kinetic energies $f(T)$ released upon fragmentation.

The kinetic energy of the reactant ion, as provided by the ion accelerating voltage V, is an important instrumental variable in determining energy resolution. We consider in the following its effects on three different facets of energy resolution in ion kinetic energy spectra.

First, consider the resolution of adjacent peaks due to different CID reactions in a spectrum. For a fragmentation reaction, $m_1^+ \to m_2^+ + m_3$, the kinetic energy release T is given (cf. introductory chapter, Section 5) by

$$T = \left(\frac{m_2}{16 m_3}\right) Ve \left(\frac{\Delta E}{E}\right)^2 \quad (3.1)$$

where Ve is the ion energy before fragmentation and $\Delta E/E$ is the fractional width of the energy peak. Thus, the absolute separation of peaks due to different transitions is directly proportional to the ion accelerating voltage V, whereas the peak width due to the kinetic energy release is proportional to $V^{1/2}$ (and also to $T^{1/2}$). hence the best resolution of adjacent peaks in an ion kinetic energy spectrum (due to different transitions giving product ions with different average energies) is obtained at high accelerating voltage.

On the other hand, resolution of components of a given reaction which differ in their kinetic energy release is maximized by maximizing the fractional width ($\Delta E/E$) and thus by *decreasing* the accelerating voltage (cf. Chapter 6, Section 3). Expressing this in terms of the amplification factor arising from the conversion of center-of-mass to laboratory coordinates (see introductory chapter, Section 7) one notes that the best resolution of fine structure in an energy distribution due to fragmentation (including CID) will occur when the amplification factor is *minimized*. This follows because the amplification is inversely dependent on $T^{1/2}$ so the relative difference in center-of-mass energies associated with two discrete energy releases T_l and T_s is greater than that between the corresponding laboratory energies. In practice the advantage of low ion energies will be offset by the finite energy distribution of the stable ion beam and/or by considerations of sensitivity and angular resolution.

The third aspect of energy resolution in collision-induced dissociation involves resolution of processes which differ in kinetic energy loss. Since the energy loss is independent of ion-accelerating voltage, while the absolute peak width due to kinetic energy release increases with $V^{1/2}$, this measurement is best made at low ion energy.

The foregoing does not include any consideration of the energy spread of the reactant ion beam which can be considered to contain both accelerating voltage-dependent and -independent terms. We have also assumed that the component of peak width due to energy release predominates over that due to energy loss. The situation of very small energy releases and low ion energies is the only one in which this assumption fails. (Energy-loss distributions can be expected to have halfwidths comparable to their most probable values, which are ~3 eV for typical reactions—somewhat larger than the corresponding activation energies.) Energy loss measurements made from the position of the center of the peak are therefore better determined at low energy when the energy loss is a relatively larger fraction of the ion energy. However, the greater width of the peak at low energy due to the energy release makes the peak center difficult to locate so somewhat higher energies may be optimum for this determination.

Although kinetic energy measurements are most accurately made by the energy or momentum analysis methods outlined in Section 3.1, there are alternatives to these procedures For example, energy loss measurements have been made using retarding potentials.[19] This principle is incorporated in the Daly scintillation detector which has the additional advantage of high sensitivity. The Daly detector[76] has also been applied for energy analysis of metastable ion dissociations and it could be extended to CID studies.

3.3. Other Considerations

The generation of ion beams for CID work has largely employed standard mass spectrometric methods including such newer ionization techniques as field ionization and chemical ionization. Ion-abundance measurements on the product species have also employed standard analog recording methods in many cases. Since relative cross sections have been of primary interest little attention has been given to such questions as detector gain as a function of the nature of the product (pulse counting has not generally been used) or to accurate pressure measurement. In some studies, McLeod gauges or capacitance manometers attached to the collision chamber have been used to record pressure but more frequently ionization gauges have been used, the pressure at the measuring point being related to that in the chamber through the conductance of the system. Ion-gauge readings are corrected for the ion yield from the target gas in question.

There has been considerable variation in the length of the collision chamber and in the optimum pressure employed. For analytical applications of collision-induced dissociation pressures can be gauged conveniently and sufficiently accurately from the attentuation of the stable ion beam. Attentuation to 10% of the original value has been used frequently; other workers have attentuated to 30% or to only 95% of the initial beam current. The possible consequences of these different conditions have been noted (Section 2.4.1) in connection with the discussion of single-collision experiments. Although much of the work done on modified mass spectrometers has employed long collision regions, short collision chambers operated at higher pressure have also been used.[77] Comparisons between the performance of the two systems have been made, for example, using the Hitachi RMH-2 in which collision gas is admitted either into the entire first field-free region or into a short (0.6 cm) chamber. The short chamber gives up to an order of magnitude improvement in sensitivity for collision-induced dissociation over that obtained in the entire field-free region. The differences between the lengths of the two regions, even when as large as a factor of 100, is probably not significant in terms of the range of ion lifetimes sampled, as this will range from 10^{-14} sec on the one hand to either 10^{-5} or 10^{-6} sec on the other. What is more significant is the fact that the best results are to be expected if all fragmentations are confined close to the point of focus of the analyzer. If this is not done, energy resolution is degraded and cross-section measurements are liable to be in error since only a fraction of the collisionally excited ions will fragment at positions from which their products can be collected. Most reactions of collisionally activated polyatomic ions will occur with rate constants of $10^7 \sec^{-1}$ or more, so that fragmentation occurs very near the focal point. To ensure that for slower reactions only processes occurring in this region are sampled the collision chamber may be at a slightly different potential from ground.[78] Typically the ions experience a slight deceleration as they approach the focal point and they traverse a short region of constant potential before reacceleration. Fragmentations occurring in the region of small but constant field about the focal point then give rise to product ions of different kinetic energies than those due to fragmentations in a region which is entirely field-free. The resulting kinetic energy distributions refer to much better conditions of angular resolution than conventionally achieved.

A further undesirable consequence of long collision chambers is the backscattering and detection of ions which have kinetic energy releases of such magnitude and direction that they would otherwise be discriminated against. This leads to a loss of energy resolution.

Studies on collision-induced dissociation of negative ions have employed methods developed for positive ions. Bowie and Hart,[79] for example, have employed a modified Hitachi RMU-7 mass spectrometer, equipped with a collision gas introduction system and differential pumping between source and analyzer. Nitrogen and toluene are most frequently employed as target gases in this work at pressures up to 5×10^{-4} torr, results being obtained under conditions of limited main-beam attenuation. This work on negative ions has employed reactions occurring in either the second or more generally the first field-free region of the instrument. Ion accelerating voltage scans have been used or else complete ion kinetic energy spectra obtained at the intermediate detector without mass analysis. This procedure is a satisfactory substitute for mass-analyzed ion kinetic energy spectrometry in negative ion mass spectrometry because the limited number of ions present makes misassignment unlikely.

4. Applications

4.1. Ion Structure Determination

4.1.1. CID Cross Sections and Ion Dissociation (CA) Spectra

A spectrum showing all the products of CID reactions undergone by any selected ion can be recorded on a reversed geometry double-focusing mass spectrometer or on a double mass spectrometer as described in Section 3.1. This MIKE spectrum or collisional activation (CA) spectrum represents quite simply *the mass spectrum of an ion*. It is from this fact that several of the applications of CID reactions follow. In particular, such a mass spectrum can be expected to provide structural information on the ion just as conventional mass spectra provide fragmentation patterns which allow structural conclusions to be drawn regarding neutral molecules introduced into the ion source. This expectation has been amply justified, as shown later.

In spite of the difference in the excitation methods used to obtain the mass spectra of ions and conventional mass spectra of molecules, there are marked similarities between the two phenomena. Both depend on essentially vertical electronic transitions and the average excitation energies of the excited ions are in the range of a few electron volts. The times during which the excited ions can undergo unimolecular fragmentation are also similar. (Ion source residence times are typically 1 μsec and field-free region transit times typically 10 μsec.) In both procedures individual reaction steps cannot always be uniquely assigned, rather an integrated view of the network of competing and consecutive unimolecular reactions is represented in the product distribution, namely in the mass spectrum.

The ions subject to CID will all be of one mass (in MIKES experiments) but a variety of isomeric structures and internal energies might be represented. For organic ions formed by electron impact most of the ions entering the field-free region will be stable to unimolecular fragmentation although a small fraction (typically 10^{-3}–10^{-4} as judged from metastable ion abundances in mass spectra) will have lifetimes which correspond to fragmentation in the field-free region or at some later point prior to detection. Most ions of higher internal energy will already have fragmented. Thus, CID provides structural information on ions which are stable to unimolecular fragmentation but which possess a range (generally unknown) of internal energies. Almost all applications of CID to the structural determination of organic ions have assumed that the resulting spectra are independent of the internal energy distributions of the ions; we return to this point in Section 4.1.2. This approach has proved a fruitful source of structural data on ions.

As an example of the use of collision-induced dissociation in characterizing ion structures consider the ion $C_2H_5O^+$. Possible structures include the oxonium ions *a*–*c*

$$CH_3-O^+=CH_2 \qquad CH_3CH=\overset{+}{O}H \qquad \underset{\underset{H}{O^+}}{H_2C-CH_2}$$

$$\textit{a} \qquad\qquad \textit{b} \qquad\qquad \textit{c}$$

as well as carbonium ion and monovalent oxygen structures. The CA spectra of $C_2H_5O^+$ ions, taken at low-energy resolution with a reversed geometry instrument and corrected for metastable peak contributions, divide into three groups.[80] For precursor molecules having a structure of the type CH_3OCH_2Y the spectrum of the $C_2H_5O^+$ ions always falls into one class, distinguished by abundant ions at m/e 15 and 29, and for molecules of type

CH_3–CHYOH, the spectra fall into another which is characterized by more abundant m/e 19 ions. Direct protonation of the ethylene oxide gives a third structure (*c*) characterized by relatively ready loss of CH_2 on CID. Clearly, the structures *a*–*c* are distinguishable using their CID reactions. Moreover, $C_2H_5O^+$ ions generated with other structures at threshold always give spectra which correspond to those characteristic of ions *a*–*c* or to mixtures thereof. This example illustrates the strength of the CA method in that the relative abundances of a number of ions can be used to characterize structures. Unfortunately, the differences between the spectra of isomeric ions are not always large so that completely correct identifications have not always been easily made.[80]

Since the CID spectrum taken on a MIKE spectrometer is simply the mass spectrum of an ion, the rules developed for deducing molecular structure from mass spectra can be applied in deducing ion structure. In this way it is possible to do more than assign ions with the same spectrum to a particular structure. Instead, the nature of the fragmentations can sometimes be used to *deduce* the ion structure when this is unknown. For example, the ion $C_2H_3^+$ is formed much more readily from $C_2H_5O^+$ ions in compounds of the type CH_3CHYOH than in CH_3OCH_2Y compounds, confirming the $CH_3CH=\overset{+}{O}H$ assignment. Similarly, the ion formed by loss of CH_3 from 2-hexanone may have either structure *d* or *e* depending on which methyl group is lost.

$$CH_3CH_2CH_2CH_2CO^+ \qquad \begin{array}{c} CH_2-O \\ | \quad\quad\;\; \diagdown \\ \quad\quad\quad C-CH_3 \\ | \quad\quad\;\; \diagup \\ CH_2-CH_2 \end{array}^+$$

$$\textit{d} \qquad\qquad\qquad\qquad \textit{e}$$

Using deuterium labeling to distinguish these ions it was found[80] that one fragments by loss of CO and by loss of CH_2CO (hence assigned as structure *d*) whereas the other fragments by loss of C_2H_4, C_3H_6, and CH_3 (consistent with structure *e*).

McLafferty and co-workers[81–85] have studied numerous other sets of isomeric ions by procedures analogous to those just described. These studies have allowed ions to be assigned to particular structures or mixtures of structures. Other workers have also made use of collisional activation spectra in this way. Levsen,[86] for example, has used the method to show that alkane and alkene molecular ions do not isomerize under typical mass spectrometric conditions whereas their even-electron fragment ions do.

Double magnetic sector mass spectrometers have also been used to measure the dissociation spectra of organic ions and thus to characterize their structures. For example, $C_6H_7^+$ ions from 1,4-cyclohexadiene and from

1-methyl-1,4-cyclohexadiene have been compared[29b] and found to represent different sets of ions in agreement with threshold thermochemical data.

A variation on the determination of ion structure by CID has been described by Giardini-Guidoni et al.[69] These authors studied collision-induced dissociation of the ion $C_2H_4O^{+\cdot}$ formed from acetaldehyde and ethylene oxide over a range of energies. Both the dissociation spectrum of the ion at 300-eV ion energy and the energy threshold for the various reactions could be used to distinguish these ion structures (cf. Section 4.2.).

Ion-structure determination by examining selected CID reactions using a double-focusing mass spectrometer of conventional geometry provides further examples of the utility of this method, particularly when the results are used in conjunction with abundance and kinetic energy release data on metastable ions.[87]

Examples of the use of collision-induced dissociation to characterize ion structures formed by low-energy ion–molecule reactions (chemical ionization) are given in Section 4.6.

4.1.2. Internal Energy Effects

As discussed in Section 3.3 the internal energy of an ion subject to collision-induced dissociation can be expected to affect the reaction in several ways, including controlling the position reached on the potential energy surface for the upper state and in rare cases the nature of the upper state. This can lead to a variation in the kinetic energy release.[88] Variation in the internal energy of the ion also changes the energy defect for the reaction and hence the cross section. If the lowest-energy dissociation of an ion is being studied, then stable ions which are subject to collision might have internal energies anywhere in the range from zero to just below the activation energy. For a typical activation energy of 2 eV, the energy required to cause dissociation would vary from 2 eV down to almost zero. The reaction cross section would be expected to vary too (see Section 2.4), and one must conclude that there will be some selectivity in the ions sampled by collision-induced dissociation. In the low-kilovolt-translational-energy range the maximum cross section lies at or below 1 eV for typical polyatomic ions (Section 2.3) so that it is expected that higher energy ions would be sampled in preference to ground state ions in CID processes (see Fig. 15).

There are data in the literature which appear to support this argument. McLafferty and co-workers[43] have observed in a number of cases that the cross section for collision-induced dissociation leading to the fragment ion of lowest appearance potential shows a large dependence on the internal energy distribution of the reactant ion. For example, collision-induced dissociation of the phenol molecular ion yields $C_5H_6^{+\cdot}$ ions and the cross section for this reaction increases by a factor of more than 2 as the ionizing electron energy is increased from 9 to 20 eV. Other products formed from the phenol molecular ion showed smaller variations, consistent with the fact that the differences in energy requirements for reactant ions of different internal energies are relatively smaller for higher-activation-energy reactions. In other compounds even larger changes in cross section were observed. Nevertheless, the effects upon the relative abundances of CID reactions which are used in assigning ion structure are smaller than the absolute effects. Moreover, if the lowest-activation-energy processes are excluded, the resulting spectra are sensibly constant within the fairly limited variations in internal energy distribution commonly encountered in ions whose structures are being compared.

To further minimize internal energy effects CA spectra can be obtained at low-ionizing electron energies. Differences from 70-eV spectra are often small, except when the ion population includes a mixture of structures. In such cases marked changes in CA spectra are observed on changing the electron energy and these changes have been associated with changes in the composition of the mixture of isomeric ions.[82]

There is a further consequence of the control of energy defect upon reaction cross section in isomeric ions. If the activation energies for fragmentation are large, the selectivity of CID in favor of higher energy ions in the distribution will be maximized. At the same time the probability of there being molecular ion isomerizations whose activation energies are below that for the lowest-energy fragmentation is also expected to be large. Under these conditions, CID might well sample predominantly the isomerized ions. Alternatively, all the ions in the beam may exist in a low-energy structure m_1^+, but the reaction of lowest free energy may be isomerization to $m_1'^+$ from which all fragmentation proceeds. Upon energization of the ions m_1^+ they would isomerize and fragment from the isomeric structure $m_1'^+$ even though this is not present in the beam. In such a situation the CID technique can only recognize the system $(m_1^+, m_1'^+)$, not the individual species.

Apparently an instance of the former type exists in the case of $C_6H_6^{+\cdot}$ ions generated from benzene and from its linear isomers, 2,4- and 1,5-hexadiyne.[89] Collision-induced dissociations, studied on a MIKE spectrometer, reveal no differences between the three populations of ions, the major reactions leading to formation of $C_4H_4^{+\cdot}$ and $C_3H_3^{+\cdot}$ occurring to similar relative and absolute extents. However, if an ion–molecule reaction with a large energy defect is studied, pronounced differences in behavior are observed. Thus, charge stripping occurs with seven times the abundance in benzene as its does in 2,4-hexadiyne. The energy defect for stripping falls in the range 16–12 eV (the difference between the single- and double-ionization potentials of benzene is 16 eV and the activation energy for the lowest-energy reaction is ~4 eV). This is to be contrasted with the energy-loss requirements (4–0 eV) for the same ions undergoing collision-induced

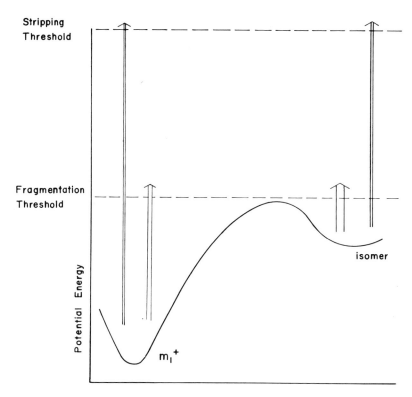

Fig. 15. Selectivity of collisional activation and charge stripping for ions of different internal energy. The heavier arrows represent larger cross sections. Collisional activation is relatively more likely to sample the isomerized ions.

dissociation. Thus, charge stripping is less susceptible to variations in the internal energy distribution of the ions and it should provide a more uniform sampling of nonfragmenting ions. Figure 15 illustrates the relationship between the cross section for an ion–molecule reaction and the internal energy of the ion.

These conclusions indicate that CID may have a role in the study of ions in excited electronic states and in the related phenomenon of ion isomerization (cf. Section 4.3.2). Isomers differing in their skeletons (carbon and heteroatoms) are usually distinguishable using CA spectra but isomers which can interconvert via hydrogen shifts are not. The lower activation energies expected for the latter processes explain this generalization. In particular cases, such as $C_6H_6^{+\cdot}$ where the critical energy for fragmentation is unusually high, isomerization can be more extensive.

Thus, in spite of its success in ion structure determinations, as discussed in Section 4.1.1, it is perhaps unfortunate that other ion–molecule reactions at high translational energy have not been more extensively used in conjunction with collision-induced dissociation. The similarity in procedures and instrumentation makes this a simple matter, and a combination of other high energy ion–molecule reactions with CID would appear to offer a deeper insight into the complex distributions of and equilibria between ion structures generated in various ways. The great advantage of collision-induced dissociation over the other methods is the fact that a large number of ionic products, not just one, are generated.

4.1.3. High Energy Resolution

If adequate energy resolution is available, the kinetic energy loss which accompanies the excitation step can be measured from the shift in the peak position, and the kinetic energy release accompanying fragmentation can be measured from its width (Section 2.1). The use of kinetic-energy-release data in characterizing the structures of metastable ions is an established procedure. In the general case, this energy can be considered to be due to contributions from both the internal energy of the ion after collision and the activation energy for the reverse reaction (Fig. 7). The kinetic energy release accompanying metastable ion dissociations has been shown[90,91] to provide a useful guide to the structures of ions formed from different sources, the fixed lifetimes associated with their observation as metastables controlling the internal energy of the ions and so allowing their characterization. The internal energies of ions which fragment by collision-induced dissociation are not similarly restricted. Nevertheless, differences in the internal energy distributions of the stable ions which are subject to CID may produce smaller variations in kinetic energy release data than those associated with isomeric structures. Some use has been made of these measurements, for example, $C_3H_7^+$ ions apparently lose H_2 via the same transition state when fragmenting as metastable ions as they do in CID reactions as judged from the similarity in the shapes of the wide peaks recorded.[32] Similarly, in a comparison of $C_3H_6O^{+\cdot}$ ions from different sources, the acetone molecular ion was readily distinguished from enolic $C_3H_6O^{+\cdot}$ ions by the fact that it alone showed a composite peak for loss of $CH_3^{\cdot}$ on collision.[87]

In some cases the method of characterizing ion structures through kinetic energy releases associated with metastable ions has been applied to particular problems also studied by collisional activation spectra. The two methods give concordant results for $C_2H_4O^{+\cdot}$ ions[92] and for $C_{12}H_{10}O^{+\cdot}$ ions[81] in spite of the differences in internal energy of the ions being sampled. The interesting $C_7H_7^+$ ion has also been studied both by kinetic energy release methods[93,94] and by collisional activation spectra.[83]

It would be desirable to make more use of comparisons between the kinetic energy releases associated with collision-induced dissociations and unimolecular reactions. Particularly if the two tests of structure gave different results, a comparison of energy releases associated with ions of different internal energy might serve to confirm the presence of different structural isomers at different internal energies. Just such a situation apparently exists in the equilibrium between the $C_3H_8N^+$ ions f and g. Kinetic-energy-release measurements[95] show the metastable ions to behave similarly; however, the same measurements for the CID reaction show a composite metastable peak in one case and not in the other (Fig. 16). It seems that the lower-energy ions examined by CID retain, at least in part, their distinct structures. In a similar case, the $C_3H_3^+$ ions generated from allene on the one hand and from propargyl or propyl halides on the other were compared.[96] The two sets of structures are indistinguishable from the kinetic energy release accompanying collision-induced loss of $2H(H_2)$. The energy releases calculated from the half-heights of all the peaks are 0.33 ± 0.01 eV. The corresponding metastable ions differ both from the CID behavior and from each other, showing energy releases of 0.02 and 0.07 eV, respectively.

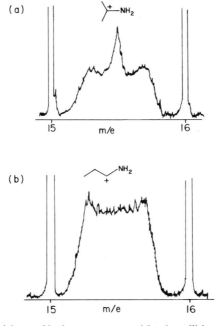

Fig. 16. Peak shapes (observed in the mass spectrum) for the collision-induced elimination of ethylene from isomeric $C_3H_8N^+$ ions. The structures of the $C_3H_8N^+$ ions are shown above the spectra.

$$\overset{+}{\diagup}\!\!\!-\text{NH}_2 \qquad\qquad \diagup\!\!\diagdown\!\!\underset{+}{\diagup}\!\text{NH}_2$$

      f   g

4.1.4. CID in Conjunction with Other Methods of Determining Ion Structure

Collision-induced dissociation in the form of collisional activation spectra has become one of the major methods of determining ion structure. We therefore close this section by considering a particular problem in which CID results are considered in conjunction with data obtained by other methods. The determination of the structures of gaseous polyatomic ions is both complex and important in many aspects of ion chemistry and mass spectrometry. The species of interest are often isolated and usually not in their ground vibrational states; indeed, they often have as much as several electron volts of energy which is sufficient for many types of structural reorganization. Structure therefore becomes a function of this internal energy and hence of the time scale of the analyzing technique. With organic ions, mixtures of structures are possible in many cases and this mixture can vary with time, isomerization proceeding during ion flight through the spectrometer. Isomerization thus complicates the interpretation of ion structure data taken by any method. The consequences for collisional activation spectra have already been noted (Section 4.1.2) and more examples are given in Section 4.3.2.

A distinction can be drawn between methods which allow the study of reacting ions and those, such as CID, which refer to ions which are stable to spontaneous fragmentation. It is also important to distinguish techniques which provide direct information on the reaction mechanism through data on activated complex properties such as isotope effects or kinetic energy release and those which probe static systems. Fuller discussion of these topics and of the range of methods of determining ion structure is given elsewhere.[97]

In the discussion which follows the application of CID in conjunction with complementary techniques is illustrated. The use of metastable ion-abundance data in the form of a MIKE spectrum, which can be taken in the absence of collision gas when recording the dissociation spectrum of the ion in question, is a particularly suitable and accessible complementary method. Kinetic energy release data are also used; they provide information on the mechanisms of the spontaneous reactions undergone by the ion, their results usually being independent of ion internal energy.[90] Thus the metastable ion-abundance data, which are quite strongly energy dependent,[97] used in conjunction with energy release data, can provide structural and energetic data on those ions in the population which have sufficient energy to react spontaneously. CID is used to examine the structures of the stable ions and these results are compared in the example discussed below with the conclusions drawn from ion-cyclotron resonance (ICR) measurements which also refer to nonfragmenting ions. However, the selectivity of these methods to ion internal energy can differ. Note in particular that kinetic-energy-release measurements can serve as a source of information on the structures of *both* the parent and the fragment ion. Moreover, for metastable ions the information so obtained usually refers to the fragment ion at threshold. Ion cyclotron resonance and collisional activation studies on such a fragment ion would provide information on its structure over a much larger and wider range of internal energies. Thus, in contrast to the usual view, kinetic energy release data can refer to lower-energy fragment ions than either collision activation spectra or ICR data.

The problem chosen to illustrate these points concerns the structures of ions m/e 84 formed from compounds **1–3**.[99] The reactions leading to 83^+ from **2** and **3** are single and double alkene eliminations, while 84^+ is the molecular ion of compound **1**. Figure 17 shows the MIKE spectra of 84^+ taken both with and without added collision gas. In each case the left-hand spectrum represents metastable ion transitions and the right-hand spectrum gives the relative abundances of the products of collision-induced dissociations corrected for unimolecular reactions.

The CID spectra of the two fragment ions are almost identical, a result which indicates that the stable ions sampled by CID have identical structures. (However, they may have different internal energy distributions given that CID is relatively insensitive to these.) The CID spectrum of 84^+ generated by ionization of **1** is quite different from the other two. One concludes that the ions which do not spontaneously fragment are structurally distinct from the rearrangement products from **2** and **3**. It will be noted, however, that the same ions are present in all three 84^+ fragmentation spectra although in quite different proportions. This is the first indication that mixtures of the same structures, in very different proportions, may be involved. ICR reactions selected to test for keto and enol functional groups show that although the molecular ion of **1** behaves as a typical ketone, the ions of mass 84 from **2** and **3** do not undergo ion–molecule reactions typical of either functionality. The fact that most of the molecular ions of **1** are

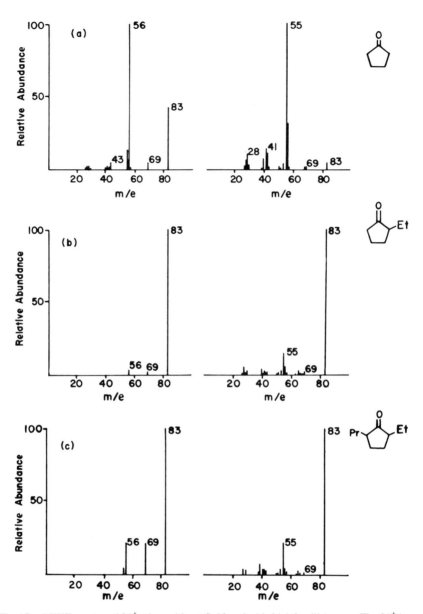

Fig. 17. MIKE spectra of 84^+ taken without (left) and with (right) collision gas. The 84^+ ions were generated (a) by ionization by cyclopentanone, (b) by fragmentation of 2-ethylcyclopentanone, and (c) by fragmentation of 2-ethyl-5-propylcyclopentanone. The collisional activation spectra (right) are corrected for unimolecular contributions. The reactant ions (84^+) are not shown.

structurally distinct from the 84$^+$ ions formed from **2** and **3** is thus confirmed. The main structure formed from compound **1** is the keto ion **h**; the chief structure contributing to m/e 84 in **2** and **3** is neither **h** nor the enol ion **i**.

$$\underset{h}{\overset{O^{+\cdot}}{\bigtriangleup}} \qquad \underset{i}{\overset{OH^{+\cdot}}{\bigtriangleup}} \qquad \underset{j}{\overset{O^+}{\diagup}\!\!\!\diagdown CH_3}$$

The unimolecular reactions of 84$^+$ from compounds **2** and **3**, examined by taking MIKE spectra (Fig. 17), show the same three processes with 84$^+$ → 83$^+$ the dominant reaction. The metastable ion abundances referred to each other are quite different but the differences are approximately in the range that is observed and predicted for ions which differ in internal energy but not in structure.[98] However, the ratio 56$^+$/83$^+$ in **1** differs from that in **2** by almost two orders of magnitude which is *prima facie* evidence for structural dissimilarity in the metastable 84$^+$ ions, corresponding to the structural dissimilarity inferred by CID in the nonfragmenting ions. The kinetic energy releases associated with the metastable ion reactions

$$84^+ \rightarrow 56^+ + 28 \tag{4.1}$$

$$84^+ \rightarrow 83^+ + 1 \tag{4.2}$$

were observed to be similar in all three compounds. This means that each population of 84$^+$ ions contains the same ions, present in different proportions. Thus, the dominant ion in **1** is the keto structure **a** which eliminates 28 mass units, and this is the minor component in **2** and **3** where the dominant ion structure(s) eliminates H· and CH$_3$. These latter modes of fragmentation suggest for the second ion an acyclic acylium structure such as **j** or a closely similar form.

Thus what appears to be a McLafferty rearrangement in **2**, for example, apparently occurs from an acyclic form of the molecular ion with hydrogen transfer to carbon not to oxygen. This is consistent with the CID spectra which show H· loss as the dominant process in **2** and **3** and with the absence of a keto or an enol function as inferred by ICR. A mixture of structures **h** and **i** in different proportions explains the data accrued on both the fragmenting and the nonfragmenting ions. This example thus illustrates the use of several techniques in ion-structure determination.

4.2. Thermochemical Determinations

The possibility of using collision-induced dissociation to obtain thermochemical data follows from the fact that an isolated chemical system is involved and energy measurements on reactants and products can be made. Studies on simple ions have emphasized this type of application (Chapter 6) and although less has been done on polyatomic ions in this connection, some information is available.

The enthalpy of the collision-induced dissociation

$$m_1^+ + N \to m_2^+ + m_3 + N \qquad (4.3)$$

can be measured in two quite different ways. The reaction can be carried out at high relative translational energy and the energy loss accompanying the reaction measured. Allternatively, the collision-induced dissociation can be studied at threshold and the minimum relative kinetic energy necessary to promote the reaction measured. Under these low-energy conditions the reaction is assumed to be adiabatic with complete availability of all forms of energy of the reactants, through the intermediacy of a long-lived collision complex.

This latter procedure has been quite widely used in tandem mass spectrometry,[100] but it is chiefly of interest here because it provides an independent method of checking thermochemical and ion structural data obtained from the study of high-energy reactions. For example, both collision-induced dissociation results[92a] and kinetic-energy-release measurements[92b] indicate that the $C_2H_4O^{+\cdot}$ ions formed from acetaldehyde and ethylene oxide are structurally distinct. In addition, ionization potential measurements[101] show that the $C_2H_4O^{+\cdot}$ ions formed at threshold from ethylene oxide and acetaldehyde differ in enthalpy by 1.5 eV. The $C_2H_4O^{+\cdot}$ ions formed by impact of 70-eV electrons and examined some microseconds after formation might or might not retain their distinct structures and enthalpies. To test this point an appropriate collision-induced dissociation

$$C_2H_4O^{+\cdot} + He \to CHO^+ + CH_3^\cdot + He \qquad (4.4)$$

was examined[69] over a range of kinetic energies up to 30 eV in the center of mass. At these low energies it is assumed that the entire relative kinetic energy of motion is converted into internal energy and that the reaction products are formed in the ground state. Thus, a plot of CHO^+ abundance versus relative translational energy of the reactants, extrapolated to zero intensity, gives the endothermicity of the reaction. From the known enthalpies of the products this allows the determination of the enthalpy of the reacting ion. For the case of the ethylene oxide molecular ion, the enthalpy of reaction was found to be 0.9 ± 0.2 eV, giving ΔH_f $(C_2H_4O^{+\cdot}) = 10.2 \pm 0.2$ eV. The corresponding value using acetaldehyde as the source of the ion was 8.5 ± 0.2 eV. The difference between these values agrees with that obtained from ionization potentials. However, such agreement may not always be expected since the two methods of ion enthalpy determination refer to different populations of ions. Ionization potential measurements give enthalpies at the threshold for ionization whereas the present method refers to those ions generated at higher electron energies which are still present in the beam some 10^{-5} sec after formation. The presence of long-lived excited states in the beam could also be shown using this type of ion–molecule reaction measurement.

The main alternative to this method of thermochemically characterizing long-lived ions is the measurement of energy losses associated with ion–molecule reactions, including both charge-transfer processes and collision-induced dissociation. In an experiment of this type the molecular ions of acetaldehyde and ethylene oxide were activated by collision with helium. The energy losses associated with H_2 elimination were measured and their difference was found to be 1.6 ± 0.6 eV, in good agreement with the ionization potential results. This type of experiment is done in an energy range in which signal strengths are normally adequate but the measurement involves the difference between two large quantities and is correspondingly inaccurate. It also requires the assumption or demonstration that target excitation is not involved (cf. Section 2.4) and some knowledge of the internal energy of the reactant ion. If this latter is assumed to be in its ground state, then the error in the determination could be as much as the activation energy (AP − IP) for the lowest-energy fragmentation reaction of the ion. If the energy loss is measured over a range of ionizing electron energies, it can be extrapolated to the onset where the assumption of ground-state ions can be made with much more confidence.

A study of this latter type on acetylene gave the results[52] shown in Fig. 18. The decrease in energy loss with increasing electron energy is expected to level out above 20 eV, energy-transfer functions between 20 and 70 eV being similar. The energy loss extrapolated to the ionization potential (11.4 eV) is measured as 5.9 ± 0.5 eV and this fixes the energy of the

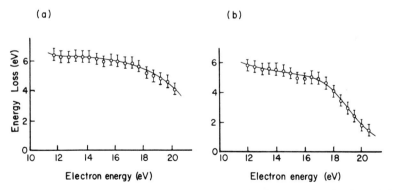

Fig. 18. Variation with ionizing electron energy of the kinetic energy loss for the collision-induced dissociation $C_2H_2^{+\cdot} \rightarrow C_2H^+ + H^\cdot$. The targets were (a) helium and (b) xenon. (Data taken from Ref. 52.)

dissociating state as 17.3 ± 0.5 eV relative to the molecule. This reaction is also interesting for the upward break in the C_2H^+ ion yield seen at the same energy. The formation of C_2H^+ ions by predissociation of an excited state of $C_2H_2^{+\cdot}$ is indicated.

A direct measure of differences in average internal energy in ions from different sources can sometimes be obtained by measuring the energy loss associated with a particular CID reaction. This procedure has been applied to the determination of the internal energy of $CO^{+\cdot}$ ions formed from $CO_2^{+\cdot}$.[102] In larger ions the mere *detection* of a particular fragment ion can fix the internal energy of the fragmenting ion within a certain, often quite narrow, energy range. At particular internal energies, characteristic products are formed and all of this information can be summarized as a breakdown curve for a particular ion showing product yield versus internal energy (Fig. 6). These data can be obtained experimentally or by statistical unimolecular rate calculations. These facts suggest an alternative to the use of energy-loss measurements in evaluating the energy transferred upon collision. In this approach the fragmentation spectrum of the ion could be used to provide a picture of the distribution of energy losses rather than just the losses (necessarily) associated with a chosen fragmentation. Variations in the ionizing electron energy would further extend the usefulness of the method by providing information on the degree of excitation of the reactant ions.

Collision-induced dissociation can be utilized in the determination of gas phase basicities.[103] The relative rates of the two competing CID reactions of the proton-bound dimer $B_1HB_2^+$ [Eq. (4.5)], is a measure of the relative proton affinities of the two bases B_1 and B_2:

$$B_1HB_2^+ \nearrow\searrow \begin{matrix} B_1H^+ + B_2 \\ B_1 + B_2H^+ \end{matrix} \qquad (4.5)$$

$$\underset{H_3C}{\overset{H_3C}{>}}C=O \cdots \underset{59}{\overset{73}{\{\cdots H^+ \cdots\}}} O=C\underset{CH_2CH_3}{\overset{CH_3}{<}}$$

k

Figure 19 shows MIKE spectra of the ion **k**, that is, for the case in which the bases are acetone and methyl ethyl ketone. The major reactions are those shown in Eq. (4.5). Clearly the larger ketone is the stronger base. Figure 19 includes results taken both in the presence and in the absence of collision

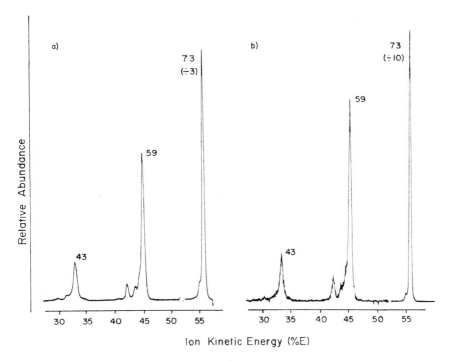

Fig. 19. Partial MIKE spectra of the acetone/methyl ethyl ketone proton-bound dimer (k, m/e 131) taken (a) in the presence of collision gas to show CID reactions and (b) in the absence of collision gas. In both cases the formation of the protonated ketones at m/e 73 and 59 constitutes the major reactions, with the larger ketone preferentially acquiring the proton. The metastable ion spectrum (b) is recorded at three times the sensitivity of the CID spectrum.

gas, showing that the method can be applied by studying either metastable ions or collision-induced dissociations. The lower internal energies of fragmenting metastable ions make this the more sensitive criterion. However, the CID signals are more intense, and these data provide a valuable check on the metastable ion results, especially if isomerization prior to fragmentation is a possibility.

4.3. Fragmentation Mechanisms

Determinations of ion structure by the methods described in Section 4.1 have immediate mechanistic implications. This is a subject in which considerable activity is underway.[104] Strong parallels with conventional mass spectrometry are expected and found. Rather than catalog this effort we concentrate on some more tentative but potentially important methods and concepts.

4.3.1. Comparisons with Metastable Ions

The comparison of CID reactions with the corresponding reactions of metastable ions can be helpful in revealing unusual fragmentation mechanisms. Since both processes yield the same fragments, with virtually the same average kinetic energy, the peaks are superimposed. That due to CID is shifted slightly to lower energy, because of the energy loss on collision, so that the combined peak will normally appear asymmetrical and will change in shape as the target pressure is varied.[105] By working at low pressure the metastable ion reaction can be isolated for study and this is usually readily done, the CID contribution being negligible at background pressures of 10^{-6} torr in typical instruments. In a few cases, however, even at these pressures there is a substantial CID peak as seen from pressure plots or from the peak shape. This is the case for H˙ loss from the acetylene molecular ion where the peak width is found[106] to decrease steadily as the pressure is lowered, the change in shape indicating a varying contribution from two processes. Even at the lowest pressures obtainable in the RMH-2 mass spectrometer (1×10^{-7} torr) the CID peak had not been entirely removed. This is interesting in view of the report[52] that this CID reaction occurs by two mechanisms, the usual electronic excitation and by what is probably an instance of collision-induced predissociation. The latter process would require no energy loss and might therefore be expected to have a relatively large cross section at unusually low pressures. Such a persistence of a CID peak at low pressures has been observed in other cases[51,107] where, as in the acetylene example, particularly small kinetic energy releases are associated with the corresponding metastable ion reactions. The correlation between small kinetic energy release and collision sensitivity is suggestive of an underlying mechanistic relationship.

Some of the more abundant metastable peaks, such as that for HCN loss from the benzonitrile molecular ion, are associated with relatively low cross sections for the corresponding collision-induced dissociation. This interesting phenomenon bears upon the mechanism of collision-induced dissociation as well as its application in studying reaction mechanisms. It is thought to be connected with the occurrence of isomerization in competition with fragmentation.

Isomerization of an ion prior to collision can lead to new excited species with a geometry (including atom arrangement) that is quite different from the original. Metastable ions, having received more internal energy on electron impact than the ions subjected to CID, may (a) isomerize in other ways or (b) fragment faster than they isomerize. Having isomerized, a different set of excited states is reached and equilibration with other states of different geometries cannot be expected although it certainly can occur in particular cases, just as it can in lower energy (nonfragmenting) ions. The higher the excitation energy the less likely such equilibration, which is expected to be a low frequency factor process relative to fragmentation.

These considerations indicate how it might be possible for an ion to give an abundant metastable peak for a particular fragmentation and yet not undergo this reaction on excitation by collision. More generally, this is one of the factors which accounts for the fact that there is not necessarily a correlation between the abundance of a metastable peak and that of the corresponding CID peak. Figure 20 illustrates the situation just described. The lowest-energy ions m_1^+ cannot isomerize but these ions are not effectively sampled by collision-induced dissociation. Higher-energy ions may isomerize irreversibly to m_2^+ (ring opening, for example, would be essentially irreversible) and collision-induced dissociation could then lead to quite different products from those expected from ions having the structure m_1^+. As indicated in the figure, isomerization of ions with sufficient internal energy to fragment spontaneously may not occur or if it does, it might not lead to fragmentation. However, if the nonfragmenting molecular ions equilibrate between the two sets of structures, then *some* of the ions subject to collision will, after excitation, have the same structure and energy as metastable ions and a CID peak will accompany the metastable peak. This is probably the usual situation.

4.3.2. Scrambling and Other Isomerization Reactions

Fragment ions in the mass spectrometer are frequently generated from rearranged ions rather than directly from the original molecular ion. These

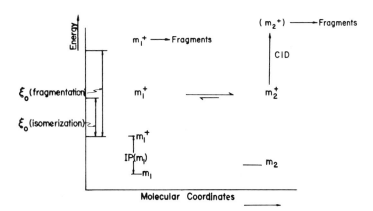

Fig. 20. Energy diagram showing how collision-induced dissociation might preferentially sample nonfragmenting ions which have isomerized. Ions of both higher and lower energies may not isomerize, the former because fragmentation competes successfully with isomerization, the latter because they have insufficient internal energy.

rearrangements may cause extensive reorganization of the molecular skeleton or they may be degenerate, only being detectable through the use of selectively labeled ions. Species which have insufficient internal energy to fragment further may nevertheless isomerize. Collision-induced dissociation provides, as just indicated, one means of detecting and characterizing these hidden rearrangments.

In applying CID to the study of isomerization reactions several principles should be recalled. First, isomerization reactions will generally have lower frequency factors than fragmentations and will therefore be relatively less favored in higher-energy ions. Second, collisional activation typically transfers a broad range of energies to the ion with a maximum probability of perhaps at most a few electron volts (Section 2.3). Third, in most experimental arrangements the ion lifetime prior to collision is at least a few microseconds whereas after collision it is at most a few microseconds but, because of the degree of excitation of the ion, typically much less than this.

Thus, ions with sufficient internal energy to isomerize will often do so prior to collision which can then serve to sample the isomerization process. On the other hand, ions formed with insufficient energy to isomerize are less likely to do so to a significant extent in competition with fragmentation after collision. Thus collision-induced dissociation can be expected to serve as a means of probing rearrangements occurring in ions whose internal energies are too low to allow fragmentation.

From these arguments one would also expect that isomerization should be at a maximum in metastable ions since these ions have the twin advantages of both time and internal energy for maximizing isomerization reactions. In agreement with this prediction it is found that when the degree of scrambling in partially labeled metastable ions is compared with that for ions which undergo the same fragmentation by CID, the latter is usually smaller. For example, in the benzoic acid molecular ion scrambling involving the carboxyl hydrogen and the two ortho ring positions is almost complete in metastable ions but only 60% complete for the corresponding CID reaction.[106] This is consistent with the fact that collision yields higher-energy ions which will then rearrange less prior to fragmentation. It is likely that scrambling is a lower-energy reaction than fragmentation with some or all of the scrambling occurring in the ion prior to collision.

The fact that isomerization, like H/D scrambling, is favored in metastable ions over those examined by CID is further exemplified by results[95] on $C_3H_8N^+$. Metastable ion abundance ratios and CA spectra as well as the kinetic energy release data described in Section 4.1.3 show that the primary amines *f* and *g* had isomerized less when sampled by CID than when the corresponding metastable ions were examined.

Exceptions to these trends can occur either when isomerization occurs after collisional activation or when metastable ions represent a minor constituent in a mixture of ions. The example of $C_3H_3^+$ (Section 4.1.3) can be noted. Another example is the $C_8H_{11}^+$ ion discussed in Section 4.6 where loss of C_2H_4 either in a metastable ion reaction or by CID occurs without H/D randomization whereas loss of $CH_3^{\cdot}$ occurs by CID with complete scrambling.

The fact that CID provides a sensitive measure of isomerization in nonfragmenting ions can be illustrated by a consideration of $C_7H_8^{+\cdot}$ formed by ionization of toluene and cycloheptatriene. Metastable $C_7H_8^{+\cdot}$ ions from both sources fragment via a common transition state as shown by the fact that the kinetic energies associated with $H^{\cdot}$ loss are identical.[108] This conclusion is supported by the fact that $C_7H_7^+$ ions formed by 70-eV electron impact behave identically upon collision.[83] Hence the activation energy for isomerization of $C_7H_8^{+\cdot}$ is lower than that for $H^{\cdot}$ loss, and some information on the extent of isomerization in ions having insufficient energy to fragment can be obtained from the collisional activation spectra of these ions. The $C_7H_8^{+\cdot}$ CA spectra[80] are very similar except that the reaction

$$C_7H_8^{+\cdot} + He \rightarrow C_6H_5^+ + CH_3^{\cdot} + He \qquad (4.6)$$

which should be diagnostic of the presence of a methyl group in the ion, was some 50% more abundant for ions generated from toluene than from cycloheptatriene. The results have been interpreted as indicating that half of the toluene molecular ions formed by 70-eV electron impact have isomerized after 10^{-5} sec. Other methods have also been used to examine the nonfragmenting $C_7H_8^{+\cdot}$ ions: in particular, a photodissociation study on longer-lived ions showed distinct $C_7H_8^{+\cdot}$ structures with no interconversion.[109] The difference in the two observations is probably associated with their sampling characteristics. By selectively examining ions of the highest internal energy, collisional activation accentuates isomerization reactions in nonfragmenting ions and allows their ready detection. In conjunction with other methods and with improved characterization of the sample characteristics of all the methods of examining nonreactive ions, it may become possible to follow isomerization as a function of the internal energy of the ion.

Collisional activation spectra have also been used to study isomerization of fragment ions which may follow or accompany their formation. Thus, simple β-bond cleavage in 2-methyl-2-phenylpropane-1,3-diol is expected to generate the benzylic ion *l*. However, in its further fragmentation, this ion behaves as if it had isomerized to *n*, perhaps via the cyclic intermediate *m*. The presence of an ion $C_2H_3O^+$ (m/e 43) in the mass spectrum was one indication of this isomerization which involves phenyl migration in the oxonium ion, *m*. The fragmentation spectrum of this ion[84] confirmed the isomerization since it was consistent only with structure *n*, the most abundant peaks being at m/e 91 ($C_7H_7^+$) and m/e 43 ($C_2H_3O^+$).

$$H_5C_6 \diagdown \atop H_3C \diagup C_+ - CH_2 \atop \diagdown :OH \quad \rightarrow \quad H_5C_6 \diagdown \atop H_3C \diagup C \underset{\underset{H^+}{O}}{\overset{}{\diagdown}} CH_2 \quad \rightarrow \quad H_3C \diagdown \atop HO \diagup C^+ - CH_2C_6H_5$$

$$\qquad \qquad l \qquad \qquad \qquad m \qquad \qquad \qquad n \qquad \qquad (4.7)$$

4.3.3. CID in Conjunction with Isotopic Labeling

The utility of isotope labeling in studying isomerizations between stable ions is evident from the preceding section. Labeling has also been employed in conjunction with CID to determine fragmentation mechanisms. Such experiments seem likely to become increasingly common because they should be greatly facilitated by the characteristic of the MIKE spectrometer that incompletely labeled compounds can be employed in the analysis.[110,111]

Consider the fragmentations of alkyl chlorides and bromides which have long been believed to involve formation of cyclic chloronium and bromonium ions, so that ethyl loss from 1-chlorohexane is proposed to occur via the following bond-forming reaction:

$$\underset{H_2C - CH_2}{\overset{C_2H_5-H_2C}{}} \overset{\overset{+\cdot}{Cl}}{\diagdown} CH_2 \quad \rightarrow \quad \underset{H_2C - CH_2}{\overset{H_2C}{}} \overset{\overset{+}{Cl}}{\diagdown} CH_2 \quad + \quad C_2H_5^{\cdot} \qquad (4.8)$$

$$o$$

The CA spectra of the $(M-C_2H_5)^+$ ions formed from 1,1-d_2- and 4,4-d_2-1-chlorohexane are identical,[85] which supports the assignment of the symmetrical structure o to the fragment ion.

Using a somewhat similar approach it should also be possible to examine parent–daughter ion pairs by CID to tell whether or not the corresponding ion source reaction is accompanied by skeletal rearrangment. For some generalized ion structure $ABCD^+$ which fragments in the ion source by loss of the neutral species D, the structure of the fragment ion can be determined by comparing its CID reactions with those of the parent, $ABCD^+$. Such a procedure should allow the comparison of the structures of molecular ions which are stable to fragmentation and of these with energies such that they fragment in the ion source.

4.4. Molecular Structure Determination

Ideally a mass spectrum should fulfill these criteria: (a) The peaks observed should be due to reactions of a specified ion; (b) the reactions involved should be of a known type and preferably simple bond cleavages; and (c) ionization of the molecule should preserve its structural integrity. MIKE spectra, taken with reverse geometry instruments, satisfy the first requirement since they make it possible to follow, in a single scan, all the reactions of any chosen ion. In structure analysis the ion selected will frequently be the molecular ion. Methods which cause ionization with minimal excitation are required to satisfy the third criterion. Field ionization and, with the appropriate reagents, chemical ionization or charge exchange serve this purpose but electron impact is less likely to do so.

Collision-induced dissociation seems to fulfill more closely the second criterion than do alternative methods which also allow product and reactant ions to be characterized. The available evidence indicates that the majority of CID reactions occur by simpler processes than do the reactions of metastable ions, which often fragment by rearrangement or elimination reactions. Since most available data on CID are for ions generated by

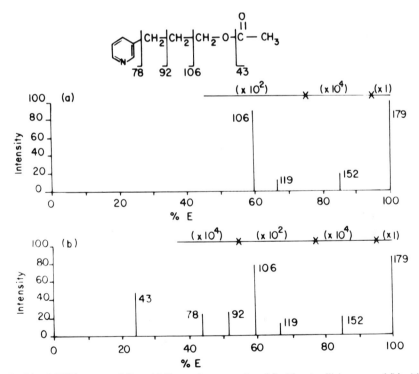

Fig. 21. MIKE spectra of (2-pyridyl)propyl acetate taken (a) without collision gas and (b) with collision gas. The simple relationship between the CID spectrum and the molecular structure facilitates structural analysis. The kinetic energy scale is given in terms of the energy E of the stable ion beam.

electron impact, rearrangement of the ion prior to collision may account for some of the cases in which simple cleavages are not involved. An example of structural analysis which employs the MIKE spectrum obtained by collision-induced dissociation is given in Fig. 21. Figure 21a shows the reactions of the metastable molecular ion. All three processes are bond-forming reactions of some type; even the formation of m/e 106 is believed to involve a spiro-fused cation. The CID reactions seen in Fig. 21b, on the other hand, are all simple bond cleavages and each reaction provides simple and direct information on the structure of the molecule.

Collision-induced dissociation is an excellent method for providing the structural information otherwise lacking in mass spectra obtained by field desorption, field ionization, and other soft ionization techniques. These methods make recognition of the molecular ion simpler than in electron impact mass spectrometry. Collisional excitation of the molecular ion, particularly in a MIKE spectrometer, then yields the fragmentation pattern and thus more detailed structural data than are given by the molecular weight alone. Beckey and Levsen[112] have employed this FI–CID–MIKES combination to good effect. They report fragmentation patterns of molecular ions induced by collision with He at 8 keV which closely resemble the corresponding 70-eV electron impact mass spectra. An example of this is shown in Fig. 22 where the spectrum of the field-ionization-produced molecular ion of methyl salicylate (molecular weight 152) is compared with its conventional mass spectrum.

Negative ion mass spectra typically show far fewer fragment ions than do positive ion spectra and this makes CID especially valuable in molecular structure studies by this technique. Some years ago an electron-addition ionization technique was developed using a plasma-type source.[113] Mass spectra were obtained for thermally labile compounds of quite high molecular weight (500 amu) and these spectra showed few fragment ions. Collision-induced dissociation might represent an approach for obtaining more stuctural information from this method. A discussion of the use of CID in studies on negative ion structure and gas phase chemistry appears in Section 5.3.

It is worth noting that structural analysis of fragment ions can be used to complement information obtained from molecular ions. In addition, useful partial structures should be obtainable even when the properties of the compound are such that it cannot be vaporized without fragmentation. Thus high molecular weight compounds, including polymers, might be examined via representative low-mass fragments which could be structurally characterized by CID. Laser ionization techniques might be used in conjunction with collision-induced dissociation in studying polymer structures. CID might also prove most useful in combination with pyrolysis as a means of studying complex biological molecules. An example illustrating this application to DNA has recently appeared.[114]

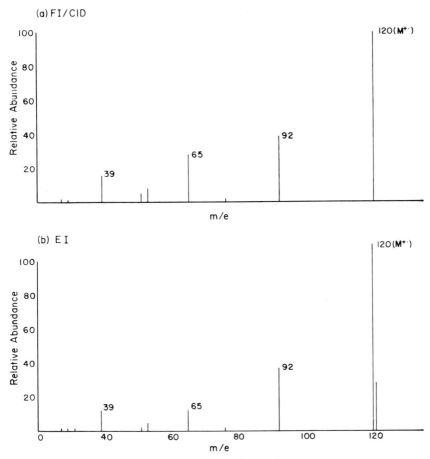

Fig. 22. Comparison of (a) the collision-induced dissociation spectrum of molecular ions of methyl salicylate produced by field ionization (FI) with (b) the conventional electron-impact (EI) spectrum. The similarity of the two spectra suggests that electron-impact fragmentation patterns may be useful in elucidating structure by the combination of field ionization with collision-induced dissociation using a reversed-sector (MIKES) instrument. (Data taken from Ref. 108.)

4.5. Analysis of Mixtures and Isotope Incorporation

The analysis of mixtures using mass spectra, although one of the first analytical applications of the technique, remains a difficult task. It requires that spectra be taken under constant ion source conditions and that the pure components be available. Each component is identified by its mass spectrum which will overlap with those of the other components. The MIKES technique provides an alternative since it enables a single ion to be used to characterize each component of the mixture. This will usually be the molecular ion. After selection of this ion by mass, collision-induced dissociation can be used to provide a fragmentation pattern and hence structural information.

Levsen and Beckey have used CID in combination with field ionization (FI) in the study of some simple mixtures.[112] The chief virtue of FI in this regard is that it is possible to associate one and *only one* peak in the mass spectrum with each component in the mixture because of the virtual absence of fragment ions. This simplifies the determination of the molecular weight on each component of the mixture. It also has the advantage that the molecular ions formed by field ionization are much less likely than those formed by electron impact to have rearranged to an isomeric structure. The absence of fragment ions in the mass spectrum is not a disadvantage in this technique since, when needed as a source of information on the structure of a component of interest, fragmentation is induced by collision and the fragmentation pattern determined by energy analysis.

A variation on this approach has been used in the analysis of steroid mixtures where metastable peaks rather than CID peaks were used to characterize the individual components; ionization was by electron impact.[115] It seems likely that this method of analyzing mixtures will be further facilitated by the use of CID since the fragmentation spectra of the individual ions can be interpreted using, as a guide, standard electron-impact mass spectra.

A promising approach to mixture analysis involves the use of chemical ionization, MIKE spectrometry, and collision-induced dissociation.[111] The method is shown schematically in Fig. 23. This procedure is capable of quantitative analysis and has a detection limit of $\sim 10^{-11}$ g. It has been applied to alkaloids and barbiturates, illustrating the fact that involatile samples can be handled. The method is complementary to GC/MC and has the advantage that the separation step is performed mass spectrometrically rather than chromatographically. Chemical noise is thus reduced and the analysis is potentially much faster although there is some loss in sensitivity vis-à-vis GC/MS.

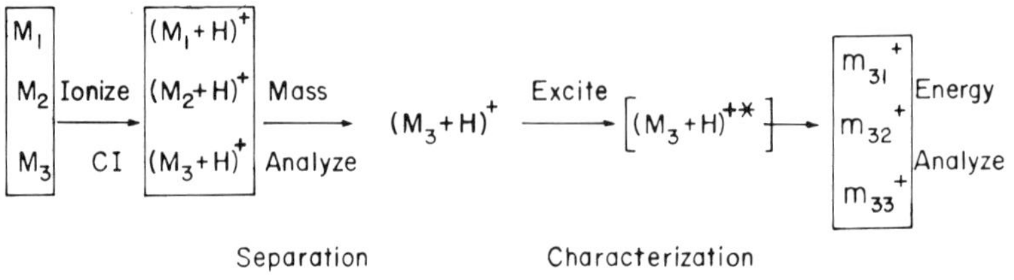

$$\boxed{\begin{array}{c}M_1\\M_2\\M_3\end{array}}\xrightarrow{\text{Ionize}\atop\text{CI}}\boxed{\begin{array}{c}(M_1+H)^+\\(M_2+H)^+\\(M_3+H)^+\end{array}}\begin{array}{c}\text{Mass}\\\text{Analyze}\end{array}\xrightarrow{(M_3+H)^+}\xrightarrow{\text{Excite}}\left[(M_3+H)^{+*}\right]\longrightarrow\boxed{\begin{array}{c}m_{31}^+\\m_{32}^+\\m_{33}^+\end{array}}\begin{array}{c}\text{Energy}\\\text{Analyze}\end{array}$$

Separation   Characterization

Fig. 23. Analysis of mixtures using collision-induced dissociation. A mixture of protonated ions is formed in the chemical ionization source and the desired component is mass-selected. It is identified by its MIKE spectrum taken to record collision-induced dissociations.

A partially labeled compound is, of course, just a particular type of mixture. MIKE spectra can therefore be used to study specifically labeled compounds even when they are present in relatively low abundance in such mixtures. This has been shown for both compounds isotopically enriched at a particular site,[115] an experiment which may yield a low isotopic incorporation, and for the selective study of specifically labeled ions using a compound which contains only the natural level of incorporation of ^{13}C or ^{34}S.[110].

A method of *determining* label incorporation based on the principles just noted can also be suggested. The essential feature of using energy spectrometry in this connection is that each ion is uniquely characterized. This removes the ambiguity met in mass spectra where a particular mass peak could be due, for example, to both the unlabeled molecular ion and the ^{13}C isotope of the $(M-H)^+$ ion. Thus, some particular reaction of the molecular ion occurring in the analyzer is chosen and the relative intensities of this peak and the corresponding peak due to the unlabeled analog are measured. With appropriate assumptions regarding isotope effects (usually that they are unity), this intensity ratio gives directly the degree of label incorporation in the molecule. Using the metastable peak for loss of H˙ from the molecular ion of toluene, this procedure has been used to determine the degree of label incorporation in d-toluene.[116] Since H˙ loss is subject to large isotope effects in metastable ions, a known mixture of labeled and unlabeled compounds also had to be run. An improved procedure would be to use a CID reaction which does not involve primary isotope effects. Isotope effects are expected, in any event, to be smaller for collision-induced dissociations which involve ions of higher internal energy than for metastable ions. By choosing a CID reaction in which the product ion does not contain the label it is possible, using a mass spectrometer of conventional geometry, to obtain a single accelerating voltage scan which includes the peak for both the unlabeled compound and that for the various labeled compounds. The CID approach to label incorporation should also be more readily applicable to determining the site as well as the degree of labeling.

4.6. *Product Characterization in Ion–Molecule Reactions and Other Applications*

Ion–molecule reactions which involve reactive collisions occur in ion cyclotron resonance cells, mass spectrometer ion sources, and the flowing afterglow apparatus, to cite some instances. Collision-induced dissociation of the ionic products after acceleration to high translational energies offers a promising new method for determination of their structures.[117-119] Consider, for example, the ion $C_8H_{11}^+$ present in the chemical ionization (CI) spectrum of benzene.[120] The ion corresponds to ethylated benzene and has been formulated as the σ complex. If a MIKE spectrometer is employed, this CI ion can be selected for analysis and then be characterized by the collision-induced dissociations it undergoes. Figure 24 shows the MIKE spectrum, taken in the presence of collision gas, of ethylated d_6-benzene (m/e 113) formed by chemical ionization of d_6-benzene with methane. By employing the deuterated benzene improved structural information is available. The spectrum is consistent with the σ-complex structure in that it shows loss of a methyl radical (m/e 95–98), ethylene (m/e 85), H˙ (not shown), and formation of the ethyl cation m/e 29, among other processes. It is noteworthy that a cluster of fragment ions corresponding to each of the carbon numbers C_2–C_7 is represented in the spectrum. The only abundant metastable peak is that associated with loss of ethylene, and even in this case the collision-induced dissociation is of comparable importance at the target pressures employed. The distribution of hydrogen and deuterium in the fragment ions is of considerable interest since complete scrambling of the ethyl and ring hydrogens has occurred in some reactions such as methyl radical loss (m/e 95–98), whereas loss of ethylene and formation of the ethyl cation show little or no hydrogen randomization. The peak (not shown) due to loss of ethylene from the metastable molecular ion shows no scrambling at all. Thus, either scrambling occurs after collisional activation (Section 4.3.2) or there is more than one form of the ethylated benzene reactant ion.

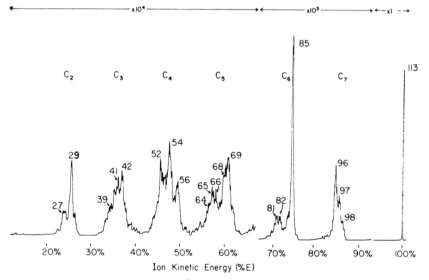

Fig. 24. Energy spectrum of $C_6D_6 \cdot C_2H_5^+$ ions (m/e 113) formed by chemical ionization of benzene using methane as reagent gas. The spectrum was taken on a MIKE spectrometer by mass selecting m/e 113 and inducing fragmentation using N_2 as collision gas. The abscissa is calibrated in terms of E, the energy of the stable ion beam. Groups of fragment ions having two to seven carbon atoms are observed.

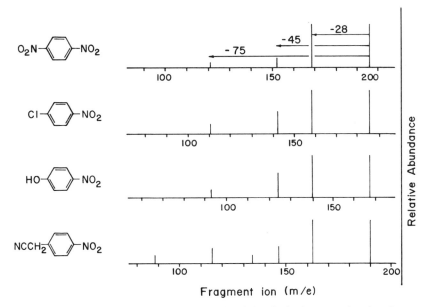

Fig. 25. The collision-induced dissociations of ethylated nitrobenzenes bearing the para substituents indicated. From the analogous reactions exhibited it is concluded that the nitro group is ethylated in the CI source.

An ion structural question of wide interest concerns the site at which an ion will bond to a polyfunctional molecule in the gas phase. Specific cases include the protonation, alkylation, and nitrosylation of aromatics and substituted aromatics. Collision-induced dissociation can provide definitive information. The ethylation of para-substituted nitrobenzenes occurs on the nitro substituent as the results of Fig. 25 illustrate.[118] The analogous fragmentation patterns of the four compounds and the characteristic losses of 45 ($C_2H_5O^.$) and 75 ($C_2H_5NO_2$) demonstrate this. Ring-ethylated aromatics show other equally characteristic fragmentations, such as $H^.$, $NH_2^.$, $C_2H_5^.$, and $Cl^.$ loss in the case of p-chloroaniline.

In the remainder of this section some suggestions are made regarding other possible applications of collision-induced dissociation at high energy.

The nature of the energy-transfer function determines the subsequent behavior of an ion excited by collision. In addition to energy-loss measurements, which provide information on this parameter, the actual distribution of products can be used provided the relationship between product yield and internal energy of the ion is known. This information, conveniently summarized in the form of a breakdown graph, is available experimentally (Section 2.2.5). This suggests that one of the more interesting aspects of energy transfer upon collision, i.e., its dependence on the nature of the target, might be explored by comparing fragmentation spectra. In particular, the abundance of a fragment ion which is formed only within a narrow range of ion energies would form a sensitive measure of the similarity in energy-transfer functions for different target species. This follows since the energy loss will be independent of the energy-transfer function whereas the fragment-ion abundance will be a sensitive measure of this function.

A possibility not yet mentioned in this chapter is collisional deactivation. Ion–molecule reactions which involve charge transfer can be endo- or exoergic or thermoneutral. Similarly, reactions which involve exchange of internal energy but not alteration in the charge on the ion can, in principle, proceed in either direction if the necessary energy balance is provided by the translational energy of the fast species.

These exothermic ion–molecule reactions should be inherently as likely as their endothermic counterparts but they may be much more difficult to detect. Collisional deactivation, if its occurrence is confirmed in polyatomic ions, may be an important factor in accounting for the pressure dependence of CID peaks (Section 2.4.1). The apparent insensitivity of some ions to CID, for example, may be explicable in terms of effective competition between activation and deactivation. This possibility may also be one of the factors responsible for the pronounced decrease in some metastable negative ion abundances observed on addition of collision gas (Section 5.1).

An area of application of collision-induced dissociation which requires more sensitive instrumentation than currently available would involve the use of these reactions in preparing ions the properties of which would then be of interest in their own right. Such experiments could be best performed by employing a second field-free region for the subsequent investigations. Comparisons with ions of known structures and energies formed, for example, directly by electron impact in the ion source could be undertaken. Some progress in this direction is to be found in the studies already made on combined CID and charge-transfer reactions discussed in the following section.

5. *Related Reactions*

This chapter has so far been concerned with the collision-induced dissociation of singly charged positive ions. The sections which follow cover the corresponding reactions of negatively charged ions and processes in which collision-induced dissociation occurs in conjunction with charge transfer. Also included are short sections on high-energy ion–surface and ion–photon reactions which are analogous to the processes occurring with gaseous targets.

5.1. Negative Ions

Negative ion mass spectrometry[121-123] is a useful complementary procedure to the far more widely used positive ion spectra. For the analysis of certain classes of compounds it is superior, giving abundant molecular ions when these may be entirely absent in positive ion spectra.

Some molecular anions show no fragmentation in negative ion mass spectrometry. In such cases, the possibility of inducing fragmentation by collision so as to be able to obtain structural information in addition to the molecular weight data is particularly valuable. The negative ion case is therefore analogous to that which obtains when field ionization is used to obtain positive ions, the low internal energies of which also greatly circumscribe fragmentation. In negative ion spectra, even when fragmentations do occur, they are often due to low-frequency factor reactions, including skeletal and hydrogen rearrangements.[79] In such a situation, molecular structure elucidation is more complex than it would be if simple bond cleavages dominated as would be expected in ions of greater internal energy.

Collision-induced dissociation of molecular anions can be illustrated using carboxylic acids, some of which show no spontaneous fragmentation. Using toluene as collision gas and techniques analogous to those employed for positively charged ions (Section 3), the loss of $CO_2H^.$ from these compounds is observed as a characteristic reaction[124] which serves to identify the functional group. In anthraquinone-2-carboxylic acid (**4**) ionization efficiency curves (Fig. 26) were obtained for the molecular anion and for the products due to CO_2 and $CO_2H^.$ loss on collision. These results indicate the occurrence of at least two distinct processes for formation of the molecular ion, direct electron capture at low energy and a process or processes involving secondary electrons at higher energy. The molecular anion populations which result from these distinct modes of formation are nonequivalent, elimination of $CO_2H^.$ being associated only with those molecular ions formed by the secondary process. These results affirm the appropriateness of collision-induced dissociation in characterizing ions, whether positively or negatively charged.

As a further illustration of the information on molecular structure provided by collision-induced dissociation of anions consider the spectrum shown in Fig. 27. This is an ion kinetic energy spectrum, taken without mass analysis, using a double-focusing mass spectrometer of conventional geometry. It is worth noting that since negative ion mass spectra are frequently very simple, energy spectra can be taken without mass analysis with little possibility of mistaken assignments. This makes the reversed-sector instrument unnecessary for work of this type. Two peaks in the spectrum, those due to loss of $NO^.$ and loss of C_6H_5OH, are due to

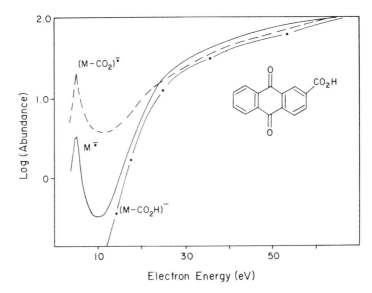

Fig. 26. Ionization efficiency curves for fragment anions produced by collision-induced dissociation compared with that for molecular anions of anthraquinone-2-carboxylic acid. Loss of CO_2 on collision occurs both from $M^{-\cdot}$ ions generated by electron capture at low electron energies and from those formed at higher energies. However, it is only the latter class of molecular anions which fragment by loss of $CO_2H^{\cdot}$.

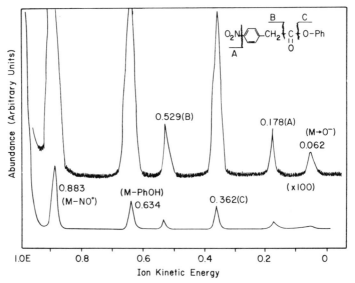

Fig. 27. Ion kinetic energy spectrum of negative ions taken without mass analysis. The peaks in this spectrum correspond to fragment anions formed both spontaneously and by collision-induced dissociation from the molecular anion of the ester shown. Spectrum retraced.

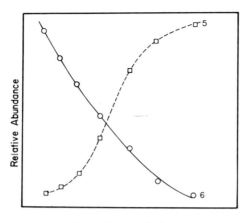

Fig. 28. Effect of target pressure on the intensity of the peak due to loss of C_6H_5OH from the molecular anions of the isomeric compounds **5** and **6**. At low pressures the observed signal is due to metastable ion reactions in both cases. Collisional activation promotes this reaction in **5** but depletes the fragmenting ion population in **6**.

spontaneous reactions, the remainder to collision-induced dissociations. All of the latter correspond to simple bond cleavages and, considered together, provide sound evidence for the molecular structure shown.

Studies on ion chemistry have long employed the ability to vary the internal energy of the population being sampled, most frequently by the simple method of varying the ionizing electron energy. This procedure is not applicable to negative ions so the alternative method of increasing internal energy by collision[79] would appear to be uniquely useful. Plots of the intensity of product ions against pressure of target gas can show sharp contrasts for different ions. For example, Fig. 28 illustrates the results[125] for a particular reaction, loss of C_6H_5OH from the isomeric molecular anions **5** and **6**. Both reactions give metastable peaks but in one case the corresponding collision-induced dissociation occurs, supplementing the peak at this position, whereas in the other the metastable peak is rapidly depleted by collision. An increase in pressure involves an increase in the average internal energy of the ion. Since ions of higher energy fragment preferentially via high-frequency factor, high-activation-energy processes, the C_6H_5OH elimination reactions are ascribed to a two-step simple cleavage process in **5** and to a rearrangement process with a tight activated complex and a low-activation energy in **6**.

The sharp fall of metastable ion abundances for rearrangement reactions with increasing collision gas pressure observed in this and other cases is

                    OH   OH
                    | |
    O₂N──⟨ ⟩──CO₂C₆H₅   O₂N──⟨ ⟩
                                              ╲CO₂C₆H₅
         5   **6**

in contrast to the behavior of positive ions in undergoing analogous reactions. In positive ions under comparable conditions, only a slight decrease in the peak height is observed. It seems likely that this difference is in part due to the nature of the potential energy surfaces for negatively and positively charged ions. The former typically show much shallower minima.

5.2. Dissociative Charge Transfer

In this section we consider reactions in which alteration of the charge of the high velocity reactant occurs in conjunction with fragmentation. Most interest attaches to single-collision events in which the collision causes charge transfer which is followed by spontaneous fragmentation. Processes in which both steps require a collision are potentially more various but are unlikely to provide the data on state assignments and energy levels for otherwise inaccessible species which the one-step reactions seem to promise.

The occurrence of charge-changes in conjunction with collision-induced dissociation has long been known. Henglein's elegant studies[16] using a parabola mass spectrograph showed the occurrence of dissociative charge exchange of doubly charged ions:

$$m_1^{2+} + N \rightarrow m_2^+ + m_3^+ + N \qquad (5.1)$$

and dissociative charge inversion of positive ions:

$$m_1^+ + N \rightarrow m_2^- + m_3 + (N^{2+}) \qquad (5.2)$$

Although the overall processes were established, the sequence of steps was not.

For each of the three most common charge-transfer reactions observed in mass spectrometry (i.e., charge stripping, charge inversion, and charge exchange of doubly charged ions) there is a corresponding dissociation process. These single-collision two-step reactions are given in Eqs. (5.3)–(5.5) together with that [Eq. (5.6)] associated with the charge stripping of negative ions:

$$m_1^+ \xrightarrow{N} [m_1^{2+}] \rightarrow m_2^{2+} + m_3 \quad \text{or} \quad m_2^+ + m_3^+ \qquad (5.3)$$

$$m_1^+ \xrightarrow{N} [m_1^-] \to m_2^- + m_3 \tag{5.4}$$

$$m_1^{2+} \xrightarrow{N} [m_1^+] \to m_2^+ + m_3 \tag{5.5}$$

$$m_1^- \xrightarrow{-N} [m_1^+] \to m_2^+ + m_3 \tag{5.6}$$

It will of course be recognized that there exist numerous other reaction types involving high-velocity neutrals, charge-separation reactions, and multicollision sequences. Even as regards reactions (5.1)–(5.4) only a beginning has been made in the study of this field of chemistry.

The variety of reactions possible can be illustrated by reference to the fragmentations of multiply charged ions, processes which have been studied extensively by conventional mass spectrometric techniques. Both unimolecular charge-separation reactions, e.g.,

$$m_1^{2+} \to m_2^+ + m_3^+ \tag{5.7}$$

and dissociation by loss of a neutral fragment, e.g.,

$$m_1^{2+} \to m_2^{2+} + m_3 \tag{5.8}$$

are well known. However, all the known cases of collision-induced dissociation of multiply charged ions appear to involve charge transfer as well as dissociation. In $C_2H_3^{2+}$, studied by Kupriyanov and Perov,[126] the cross section for charge exchange of doubly charged ions was found to be somewhat larger than those for the dissociative charge-exchange reactions, which is consistent with these reactions occurring by charge exchange followed by spontaneous fragmentation. In general, the relative cross sections for competitive collisional reactions are expected to be controlled by the magnitude of the energy loss. This is typically small (several electron volts) for CID, large for stripping (10/20), and variable but often very small for charge exchange of doubly charged ions (20/11)—hence the importance of this last reaction.

Charge-transfer reactions provide a means of forming species in states which might be inaccessible if electron impact, electron attachment, photoionization, photodetachment, or other methods are used. Chapters 4 and 5 deal with the rich chemistry which follows from the study of high energy charge transfer reactions. The type of data used to deduce this chemistry comes primarily from measurements of various energy terms but, particularly for polyatomic systems, the crucial information on the nature of the bonding is more accessible if the species of interest can be made to undergo some chemical reaction. In particular, fragmentation of the charge-transfer product ion can provide valuable information on its structure. Experiments in which the products of charge-transfer reactions are excited by collision and their fragmentation examined would require, to avoid ambiguity in characterizing the process, that each collision occur in a separate reaction chamber of field-free region. Experiments of this type are technically difficult and few have been done. On the other hand, single-collision reactions such as (5.3)–(5.6) can be studied in one collision chamber, and they provide unique data on species in unusual states. In such experiments, it is usually not possible to distinguish products of the ion-molecule reaction which are truly unstable (lifetimes less than one bond vibration) from those which are stable but formed with internal energy sufficient to allow fragmentation within the collision chamber in a time on the order of 10^{-5} sec.

Since ion–molecule reactions tend to occur so as to minimize the change in kinetic energy of the system, the internal energy of the high velocity product and hence its fragmentation behavior should be a function of the nature of the target species. The operation of this effect is shown later. Based on similar reasoning one can predict that exothermic processes should be more likely to lead to products with sufficient internal energy to fragment spontaneously than those which involve energy loss. Thus reactions (5.5), which will typically involve an energy gain or a smaller energy loss than reaction (5.6), should be more likely to result in fragmentation of m_1^+ given that the structures of the m_1^+ ions formed by the two methods are the same. This would also lead to the prediction that reaction (5.5) should be very much more common than reaction (5.3), which is much more endoergic. The validity of the prediction is evident from results on $E/2$ and $2E$ mass spectra of a large number of compounds taken on a conventional geometry mass spectrometer (Chapter 5). These spectra record, respectively, all ions which undergo charge exchange $(2+ \rightarrow +)$ and charge stripping $(+ \rightarrow 2+)$ in the first field-free region of the instrument. Since complete mass spectra are plotted, any subsequent delayed fragmentation of the charge transfer products occurring in the second field-free region appears as a metastable peak. Such peaks are more common in $2E$ mass spectra than in $E/2$ spectra plotted at comparable sensitivity.

For example, Fig. 29 shows a portion of the $2E$ mass spectrum of naphthalene. The peaks at integral mass are due to singly charged ions produced by charge exchange of doubly charged ions. Peaks at nonintegral mass are due to the further fragmentation of these ions and the figure shows such a peak due to the spontaneous fragmentation:

$$C_{10}H_8^+ \rightarrow C_8H_6^+ + C_2H_2 \tag{5.9}$$

If collision gas is now added to the second field-free region as well, collision-induced dissociation of the products of charge exchange can be observed. It is of note that the kinetic energy release associated with the spontaneous reaction (5.9) is 0.36 eV which is equal within experimental error to the value of 0.39 eV recorded for the same reaction when the

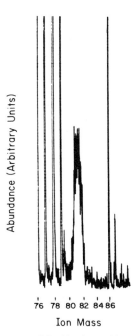

Fig. 29. Portion of the mass spectrum of doubly charged ions after charge exchange ($2E$ mass spectrum) showing a broad peak due to the fragmentation together with more abundant ions due to charge exchange. The fragmentation reaction is due to the elimination of acetylene from 102^+ in naphthalene by the overall sequence $C_{10}H_8^{2+} + C_6H_6 \rightarrow C_{10}H_8^{+\cdot} + (C_6H_6^{+\cdot}) \rightarrow C_8H_6^{+\cdot} + C_2H_2$.

$C_{10}H_8^{+\cdot}$ ion is prepared directly by electron impact. Clearly the reacting form of this ion has the same structure when prepared by either route. In other cases charge-changing collisions have been specifically selected as a means of preparing ions with particular structures. Thus, charge exchange of the linear doubly charged ion of benzene has been used[127] to prepare linear $C_6H_6^{+\cdot}$. Comparison of the cross sections and energy releases associated with the further fragmentation of this ion with that of the metastable molecular ion of benzene provides evidence that the latter is acyclic.

Another general factor which controls the reactivity of charge-transfer products is the relationship between the potential energy surfaces for the ions in different charge states. Since the ion–molecule reactions in question involve essentially vertical transitions, formation of an excited or unstable product will result when the minima in the curves are displaced relative to each other. Such displacements are particularly likely in charge inversion reactions and this may be part of the reason why these reactions show large cross sections for subsequent dissociations.

Frequently the charge inversion reactions *only* occur with dissociation, stable forms of the cation or anion in question apparently being inaccessible.

Examples from the charge inversion of positive ions include the case of anisole, the molecular ion (108^+) of which does not yield a detectable ion at 108^-. An abundant ion (93^-) corresponding to loss of a methyl group from the molecular anion is observed. A related phenomenon is that a particular ion may be inaccessible by a simple charge transfer reaction but accessible in a dissociative process. For example, the species $(M-NO)^-$ in dinitrobenzene is not formed in detectable quantity on charge inversion of the $(M-NO)^+$ ion. However, dissociative inversion of the molecular ion does yield this anion. These observations are related to those reported elsewhere[128] concerning the accessibility of certain doubly charged ions from the corresponding singly charged ion but not from the neutral molecule. Although the phenomenon is understood in general terms, detailed information on the relative positions of the surfaces is probably obtainable only in the simplest cases.

Dissociative charge inversion of the type in which fragmentation precedes inversion potentially provides a means of characterizing, through the charge inversion process, the products of metastable ion or collision-induced reactions. Even more interesting are cases in which inversion precedes fragmentation. If fragmentation of the negative ion is spontaneous, this allows the study of ions which are otherwise inaccessible and which may be quite unstable. Clearly it is important to be able to characterize not only the number of collisions involved but also the order in which they occur. If fragmentation precedes inversion, the kinetic energy release will be that for the corresponding positively charged ion reaction. Examination of the kinetic energy peak shape can thus indicate the type of mechanism. For example, the overall reaction

$$H_2O^{+\cdot} \rightarrow HO^- + H^\cdot \qquad (5.10)$$

has been shown[62] by pressure studies to involve one collision. The kinetic energy release was almost an order of magnitude larger than that for the reaction

$$H_2O^{+\cdot} \rightarrow HO^+ + H^\cdot \qquad (5.11)$$

hence the process must occur in the sequence

$$H_2O^{+\cdot} \xrightarrow{N} [H_2O^{-\cdot}] \rightarrow HO^- + H^\cdot \qquad (5.12)$$

These observations set the stage for accumulation of experimental information on $H_2O^{-\cdot}$ as formed in a vertical transition from $H_2O^{+\cdot}$. Highly dissociative states of $H_2O^{-\cdot}$ are reached which correlate with $HO^- + H^\cdot$ on the one hand and with $O^{-\cdot}$ and excited hydrogen atoms on the other.[62]

Similarly it has been found that the reaction

$$H_2S^{+\cdot} + N \rightarrow HS^- + H^\cdot \qquad (5.13)$$

occurs via the anion $H_2S^{-\cdot}$, the kinetic energy release accompanying its fragmentation being substantially greater than that for formation of HS^+

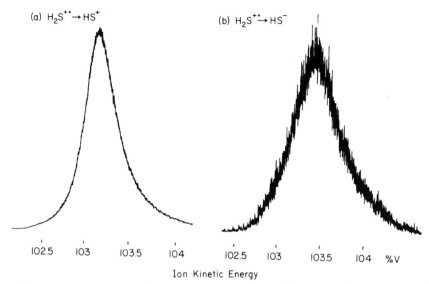

Fig. 30. Comparison of the ion kinetic energy peaks associated with (a) collision-induced dissociation of $H_2S^{+\cdot}$ to give HS^+ and (b) dissociative inversion of $H_2S^{+\cdot}$ to give HS^-. The positions and widths of these peaks provide kinetic-energy-loss and kinetic-energy-release measurements from which an inversion–dissociation sequence is deduced for (b). The abscissa is given in terms of the accelerating voltage (V) necessary to transmit the product ion taking the stable ion beam as 100%.

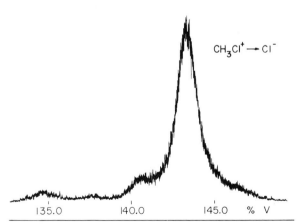

Fig. 31. Illustration of signal-to-noise characteristics of peak due to inversion of charge accompanied by dissociation in chloroform. The main peak at 143% V is due to the overall reaction $CH_3Cl^{+\cdot} \rightarrow Cl^-$.

from $H_2S^{+\cdot}$. Figure 30 shows the peak shapes for the formation of HS^+ and HS^- from $H_2S^{+\cdot}$.

The high cross section for dissociative charge inversion of organic ions[129] is illustrated by the accelerating voltage scan in Fig. 31 which shows

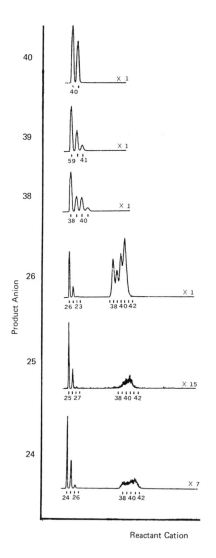

Fig. 32. Peaks due to dissociative charge inversion of some positive ions in acetonitrile. The peaks at the left of each scan have the same kinetic energy as the reactant ions. Read vertically, they represent the charge inversion $(-E)$ mass spectrum of the compound. Associated with each of these peaks are others due to ions of lower energy. Any of these peaks is due to the conversion of a positive ion of the indicated mass to a negative ion of the mass shown on the left of each scan.

peaks due to CH_nCl^+ ions forming Cl^- as product. Other examples of this phenomenon in polyatomic ions have been noted,[130] each peak in the $-E$ (charge inversion) spectrum of acetonitrile, for example, showing a number of fragmentation reactions as illustrated in Fig. 32. Only a beginning has been made in using such data in correlating the structures of negative and positive polyatomic ions.

Since the methods used to study collision-induced dissociation specify only the reactant and the product ion, it is possible that a charge-changing collision may occur in the reaction sequence but not be recognized. Thus the intermediate m_1^{2+} in

$$m_1^+ + N \rightarrow [m_1^{2+}] \rightarrow m_2^+ + m_3^+ \qquad (5.14)$$

might not be distinguished from simple collision-induced dissociation in an experiment which monitors m_2^+ fragment ions. Indeed, the results for $C_3H_7^+$ shown in Fig. 1 may well be due to the operation of both a simple collision-induced dissociation, associated with the small energy release, and the more complex sequence (5.14) associated with the very broad peak with sharply defined wings and an energy release of several electron volts. The fact that the experiment was done with very-high-energy ions can be expected to promote reaction (5.14), which requires a relatively large energy loss for the charge-stripping step.

5.3. Related Ion–Surface Reactions

Studies of ion–surface phenomena have dealt almost exclusively with atomic ions. The energy spectra of kilovolt-energy ions scattered from surfaces, and the nature of inelastic events including neutralization and stripping, have been extensively studied for these systems. It would seem that a surface might be substituted for a target gas in the collisional excitation of high-energy ions. Experiments in which high-energy ion beams are allowed to pass through thin foils of metal or organic polymers (beam-foil experiments) rely on the similarity between gaseous and solid target. A few such studies have employed polyatomic ions, as in the dissociation of $CO_2^{+\cdot}$ at 45 keV to give both positive and negative atomic ions.[131]

An alternative experimental arrangement which circumvents the difficulty of preparing the necessary thin films is that in which the ion beam incident at grazing angles is scattered from the surface[132] (Fig. 33). This arrangement maximizes the time the ion spends near the surface and can lead to high cross sections for surface-induced dissociation. Energy losses, which are readily measured, are almost entirely inelastic and can therefore be related to the nature of the species involved; the method shows promise of experimental convenience compared with the conventional gaseous

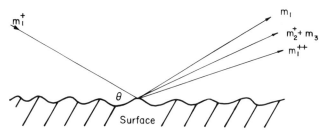

Fig. 33. Ion–surface interactions at small scattering angle leading to dissociation, charge stripping, or charge inversion. The kinetic energy change of the projectile approximates the heat of reaction Q at small θ.

target procedures. It should also provide information on the nature of the surface.

5.4. Related Photodissociations

Processes in which an ion is excited by absorption of radiation prior to fragmenting are analogous to those in which the excitation is by collision. They may even be simpler to explain, given the well-known nature of optical selection rules. A practical difficulty, however, is the high photon fluxes required. Some studies of this type have been made in the ion source of the mass spectrometer.[133] Ion cyclotron resonance is particularly promising in this connection since it allows ions to be trapped for several seconds, so facilitating photodissociation.[134] In this way it is possible to record photodissociation spectra of polyatomic ions as a function of photon wavelength and thus to distinguish isomeric ions.[109] Since either the disappearance of the reactant ion or the appearance of the dissociation product can be followed, the method provides direct structural data on long-lived ions in the ICR cell. Correlations observed[135] between spectral features in the photodissociation spectra of aromatic ions and the photoelectron spectra of the corresponding molecules allow assignments of the upper states of the ions in the excitation process and provide evidence for the similarity of the ion structures generated in the two techniques.

A sensitive method of studying photodissociation of low energy ions in a tandem quadrupole instrument has also been described.[136] In this experiment one quadrupole is employed for reactant ion selection and another for product ion analysis.

Reactions in which *fast* moving ions are photodissociated[137,138] are most directly comparable with collision-induced dissociation. Because of the large cross sections for CID, the resulting signal may be much greater than that due to photodissociation at typical background pressures. For this reason ultrahigh vacuum has been employed in conjunction with a pulsed irradiating source. Experiments of this type have given kinetic energy spectra which show vibrational structure in simple species. Extensions to the study of more complex ions are anticipated.

5.5. CID Studied by Ion–Cyclotron Resonance Spectrometry

We have already indicated in some detail (Section 4) how the excitation of ions by collision can be analytically useful. Such processes are most significant when the reaction involved can be specified. The two general mass spectrometric methods for achieving this are the study of field-free region reactions in a dual-analyzer instrument, as detailed in Section 3.1, and the use of double-resonance techniques in ion-cyclotron resonance spectroscopy. Thus, by irradiating one ion and observing another (at a different frequency) it is possible to demonstrate coupling between two ions through chemical reactions.

Ion-cyclotron resonance spectrometry has been extensively employed in recent years for studies on ion-molecule reactions but in only a few cases were fragmentation reactions specifically studied. This is a little curious since ion kinetic energies range up to several hundred electron volts. One reaction interpreted[139] as proceeding by collision-induced dissociation of the molecular ion was the process

$$\text{4-Cl-C}_6\text{H}_4\text{-CH}_2\text{CH}_3{}^{+\cdot} \rightarrow \text{4-Cl-C}_6\text{H}_4\text{-}{}^+\text{CH}_2 + \text{CH}_3^{\cdot} \qquad (5.15)$$

The double-resonance spectrum was obtained in the presence of a large excess of nitrogen at an ICR cell pressure of 1×10^{-5} torr, although no data were given on the double-resonance spectrum in the absence of collision gas. The observed effect is due to an increase in the molecular ion translational energy through absorption of radio-frequency energy, with subsequent collisions resulting in the conversion of translational to internal energy. There are two disadvantages of the ICR approach—first the ion kinetic energy is not known, and in fact it varies as the ion spirals. Hence the collision energy depends on the target gas pressure. Second, as in the usual mass spectrometric techniques, any intermediates occurring between reactants and products are not defined. This is the more severe in ICR because here we can have ready atom- and group-transfer reactions so that the sequence between m_1^+ and m_2^+ is potentially very complex. Nonetheless, exploration of this method and particularly experiments in which comparisons are made with CID in field-free regions might be very worthwhile.

5.6. *CID of Neutral Molecules*

This discussion of collision-induced dissociation has so far been restricted to the fragmentation of ions. Considerable interest would attach to the dissociation of polyatomic molecules in this way, direct comparisons with photochemistry being just one area open to enquiry. Experiments with molecular beams in this connection are extremely difficult and apparently have not been done using polyatomic species at the energies of interest here. An alternative experiment which is conceptually related to much of the work detailed in this book is also possible. Instead of using a fast molecule and a slow target, a fast target and a slow molecule could be employed. The processes which ensue are equivalent. In the usual mechanism of collisional activation involving electronic excitation it is the electrons of the target which are involved in the excitation, and the presence or absence of an overall charge on the target can be viewed as a minor perturbation. Hence the target might be made a fast-moving ion, and this allows the study of CID of neutral molecules. In one application of this principle, direct analysis of the slow secondary products of collision has been effected.[140]

As an alternative to analysis of the slow species (before or after acceleration in the case of ionic products) the fast product can be analyzed, information on the reaction being derived from its kinetic energy spectrum. This latter type of study therefore employs an inelastic scattering process of the 10/10* type and it has been applied in studying H_2 dissociation.[141] By varying the ion or by using species where the first excited state lies above the ground state by an energy which exceeds the energy loss in typical experiments, it can be arranged that the measured inelastic energy loss be associated entirely with the target molecule. Separation of the elastic energy loss from the measured loss requires a knowledge of the scattering angle.

These considerations suggest that a conventional mass spectrometer might be adapted to the study of the molecular CID through use of the following reaction:

$$m^+ + N_1 \rightarrow m^+ + N_2 + N_3 \tag{5.16}$$

where m^+ could be H^+ or some other simple ion and N_1 is a polyatomic molecule. The measurement of the kinetic energy change associated with reaction (5.16) is facilitated if it is studied at a nonzero scattering angle. If this is not done, the intense peak due to elastic scattering or unreacted ions may completely obscure the signal due to inelastic reactions. Nonzero angles can be selected by offsetting the ratio of electric sector to ion-accelerating voltages in a conventional mass spectrometer.[143,144] Such an experiment would directly yield the excitation energy distribution acquired by the molecule but further information would be required to characterize the reaction channel(s) involved. The study of excitation processes in organic ions using energy-loss measurements of this type has been described.[145]

An alternative experiment would be to study small impact-parameter reactions at zero scattering angle as has been done for H_2 by Los and co-workers[146] in the kilovolt energy range. The processes which occur involve direct hard-sphere collisions and include impulsive stripping. The energy loss associated with such backscattering reactions at 0° is a considerable fraction of the ion energy.

Dissociative charge transfer of polyatomic molecules is a related process which can be studied conveniently in modified mass spectrometric instrumentation. The molecule is converted to the ion in a high-translational-energy charge-transfer reaction and the internal energy of the ion so formed may cause it to dissociate spontaneously.[147,148] The cross section for the charge-transfer reaction

$$m^+ + N_1 \rightarrow m + [N_1^+] \rightarrow N_2^+ + N_3 \qquad (5.17)$$

has been measured as a function of the translational energy in the cases of methane and acetylene and interpreted in terms of the adiabatic maximum criterion of Hasted (Section 2.3). Experiments in which a fast beam of neutral molecules, formed by charge exchange of an ion beam in one collision chamber, is dissociated by collision in a second chamber would help to lift the present restriction of CID to systems which include ions. Experiments utilizing multiple collision chambers are already in use in the study of atomic systems.[149]

6. Prospects

Collision-induced dissociation provides a means of obtaining structural information on gaseous ions and thermochemical data on their fragmentation reactions. Several lines of work suggest that applications in molecular structure determination may also develop strongly. There is a particular compatibility between the technique of collision-induced dissociation and the use of mass-analyzed ion beams in reversed-sector mass spectrometers. This facilitates both analytical studies using collisional activation spectra and studies on individual reactions employing high energy resolution. The use of field ionization or chemical ionization, in association with the CID–MIKES combination, offers particular promise for the analysis of mixtures. In this regard it offers an alternative to the combined gas chromatography/mass spectrometry (GC/MS) technique with the advantage that separation is accomplished after ionization in the time required to scan a mass spectrum.

There is an unusual appropriateness in the fact that collision-induced dissociation which is based on high-energy ion–molecule reactions, should find some of its more promising applications in the characterization of the products of conventional low-energy ion–molecule reactions. The study of chemical ionization products in this way is in progress.

Collision-induced dissociation offers a means of studying an aspect of ion chemistry which is usually encountered only indirectly, isomerizations which do not lead to fragmentation. In this as in other areas of investigation it may profitably be employed in concert with other high-energy ion–molecule reactions. Other growth points in this young subject may be in the related surface reactions and in the more complex reaction sequences in which collisional activation occurs in conjunction with charge-changing collisions. These latter processes increase the degree of control which can be maintained over ion preparation so that highly unstable species can be studied by monitoring the energy spectra of their dissociation products.

The information which collision-induced dissociation may provide on the dynamics of polyatomic ion reactions is largely unexplored but nevertheless promising. Ions of high internal energy, not accessible by conventional metastable ion studies, can be scrutinized via the measurement of kinetic energy release. As in metastable ion reactions, an important criterion in future work may be the need to more accurately define the internal energies of the ions being studied.

Future directions of progress have been suggested in the text and special attention is drawn to the suggestions for new experiments or topics of study scattered throughout this chapter.

ACKNOWLEDGMENT

The Purdue work described in this chapter was supported by the National Science Foundation.

References

1. See, for example, L. Friedman and B. G. Reuben, *Adv. Chem. Phys.* **19**, 33 (1970).
2. K. R. Jennings, in *Mass Spectrometry: Techniques and Applications* (G. A. W. Milne, ed.), Wiley, New York, 1971, p. 419.
3. J. L. Holmes and F. Benoit, in *MTP Review of Science, Physical Chemistry*, Series One, Vol. 5 (A. Maccoll, ed.), University Park Press, Baltimore, 1972, Chapter 8.
4. R. G. Cooks, J. H. Beynon, R. M. Caprioli, and G. R. Lester, *Metastable Ions*, Elsevier, Amsterdam, 1973.
5. J. A. Hipple and E. U. Condon, *Phys. Rev.* **68**, 54 (1945).
6. J. A. Hipple, R. E. Fox, and E. U. Condon, *Phys. Rev.* **69**, 347 (1946).
7. J. J. Thomson, *Rays of Positive Electricity and Their Application to Chemical Analysis*, Longmans, Green, London, 1913, p. 94.
8. F. W. Aston, *Proc. Cambridge Phil. Soc.* **19**, 317 (1919).
9. F. W. Aston, *Mass Spectra and Isotopes*, 2nd ed., Arnold, London, 1942, p. 61.
10. H. D. Smyth, *Phys. Rev.* **25**, 452 (1925).

11. T. R. Hogness and E. G. Lunn, *Phys. Rev.* **30**, 26 (1927).
12. K. E. Dorsch and H. Kallmann, *Z. Phys.* **53**, 80 (1929).
13. H. D. Smyth, *Rev. Mod. Phys.* **3**, 347 (1931).
14. K. T. Bainbridge and E. B. Jordon, *Phys. Rev.* **51**, 595 (1937).
15. J. Mattauch and H. Lichtblau, *Phys. Zeit.* **40**, 16 (1939).
16. A. Henglein, *Z. Naturforsch.* **7a**, 165 (1952).
17. A. Henglein, *Z. Naturforsch.* **7a**, 208 (1952).
18. F. M. Rouke, J. C. Sheffield, W. D. Davis, and F. E. White, *J. Chem. Phys.* **31**, 193 (1959).
19. J. W. McGowan and L. Kerwin, *Can. J. Phys.* **41**, 316 (1963).
20. V. M. Kolotyrkin, M. V. Tikhomirov, and N. I. Tunitskii, *Dokl. Akad. Nauk SSR* **92**, 1193 (1953).
21. C. E. Melton and H. M. Rosenstock, *J. Chem. Phys.* **26**, 568 (1957).
22. C. E. Melton, in *Mass Spectrometry of Organic Ions* (F. W. McLafferty, ed.), Academic Press, New York, 1963.
23. S. E. Kupriyanov and A. A. Perov, *Zh. Tekhn. Fiz.* **34**, 1317 (1964).
24. S. E. Kupriyanov and A. A. Perov, *Zh. Fiz. Khim.* **42**, 857 (1968).
25. S. E. Kupriyanov, *Zh. Tekhn. Fiz.* **36**, 2161 (1966); *Chem. Abstr.* **66**, 41645 (1967).
26. S. E. Kupriyanov, *Elementarnye Protesessy Khim. Vysokikh. Energ. Akad. Nauk SSSR*, Inst. Khim. Fiz. Tr. Simposiuma, Moscow, 1963, pp. 23–26; *Chem. Abstr.* **65**, 11512 (1966).
27. (a) Z. Z. Latypov and S. E. Kuprianov, *Zh. Fiz. Khim.* **39**, 1572 (1965); *Chem. Abstr.* **63**, 12484 (1965); (b) S. E. Kuprianov and A. A. Perov, *Zh. Tekhn. Khim.* **33**, 823 (1963); *Chem. Abstr.* **59**, 12196 (1963).
28. S. E. Kuprianov, *Tr. Nauchn.-Issled.*, Fiz. Khim. Inst., 1963, p. 76; *Chem. Abstr.* **60**, 1215 (1964).
29. (a) S. E. Kupriyanov and A. A. Perov, *Russ. J. Phys. Chem.* **39**, 871 (1965); (b) A. A. Perov, S. E. Kupriyanov, and M. I. Gorfinkel, *ibid.* **45**, 1014 (1971).
30. K. R. Jennings, *Int. J. Mass Spectrom. Ion Phys.* **1**, 227 (1968).
31. F. W. McLafferty and H. D. R. Schuddemage, *J. Am. Chem. Soc.* **91**, 1866 (1969).
32. W. F. Haddon and F. W. McLafferty, *J. Am. Chem. Soc.* **90**, 4745 (1968).
33. J. H. Beynon and R. G. Cooks, *Res. Develop.* **22**, No. 11, 26 (1971).
34. T. Wachs, P. F. Bente, III, and F. W. McLafferty, *Int. J. Mass Spectrom. Ion Phys.* **9**, 333 (1972).
35. J. H. Beynon, R. G. Cooks, J. W. Amy, W. E. Baitinger, and T. Y. Ridley, *Anal. Chem.* **45**, 1023 A (1973).
36. D. T. Terwilliger, J. H. Beynon, and R. G. Cooks, *Proc. R. Soc. (London) A* **341**, 135 (1974), and references cited therein.
37. M. Roussel and A. Julienne, *Int. J. Mass Spectrom. Ion Phys.* **9**, 449 (1972).
38. K. M. Rafaey and W. A. Chupka, *J. Chem. Phys.* **43**, 2544 (1965); cf. Ref. 17.
39. J. Durup, in *Recent Developments in Mass Spectrometry* (K. Ogata and T. Hayakawa, eds.), Univ. Park Press, Baltimore, Maryland, 1970, p. 921.
40. H. S. W. Massey and H. B. Gilbody, *Electronic and Ionic Impact Phenomena*, 2nd ed., Vol. 4, Oxford University Press, London, p. 2932.
41. (a) S. E. Kupriyanov, *Izv. Akad. Nauk. SSR Ser. Fiz.* **27**, 1102 (1963); (b) S. E. Kupriyanov and A. A. Perov, *Russ. J. Phys. Chem.* **39**, 871 (1965).
42. L. P. Hills, M. Vestal, and J. H. Futrell, *J. Chem. Phys.* **54**, 3834 (1971).
43. F. W. McLafferty, P. F. Bente, III, R. Kornfeld, S.-C. Tsai, and I. Howe, *J. Am. Chem. Soc.* **95**, 2120 (1973).
44. R. G. Cooks, L. Hendricks, and J. H. Beynon, *Org. Mass Spectrom.* **10**, 625 (1975).
45. J. D. Rynbrand and B. S. Rabinovitch, *J. Chem. Phys.* **54**, 2275 (1971).
46. I. G. Simm, C. J. Danby, and J. H. D. Eland, *Int. J. Mass Spectrom. Ion Phys.* **14**, 285 (1974).

47. Compare J. Turk and R. H. Shapiro, *Org. Mass Spectrom.* **5**, 1373 (1971), and references cited therein.
48. See, for example, P. J. Robinson and K. A. Holbrook, *Unimolecular Reactions*, Wiley-Interscience, New York, 1972.
49. H. M. Rosenstock, *Adv. Mass Spectrom.* **4**, 523 (1967).
50. C. E. Klots, *J. Phys. Chem.* **75**, 1526 (1971).
51. B. H. Solka, J. H. Beynon, and R. G. Cooks, *J. Phys. Chem.* **79**, 859 (1975).
52. H. Yamaoka, D. Pham, and J. Durup, *J. Chem. Phys.* **51**, 3465 (1969).
53. S. J. Anderson and J. B. Swan, *Phys. Lett.* **48A**, 435 (1974).
54. R. Caudano and J. M. Delfosse, *J. Phys. B* **1**, 813 (1968).
55. F. P. G. Valckx and P. Verveer, Proceedings of the 4th International Conference on the Physics of Electronic and Atomic Collisions, Quebec, 1965, p. 333.
56. J. Hasted, *J. Appl. Phys.* **30**, 25 (1959).
57. H. S. W. Massey and E. H. S. Burhop, *Electron and Ionic Impact Phenomena*, 1st ed., Oxford University Press, London, 1956, pp. 441, 515.
58. H. S. W. Massey, *Rep. Prog. Phys.* **12**, 248 (1949).
59. J. B. Hasted, *Physics of Atomic Collisions*, 2nd ed., American Elsevier, New York, 1972, p. 621ff.
60. Compare, for example, T. Ast, J. H. Beynon, and R. G. Cooks, *J. Am. Chem. Soc.* **94**, 6611 (1972).
61. R. Kornfeld, Ph.D. Thesis, Cornell University, 1971, p. 129, Table II.
62. T. Keough, J. H. Beynon, and R. G. Cooks, *Chem. Phys.* **12**, 191 (1976).
63. S. E. Kupriyanov and A. A. Perov, *Russ. J. Phys. Chem.* **42**, 447 (1968).
64. B. H. Solka and R. G. Cooks, unpublished results.
65. J. Durup, P. Fournier, and D. Pham, *Int. J. Mass Spectrom. Ion Phys.* **2**, 311 (1969).
66. T. Baer, L. Squires, and A. S. Werner, *Chem. Phys.* **6**, 325 (1974).
67. J. W. McGowan and L. Kerwin, *Can. J. Phys.* **42**, 2086 (1964).
68. (a) S. E. Kupriyanov, *Zh. Eksper. Teor. Fiz.* **47**, 2001 (1964). (b) F. W. McLafferty and W. T. Pike, *J. Am. Chem. Soc.* **89**, 5951 (1967).
69. A. Giardini-Guidoni, R. Plantania, and F. Zucchi, *Int. J. Mass Spectrom. Ion Phys.* **13**, 453 (1974).
70. H. R. Jennings, *J. Chem. Phys.* **43**, 4176 (1965).
71. J. H. Futrell, K. R. Ryan, and L. W. Sieck, *J. Chem. Phys.* **43**, 1832 (1965).
72. A. F. Weston, K. R. Jennings, S. Evans, and R. M. Elliott, *Int. J. Mass Spectrom. Ion Phys.* **20**, 317 (1976).
73. D. Kemp, R. G. Cooks, and J. H. Beynon, *Int. J. Mass Spectrom. Ion Phys.* **21**, 93 (1976).
74. J. H. Futrell and T. O. Tiernan, in *Ion–Molecule Reactions*, Vol. 2 (J. L. Franklin, ed.), Plenum Press, New York, 1972, Chapter 11.
75. K. C. Kim, M. Uckotter, J. H. Beynon, and R. G. Cooks, *Int. J. Mass Spectrom. Ion Phys.* **15**, 23 (1974).
76. N. R. Daly, A. McCormick, R. E. Powell, and R. Hayes, *Int. J. Mass Spectrom. Ion Phys.* **11**, 255 (1973).
77. J. H. Beynon, R. G. Cooks, and T. Keough, *Int. J. Mass Spectrom. Ion Phys.* **13**, 437 (1974).
78. J. G. Maas, N. P. F. B. van Asselt, and J. Los, *Chem. Phys.* **8**, 37 (1975).
79. J. H. Bowie and S. G. Hart, *Int. J. Mass Spectrom. Ion Phys.* **13**, 319 (1974).
80. (a) F. W. McLafferty, R. Kornfeld, W. F. Haddon, K. Levsen, I. Sakai, P. F. Bente, S.-C. Tsai, and H. D. R. Schuddemage, *J. Am. Chem. Soc.* **95**, 3886 (1973); (b) B. van der Graaf, P. P. Dymerski, and F. W. McLafferty, *J. Chem. Soc., Chem. Commun.*, 978 (1975).
81. K. Levsen and F. W. McLafferty, *Org. Mass Spectrom.* **8**, 353 (1974).
82. N. M. M. Nibbering, T. Nishishita, C. C. Van de Sande, and F. W. McLafferty, *J. Am. Chem. Soc.* **96**, 5668 (1974).

83. F. W. McLafferty and J. Winkler, *J. Am. Chem. Soc.* **96**, 5182 (1974).
84. N. M. M. Nibbering, C. C. Van de Sande, T. Nishishita, and F. W. McLafferty, *Org. Mass Spectrom.* **9**, 1059 (1974).
85. C. C. Van de Sande and F. W. McLafferty, *J. Am. Chem. Soc.* **97**, 2298 (1975).
86. (a) K. Levsen, *Org. Mass Spectrom.* **10**, 43 (1975); (b) **10**, 55 (1975).
87. J. H. Beynon, R. M. Caprioli, and R. G. Cooks, *Org. Mass Spectrom.* **9**, 1 (1974).
88. W. McGowan and L. Kerwin, *Can. J. Phys.* **42**, 972 (1964).
89. R. G. Cooks, J. H. Beynon, and J. F. Litton, *Org. Mass Spectrom.* **10**, 503 (1975).
90. E. G. Jones, L. Bauman, J. H. Beynon, and R. G. Cooks, *Org. Mass Spectrom.* **7**, 185 (1973).
91. J. L. Holmes and J. K. Terlouw, *Org. Mass Spectrom.* **10**, 787 (1975).
92. J. K. Terlouw and J. L. Holmes, paper presented at the 23rd Annual Conference on Mass Spectrometry and Allied Topics, ASMS, Houston, May 1975.
93. R. G. Cooks, J. H. Beynon, M. Bertrand, and M. K. Hoffman, *Org. Mass Spectrom.* **7**, 1303 (1973).
94. J. L. Holmes, *Adv. Mass Spectrom.* **6**, 865 (1974).
95. G. Gum, G. Sindova, and N. Uccella, *Ann. Chim.* **64**, 169 (1974).
96. C. S. Hsu and R. G. Cooks, unpublished results.
97. J. H. Beynon and R. G. Cooks, *Adv. Mass Spectrom.* **6**, 835 (1974).
98. (a) R. G. Cooks, I. Howe, and D. H. Williams, *Org. Mass Spectrom.* **2**, 137 (1969); (b) C. W. Tsang and A. G. Harrison, *ibid.* **7**, 1377 (1973).
99. R. Hass, R. G. Cooks, J. F. Elder, D. G. I. Kingston, and M. M. Bursey, *Org. Mass Spectrom.* **11**, 697 (1976).
100. See, for example, M. DePas, J. J. Leventhal, and L. Friedman, *J. Chem. Phys.* **51**, 3748 (1969).
101. J. L. Franklin, J. G. Dillard, H. M. Rosenstock, J. T. Herron, K. Draxl, and F. H. Field, *Ionization Potentials, Appearance Potentials and The Heats of Formation of Gaseous Ions*, NSRDS-NBS 26, 1969.
102. T. F. Moran, F. C. Petty, and A. F. Hendrick, *J. Chem. Phys.* **51**, 2112 (1969).
103. R. G. Cooks and T. L. Kruger, unpublished results.
104. K. Levsen and H. Schwarz, *Org. Mass Spectrom.* **10**, 752 (1975).
105. J. H. Beynon, M. Bertrand, E. G. Jones, and R. G. Cooks, *Chem. Commun.*, 341 (1972).
106. R. G. Cooks and T. Keough, unpublished results.
107. R. G. Cooks, K. C. Kim, and J. H. Beynon, *Chem. Phys. Lett.* **26**, 131 (1974).
108. I. Howe and F. W. McLafferty, *J. Am. Chem. Soc.* **92**, 3797 (1970).
109. R. C. Dunbar and E. W. Fu, *J. Am. Chem. Soc.* **95**, 2716 (1973).
110. J. H. Beynon, D. F. Brothers, and R. G. Cooks, *Anal. Chem.* **46**, 1299 (1974).
111. J. F. Litton, T. L. Kruger, R. W. Kondrat, and R. G. Cooks, *Anal. Chem.* **48**, 2113 (1976).
112. K. Levsen and H. D. Beckey, *Org. Mass Spectrom.* **9**, 570 (1974).
113. M. von Ardenne, K. Steinfelder, and R. Tummler, *Electronenanlagerungs-Massenspektrographic organischer Substanzen*, Springer Verlag, Berlin, 1971.
114. K. Levsen and H. R. Schulten, *Biomed. Mass Spectrom.* **3**, 137 (1976).
115. D. H. Smith, C. Djerassi, K. H. Maurer, and U. Rapp, *J. Am. Chem. Soc.* **96**, 3482 (1974).
116. J. H. Beynon, J. E. Corn, W. E. Baitinger, J. W. Amy, and R. A. Benkeser, *Org. Mass Spectrom.* **3**, 191 (1970).
117. J. H. Beynon and R. G. Cooks, *Int. J. Mass Spectrom. Ion Phys.* **19**, 107 (1976).
118. T. L. Kruger, R. Flammang, J. F. Litton, and R. G. Cooks, *Tetrahedron Lett.* **50**, 4555 (1976).
119. F. W. McLafferty, paper presented at the 23rd Annual Conference on Mass Spectrometry and Allied Topics, ASMS, Houston, 1975.
120. M. S. B. Munson and F. H. Field, *J. Am. Chem. Soc.* **89**, 1047 (1967).
121. J. H. Bowie, in *Mass Spectrometry, A Specialist Periodical Report*, Vol. 2 (D. H. Williams, ed.), The Chemical Society, London, 1973.

122. J. H. Bowie and B. D. Williams, in *Mass Spectrometry, International Review of Science, Physical Chemistry*, Series Two (A. Maccoll, ed.), University Park Press, Baltimore, 1975.
123. P. W. Harland, K. A. C. MacNeal, and J. C. J. Thynne, in *Dynamic Mass Spectrometry*, Vol. 1 (D. Price and J. E. Williams, eds.), Heyden, London, 1970.
124. J. H. Bowie, *Org. Mass Spectrom.* **9**, 304 (1974).
125. J. H. Bowie and A. C. Ho, *Org. Mass Spectrom.* **9**, 1006 (1974).
126. S. E. Kupriyanov and A. A. Perov, *Dokl. Akad. Nauk SSR* **49**, 1368 (1963); *Chem. Abstr.* **59**, 8220 (1963).
127. T. Keough, T. Ast, J. H. Beynon, and R. G. Cooks, *Org. Mass Spectrom.* **7**, 245 (1973).
128. Ref. 4, p. 146.
129. T. Keough, Ph.D. Thesis, Purdue University, 1975.
130. K. Varmuza, R. Heller, and P. Krenmayr, *Int. J. Mass Spectrom. Ion Phys.* **15**, 218 (1974).
131. F. A. White, F. M. Rourke, and J. C. Sheffield, *Rev. Sci. Instr.* **29**, 182 (1958).
132. R. G. Cooks, D. T. Terwilliger, T. Ast, J. H. Beynon, and T. Keough, *J. Am. Chem. Soc.* **97**, 1583 (1975).
133. R. E. Ellefson, B. A. Osterlitz, J. M. Phillips, A. B. Denison, and J. H. Weber, *Chem. Phys. Lett.* **31**, 364 (1975).
134. R. C. Dunbar, *J. Am. Chem. Soc.* **93**, 4354 (1971).
135. P. D. Dymerski, E. Fu, and R. C. Dunbar, *J. Am. Chem. Soc.* **96**, 4109 (1974).
136. M. L. Vestal and J. H. Futrell, *Chem. Phys. Lett.* **28**, 559 (1974).
137. J. B. Ozenne, D. Pham, and J. Durup, *Chem. Phys. Lett.* **17**, 422 (1972).
138. (a) N. P. F. B. Van Asselt, J. Maas, and J. Los, *Chem. Phys. Lett.* **24**, 555 (1974); (b) *Chem. Phys.* **5**, 429 (1974).
139. F. Kaplan, *J. Am. Chem. Soc.* **90**, 4483 (1968).
140. E. Mitani, H. Tsuyama, and M. Itoh, in *Recent Developments in Mass Spectrometry* (K. Ogata and T. Hayakawa, eds.), Univ. Park Press, Baltimore, Maryland, 1970, p. 949.
141. J. Schottler and J. P. Toennies, *Chem. Phys. Lett.* **12**, 615 (1972).
142. D. F. Diltner and S. Datz, *J. Chem. Phys.* **49**, 1969 (1968).
143. R. G. Cooks, K. C. Kim, and J. H. Beynon, *Chem. Phys. Lett.* **23**, 190 (1973).
144. T. Ast, D. T. Terwilliger, R. G. Cooks, and J. H. Beynon, *J. Phys. Chem.* **79**, 708 (1975).
145. M. Medved, R. G. Cooks, and J. H. Beynon, *Chem. Phys.* **15**, 295 (1976).
146. H. van Dop, A. J. H. Boerboom, and J. Los, *Physica* **54**, 223 (1971).
147. J. B. Homer, R. S. Lehrle, J. C. Robb, and D. W. Thomas, *Trans. Faraday Soc.* **62**, 619 (1966).
148. R. S. Lehrle and R. S. Mason, *Adv. Mass Spectrom.* **6**, 759 (1974).
149. J. Appell, M. Durup, P. Dông, and J. Durup, paper presented at the 23rd Annual Conference on Mass Spectrometry and Allied Topics, Houston, May 1975.

Index

Acetylene, target in double-electron transfer, 244-245
Adiabatic criterion, 20, 85, 156, 157, 199
  in charge transfer, 122, 229-231
  in collision-induced dissociation, 383-384
Adiabatic maximum rule, 230
Adiabatic molecular orbitals
  correlation diagrams for, 150-154
  and inner-shell excitation, 194-196
Alkenes, excited states of, 82-84
Ammonia, states of doubly ionized, 246
Amplification factor, for kinetic energy release, 366-367
Angular distribution
  in charge exchange, 131-142
  in collision-induced dissociation, 313-314
  in collisional excitation, 63-71
  of emitted radiation, 214-217
  see also Scattering angle
Angular resolution
  effect on energy resolution, 7-8, 311-313
  effect on measured cross section, 170-171
  in energy-loss spectra, 64, 66
  and fine structure in momentum distributions, 337-345
  requirements in dissociation studies, 317-321
Angle, scattering, *see* Scattering angle
Aniline, $2E$ mass spectrum of, 262
Anisidine, charge stripping of, 278
Anisotropy
  in collision-induced dissociation, 297-300, 324-328, 333-334
  in photodissociation, 350-351

Appearance potential, measurements in studying ion dissociation, 306, 329-330
Aston bands, 305-312
  for HeH^+, 340-343
  for N_2^+, 336-337
  see also Collision-induced dissociation
Atomic hydrogen
  energy loss spectrum of, 49-53
  excitation of, 44-53
Auger electron spectroscopy, double-ionization potentials from, 244

Barat–Lichten rule, 152
Basicity, gas phase, 416-417
Born approximation
  applied to collision-induced dissociation, 297-300
  calculations for H_2^+ dissociation, 324-325
  cross sections from, 20-23, 48-49, 52, 63, 85-86
Born–Oppenheimer approximation, 4, 107, 148-150
Breakdown curves, and energy loss on collision, 374-375

Capacitance manometer, 41-42, 110-111
Charge exchange
  by coincidence measurements, 203-206
  dissociative, 434-441
  cross section
    differential, 114-115
    doubly charged ions, 230-231, 262-263
    energy dependence, 119-124
    total, 114-115

451

Index

Charge exchange (cont'd)
  of doubly charged ions, 124-131, 136-142, 227-254, 260-267
  neutral product detection, 183-186
  product internal energy in, 436-437
  in H^+-H collisions, 45-49
  of He^+ in collisions with Ne, 183-186
  mechanism, 96-107, 156-163
  multielectron transfer, 124-142, 227-228
  partial beam model for, 30-31
  resonant, 92, 100
  singlet → triplet transitions by, 81-84
  of singly charged ions, 119-124, 131-134
    organics, 284-286
  theory of, 96-108
Charge-exchange mass spectrum, 286
Charge inversion
  dissociative, 437-444
  of negative ions, 281-284
  of positive ions, 267-272
  see also general entries under Charge exchange, Charge stripping
Charge inversion mass spectrum, 270-272
Charge separation, in polyatomic dications, 265-267, 274
Charge stripping
  compared to collision-induced dissociation, 273-274
  competitive with collision-induced dissociation, 406-407
  of ions formed by chemical ionization, 281
  ot negative ions, 281-284
  partial beam for, 30-31
  of polyatomic ions, 272-281
  using a MIKE spectrometer, 279-280
Charge transfer, see Charge exchange
Chemical ionization
  charge stripping and, 281
  gas phase basicities by, 416-417
  ion structures from, 427-430
Coincidence measurements
  experimental arrangement, 200-202
  ion-electron, 217
  ion-photon, 198-219
  on molecular species, 217-219
  photon-photon, 219
  of polarization, 215-217
Collision chamber
  collision-induced dissociation in, 307-309, 401-402
  in double-focusing mass spectrometers, 260, 401-402

Collision chamber (cont'd)
  location at focal point, 308
  negative ion formation in, 232
  neutral beam formation in, 169
  potential applied to, 308-309, 313
Collision gas, see Target
Collision-induced dissociation
  adiabatic criterion in, 383-384
  of anions formed by charge inversion, 270-272
  Born approximation and, 297-300
  of cations formed by charge inversion, 283-284
  charge exchange with, 434-441
  coincidence measurements on, 217-219, 316
  collision energy and, 388-390
  in competition with stripping, 273-274, 406-407
  conversion of coordinate systems in, 317-321
  cross section for, 362-363, 380-385
  of diatomic ions, 289, 352
  dynamics of, 291-305
  energy partitioning accompanying, 375-377
  energy release in, 296-297, 361-362
  energy resolution in, 311-313
  gas phase basicity determination, 416-417
  of H_2^+, 322-331
  of HeH^+, 331-334, 340-346
  internal energy effects on, 393-395
  internal energy of ions in, 373-377
  in ion cyclotron resonance, 443-444
  ion structure determination by, 402-413
  of ions in $2E$ mass spectra, 265
  of ions in $E/2$ mass spectra, 274-279
  isolated electronic states in, 376-377
  kinetics of, 372-377
  mechanism of, 294-305, 367-380
    electronic excitation, 294-300
    Franck-Condon transitions in, 369
    hard collisions in, 326-327
    predissociation, 302-305, 379
    Russek mechanism, 379
    vibrational excitation, 295-302, 378-380
  metastable ions and, 358-359, 418-419
  molecular structure elucidation, 442-445
  multiple collisions in, 386-387
  of N_2^+, 336-340

Collision-included dissociation (cont'd)
  of neutral molecules, 444-445
  by photons, 442-443
  of polyatomic ions, 357-446
  resonance in, 383-384
  on surface impact, 441-442
  target pressure effects, 385-388
Collisional activation
  mechanism of, 294-302
  see also Collision-induced dissociation
Collisional activation spectra, 363
  energy resolution in, 399-400
  internal energy of ions and, 370-371
  ion structures from, 402-413
Collisional excitation, of molecules, 71-84
Coordinate systems, conversions between, 8-14, 289-290, 311, 317-321
Correlation diagram
  for adiabatic states, 150-155
  for diabatic states, 103, 151-157
  in inner shell excitations, 194-196
Coupling of states
  long range via Stark effect, 212-214
  see Rotational Coupling, Radial Coupling
Cross section
  charge exchange, 99-108
    of doubly charged ions, 262-263
    kinetic energy dependence, 119-124
    relation to energy loss spectrum, 97
  collision-induced dissociation
    anisotropy of, 324-328, 333-334
    of organic ions, 362-363, 380-385
  collisional excitation
    of atomic hydrogen, 44-53
    general features of, 85-86
    of helium, 53-63
    and impact parameter, 161-163
    partial beam method, 30-34
    by photon emission, 35-44
    of projectile, 31-34
    of target, 31-34
    theory of, 156-163
  differential
    for charge exchange, 124-142
    for inelastic collisions, 66-68, 171-177
  double charge transfer
    energy defects and, 230
    Franck–Condon principle and, 239
    pressure effects on, 234-235
  measurement
    effect of angular resolution, 170-171
    of total, 114-115

Cross section (cont'd)
  oscillations in, 133-137, 180
  photodissociation, dependence on vibrational population, 348-351
  reduced
    for charge transfer, 132, 140-142
    for He^+/Ne collisions, 180
  total, for excitation, 170-171
Curve crossing, charge exchange by, 99-108, 156-163
Crossed beam technique, for study of H^+/H collisions, 44-45
Cyclohexene, charge exchange mass spectrum of, 286

Detectors
  for ions
    electron multipliers, 112
    Faraday cup, 39-40
    ion counting, 174-175, 200, 232
    photoplate, 361
    scintillator, 112
  for neutrals, 40
    counting, 178-179
    efficiency of, 176
    metastable atoms, 116
  for photons, 42-44, 216
Diabatic molecular states, 103, 151-157
  in charge exchange, 98-99
  in dissociation, 299-300
  for H_2^+, 151
  for He_2^{2+}, 103
  and Steuckelberg oscillations, 207-209
Differential cross section, see Cross section, differential
Disproportionation factor, 5, 254, 365
Dissociation
  with charge exchange
    negative ions, 271-272
    positive ions, 265-267, 274-279
  with charge inversion, 437-441
  with double charge transer, 251-252
  Newton diagram for, 12-13
  with rearrangement, 419-421
  see also Collision-induced dissociation
Doppler shift, and excitation mechanisms, 47
Double-electron transfer
  cross section, 229-231
  differential, 242
  double-ionization potentials from, 236-238

Double electron transfer (cont'd)
  of doubly charged ions, 108
  Franck–Condon principle and, 238-240
  ion structure from, 250
  negative ion formation by, 235, 247-248
  pressure effects, 234-235
  radiative processes in, 247-248
  selection rules, 236-238, 241-242
  thermochemistry, 228
  see also Charge exchange
Double-charge-transfer spectroscopy, 227-254
Doubly charged ions
  charge exchange of, 124-142, 260-267
  charge separation of, 274
  dissociation of, 434-437
  in mass spectra, 260-261
Dynamics
  of collision-induced dissociation, 291-305
  of ion-molecule collisions, 8-14, 252-254

+E Mass spectra, 281-287
−E Mass spectra, 267-272
$E/2$ Mass spectra, 272-281
$2E$ Mass spectra, 260-267
Eikonal approximation, 23, 86
Elastic scattering
  He^+ + Ne, 180-185
  He^{++}, 135, 140-141
  Newton diagrams for, 11-12
  and target pressure, 234-235
Electron capture, see Charge exchange
Electron affinity, from double electron transfer, 250-251
Electron spectroscopy, in studying ionic collisions, 163-164
Electron transfer, see Double-electron transfer, Charge exchange
Emission
  cascade processes, 36, 47
    cross section measurements by, 36-38
  in charge changing collisions, 247-248
  during collisions, 108
  in excitation, 24-25, 35-44
  modulated beam experiments, 44-48
  oscillations in cross-section, 199-200
  photon flux determination, 42-44
  polarization of radiation, 43-44, 47, 213-217
  in proton-molecule collisions, 74-84
  of x-rays, 217

Energy analysis
  electrostatic
    cylindrical, 117, 396-398
    127° cylindrical, 234
    cylindrical mirror, 200
    hemispherical, 26-28
    parallel plate, 174
  of neutrals, 177-179
  time of flight, 177-179, 186
Energy-change spectrum, see Energy-loss spectrum
Energy defect
  for charge exchange, 124, 228
  for collision-induced dissociation, 309, 367, 380-384
  for inelastic collisions, 5, 92, 164-166, 252-254
  see also Energy loss, inelastic
Energy loss
  in collision-induced dissociation, 365
    and breakdown curves, 374-375
    collision energy effects, 387-390
    effect on cross section, 380-383
    and ion enthalpy, 415-416
    target pressure effects, 385-387
  disproportionation factor for, 365
  in dissociative excitation, 292-293, 309-310
  of N_2^+, 337-339
  in double-electron transfer, 228
  calibration of scale, 232-233
  of He^+ in He^+/Ne collisions, 180-183
  kinematics of, 252-254
  by momentum analysis, 232
  resolution of peaks due to, 400
  due to vibrational excitation, 189-193
Energy loss, inelastic, 126-147, 164-166, see also Energy defect
Energy-loss measurements, in neutral–neutral collisions for Ar–Ar, 189
Energy-loss spectrum, 4-6, 26-35
  of argon, 175-176
  of atomic hydrogen, 49-53
  differential in scattering angle, 63-71, 125-131
  of helium, 53-54
  for He^{2+}/Ne collisions, 125-127
  for He^{2+}/Ar collisions, 127-131
  of molecular hydrogen, 73-75
  of molecular nitrogen, 75-79
  of molecular oxygen, 79-81

Energy-loss spectrum (*cont'd*)
  partial beam contributions to, 31-33
  for protons on molecular targets, 73-84
Energy-loss spectrometry, 147-221
  energy resolution in, 117, 164-166, 233-234
Energy partitioning, in collision-induced dissociation, 375-377
Energy release, accompanying fragmentation, *see* Kinetic energy release
Energy resolution
  in collision-induced dissociation, 399-400, 408-409
  in electron and photon spectroscopy, 164
  in energy-loss spectra, 26-28, 117, 164-166, 233-234
  and placement of collision chamber, 260
  of time of flight analyzer, 179

Faraday cup, 39-40, 111
Flavanone, charge stripping of negative ions, 283-284
Fragmentation, *see* Dissociation
Franck–Condon principle
  in charge exchange of doubly charged ions, 238-240, 262
  in collision-induced dissociation, 369
  in double electron transfer, 238-240
  in electronic excitation, 296-297, 323-324
  in target excitation, 72

Glauber approximation, 23, 53, 86

H^-
  excited states of, 235
  from H^+ in double electron transfer, 235-248
He^{+2}, charge exchange, 125-130
Heat of reaction, *see* Energy defect
Helium
  energy loss spectrum of, 53-54
  excitation of, 53-63
3-Hexanone, $E/2$ mass spectrum, 273, 276
Hydrocarbons, $2E$ mass spectra of, 266-267
Hydrogen, molecular
  double charge transfer, 238-239
  excitation by protons, 73-75

Impact parameter, 6, 93, 164
  from coincidence measurements, 206-211
  and reduced scattering angle, 67, 101-102
  and rotational excitation, 161-163

Inelastic collision, Newton diagram for, 12-13
Inner shell excitation, 193-198
  state correlations for, 194-196
  vacancy in MO and, 195-196
  x-ray emission during, 217
Internal energy
  from energy-loss measurements, 416
  and fragmentation, 285-286
  of ions,
    after charge exchange, 436-437
    after collisional excitation, 371-372, 393-395, 419
    before collision-induced dissociation, 314-315, 373-377, 381, 405-408, 419
    and isomerization, 419-421
    of metastable ions, 359
    of negative ions, 433-444
    from photodissociation measurements, 348-351
Ion-cyclotron resonance, collision-induced dissociation in, 443-444
Ion energy-loss spectra, *see* Energy-loss spectra
Ion enthalpy, from energy-loss measurements, 415
Ion kinetic energy spectrometry, 258-259, 396-397
  of negative ions, 282-284, 432
  see also Energy spectra, Mass-analyzed ion kinetic energy spectrometry
Ion–molecule reactions, at low energy, 2, 357, 427-428
Ion sources
  charge exchange in, 285
  chemical ionization, 426
  for coincidence experiments, 200
  for collision-induced dissociation, 401
  excited state ions from, 39, 113, 315
  field ionization, 424-425
  Nier type, 166-168, 232, 312, 315
  photoionization, 315
  plasma, 234, 312, 314-315
Ion structure
  from charge inversion, 268-270
  from charge stripping, 277-279
  chemical ionization and, 427-430
  from collision-induced dissociation, 402-413
  from double-charge transfer, 250
  from fragmentations of doubly charged ions, 265

Ion structure (cont'd)
  isomerization and, 419-422
  from isotopic labeling, 422
  from kinetic energy release, 408-413
  negative ions, 402-413, 424, 431-434
  positive and negative ions compared, 284
  singly and doubly charged ions compared, 262
Ionic collisions
  types of, 21
  see also the individual types
Ionization, in neutral–neutral collisions, 187-189
Ionization potential, double
  and charge stripping, 406-407
  determined by charge exchange, 236
  of molecules, 243-246
  of O_2, 241-242
  SCF calculation of, 246
Isolated states, in collision-induced dissociation, 376-377
Isomerization, studied by collision-induced dissociation, 419-421, 428
Isotopic labeling, for ion structure studies, 422, 428

Kinetic energy loss, see Energy loss
Kinetic energy release, 6-7
  amplification factor for, 366-367
  angular distribution of, 324-329
  angular resolution and, 317-321
  on collision-induced dissociation, 296-297, 310-311, 361-362, 399
  for diatomic ions, 296-297, 323, 334
  distribution of, 317-321
  for doubly charged ions, 265-266
  fine structure in, 338-342
  in H_2^+ dissociation, 323
  in HeH^+ dissociation, 331-334, 340-345
  and interchange distance, 265-266
  ion structure and, 408-413
  on photodissociation, 346-351
  resolution of, 311-313
Kinetics, of collision-induced dissociation, 372-377
K-shell excitation, see Inner-shell excitation

Landau–Zener level crossings, 102-106, 120, 155-157
Lifetimes, of dissociative states, 339-340, 342-344

McLeod gauge, 40-42, 110
Mass-analyzed ion kinetic energy spectrometer, $E/2$ mass spectra, 279-280
Mass-analyzed ion kinetic energy spectrometry, 363, 397-398
  chemical ionization and, 427-430
  ion structure determinations by, 402-413
  mixture analysis by, 424-427
Mass analyzer
  inhomogeneous magnetic field, 307, 312
  magnetic sector, 396-398
  Wien filter, 234
Mass spectra
  new types using ion-molecule reactions, 257-287
  pressure effects on, 287
Mass spectrometers
  modified for charge-transfer reactions, 231-232, 258-260
  modified for collision-induced dissociation, 305-312, 363-364, 395-399
  reversed sector, 363
Massey criterion, see Adiabatic criterion
Merged beam experiments, 113
Metastable ions
  by charge exchange, 436-437
  compared to collision-induced dissociation, 358-359, 418-419
  doubly charged, 265-266
  energy release in, 318-320
  ion kinetic energy spectra and, 258-259
  negative ions, 433-434
  predissociation of, 305
Metastable peaks
  in $2E$ mass spectra, 263-264
  fine structure in, 341-343
  in mass spectra, 360-361
Metastable states, formed in ion sources, 94, 113
Mixture analysis, by collision-induced dissociation, 425-427
Molecular orbital theory
  double ionization potentials from, 246
  of ionic collisions, 96-108, 149-163
Momentum analysis, resolution, 311
Momentum distribution
  of dissociation fragments, 317-321
  fine structure in, 338-345
  in photodissociation, 346-351
  see also Ion kinetic energy spectra

Negative ions
  from charge inversion, 267-272
  collision-induced dissociation of, 402, 431-434
  from dissociative charge inversion, 251-252
  ion kinetic energy spectra of, 432
  metastable, 433-434
  structure determination of, 424, 431-434
Neutral beams, kilovolt energy, 168-169
  by charge exchange, 168-169
  in neutral−neutral collisions, 186-189
Neutral particles, detected in coincidence experiments, 204-206
Newton diagram, 10-14
Nitriles, charge inversion of, 268-271
Nitrobenzene
  $-E$ mass spectrum, 270
  mass spectrum at high target pressure, 287
Nitrogen, excitation by protons, 75-79
Nomenclature, for ionic collisions, 20-21, 230
Non-crossing rule, for adiabatic MO's, 150, 154-156, 158

Oxygen
  double-charge transfer, 241-242, 249-250
  double-ionization potential, 242
  excitation by protons, 72, 79-81

Partial beam, model for ionic collisions, 31-33
Photodissociation, 346-351, 442-443
  energy release in, 318, 347
  of H_2^+, 346-351
  selection rules for, 350-351
Photomultiplier, 42
Photon emission, see Emission
Phthalic anhydride, ion kinetic energy spectrum, 259
Polarization
  of emitted radiation, 214-217
  in photodissociation, 350-351
Polarization forces
  dissociation by, 301-302, 312
  see also Russek mechanism
Predissociation
  of HeH$^+$, 342-345
  in ion-molecule collisions, 302-305
  of N_2^+, 338-340

Projectile
  electron affinity, 250-251
  molecular structure, 251
  neutralization of, 184-186
  state composition, 249-251
Pyridine
  metastable peaks, 263-264
  $2E$ mass spectrum, 263

Quasi-equilibrium theory, applied to collision-induced dissociation, 373
Quadrupole transitions, 63

Radial coupling, of nuclear and electronic energy, 107, 126-127, 155-163
  energy-level crossings and, 155, 157
  inner-shell excitation by, 194
Radiative transitions
  in charge-exchange collisions, 247-248
  see also Emission
Radiation
  emission of, 31, 74-84, 163-164, 247-248
  polarization of, 213-217
  see also Emission
Rare gases, doubly ionized states of, 237-238
Reduced cross sections, see Cross section, reduced
Resonance criterion, in collision-induced dissociation, 383-385
Rosenthal oscillations, 211-214
Rotational coupling, of nuclear and electronic energy, 107, 126-127, 156-163
  inner-shell excitation by, 194-197
Rotational predissociation, 342-345
Russek mechanism, of collision-induced dissociation, 301-302, 331-334, 346

Scattering angle
  in charge exchange, 125-127, 131-138, 184-185
  in collision-induced dissociation, 313-314, 324-326
  in double-electron transfer, 232, 242-243
  in energy-loss spectra, 63-71, 165, 252-254
  reduced, 180
    relation to impact parameter, 67, 101-102
  selection of, 64

Selection rules
  in charge exchange
    spin selection, 236-238, 241-242
    symmetry selection, 240-242
  for electronic angular momentum, from polarization measurements, 215-216
  for transitions between diabatic states, 107-108
Semi-classical treatment, of ionic collisions, 92-93, 97-108
Singlet → triplet transitions, 81-84
Spin conservation, Wigner rule, 53, 72-73, 81-84
Spin-orbit coupling, 127-128
  in charge exchange, 123-124
State composition, from double electron transfer, 249-250
Statistical theory, of collision-induced dissociation, 373-377
Stueckelberg oscillations, 124-125, 131-138, 180-182, 206-211
Superelastic collisions, 3, 186
Surface-induced dissociation, 441-442
Symmetry rules
  for radial coupling, 158
  for rotational coupling, 158

Target
  dissociation in double-charge transfer, 251-252
  for dissociation of polyatomics ions, 372, 390-391
  dynamics of collisions, 191-193
  effect on $2E$ mass spectra, 264
  effect on energy release in dissociation, 328-331
  excitation in collision-induced dissociation, 391
  ionization to excited states, 235-236
  momentum acquired in collision-induced dissociation, 290, 326-328
  nature of
    effect on inner-shell excitation, 196-197
    polyatomic molecules as, 243-246
  polarizability leading to Rosenthal oscillations, 212-214

Target (*cont'd*)
  pressure
    determination of, 40-42, 110-111
    and emission intensity, 37-38
    and excitation cross section, 32-34, 38-39
    and mass spectrum, 287
    and mechanism of double-electron transfer, 234-235
    metastable ions and collision-induced dissociation, 386-387, 418-419
    negative ion attenuation, 433-434
    and scattered ion signal, 172-173
  recoil energy of, 2-3, 165-166, 253
Thermochemical determinations, by collision-induced dissociation, 413-417
Thermochemistry, of charge exchange reactions, 262-263
Toluene, $2E$ mass spectrum of, 261-262
Transient molecules
  alignment of, 215
  formed in ionic collisions, 96-108, 149-163, 193-196
  transitions between MO's of, 156-163
Translational energy spectrum, *see* Energy loss spectrum
Translational spectroscopy, 335-352
  of H_2, 346-351
  of HeH^+, 340-346
  of N_2^+, 336-340
  see also Ion kinetic energy spectrometry
Triplet states of organic molecules, 82-84

Vibrational excitation
  in Ar^+-N_2 collisions, 191-193
  dissociation due to, 300-302
  in He^+-H_2 collisions, 193-194
  by impulsive momentum transfer, 326-328, 334
  in ion–molecule collisions, 295-302
Vibrational population, effect on photodissociation, 348-350

Wigner spin conservation rule, *see* Spin conservation
Wigner–Witmer coupling, 127
Wien filter, for mass analysis, 234

THE LIBRARY
UNIVERSITY OF CALIFORNIA
San Francisco
(415) 476-2335

THIS BOOK IS DUE ON THE LAST DATE STAMPED BELOW

Books not returned on time are subject to fines according to the Library Lending Code. A renewal may be made on certain materials. For details consult Lending Code.

14 DAY
DEC 12 1989
RETURNED
DEC 12 1989